Stability Theory

STABILITY THEORY

An Introduction to the Stability of Dynamic Systems and Rigid Bodies

HORST LEIPHOLZ

DEPARTMENT OF CIVIL ENGINEERING
UNIVERSITY OF WATERLOO
WATERLOO, ONTARIO, CANADA

Translated by Scientific Translation Service

 1970

ACADEMIC PRESS   New York and London

First published in the German language in the series Mathematik
und Mechanik under the title *Stabilitätstheorie*.
©1968 by B.G. Teubner, Stuttgart, Germany.

ACADEMIC PRESS, INC.
111 Fifth Avenue, New York, New York 10003

United Kingdom Edition published by
ACADEMIC PRESS, INC. (LONDON) LTD.
Berkeley Square House, London W1X 6BA

LIBRARY OF CONGRESS CATALOG CARD NUMBER: 71-131388

PRINTED IN THE UNITED STATES OF AMERICA

Contents

Preface

There have been great advances in the theory of stability in recent decades due to the requirements of control theory and flight mechanics, for example. We need only mention the theory of A. M. Lyapunov. A number of specialists have given a very mathematical and abstract description of the Lyapunov stability theory. A "stability theory of motion" has been developed which can be applied to the kinetics of rigid bodies and systems. The stability theory of elastomechanics was developed independently. However, there have been a number of important developments in recent years dealing with the following problems. The concept of the "follower forces," nonconservative loads, respectively, has been introduced in aeroelasticity. A number of problems in elastokinetics that involve pulsating loads or periodically varying parameters have led to new stability questions. So-called "kinetic" methods have become necessary in elastomechanics in order to determine the stability boundaries. An evaluation of the generally valid stability criteria of elastostatics has shown that they can only be applied to a limited number of problems under special assumptions. The transition from stability to instability is a kinetic process in elastostatics. The most general and most certain method of determining stability is the *kinetic stability criterion* even in special cases in which the classical stability criteria of elastostatics hold. This will be discussed in detail in Section 2.3.

On this basis, it is possible to go beyond the original framework of the theory and treat stability problems of kinetics of rigid bodies and systems as well as of elastomechanics using a single, unified theory. One of the primary purposes of the author was to show the *common traits* of stability problems which occur in all branches of mechanics. The stability problems of hydromechanics and aeromechanics were omitted, because they are discussed in separate lectures, outside of technical mechanics lectures in our schools. The interested and knowledgable reader may readily apply the methods for determining stability to stability problems of continuum mechanics of liquid and gaseous media.

Stability theory is primarily expressed in mathematical terms. For purposes of brevity, it was assumed that the reader is familiar with these mathematical methods. There are several references which can provide the reader with the requisite mathematical knowledge. The mathematics have been discussed in detail if they represent new and modern developments. Thus, for example, in the section on mathematical approximation methods used to solve non-self-adjoint or nonlinear problems, the Galerkin method was pointed out and discussed.

The Einstein summation convention was used in most cases, unless this was not possible or not appropriate. The usual subscript notation was used in vector and tensor calculus.

I would like to express particular appreciation to Professor Dr. K. Magnus, Stuttgart, who inspired me to write the book and who aided me with many valuable suggestions. I am indebted to my assistant, Mr. Dipl.-Ing. H. J. Hoffmann for his critical remarks. Dipl. Math. W. Hauger made the drawings and performed a number of necessary calculations. Mrs. L. Döttinger and Miss G. Walter worked with great dedication in completing the manuscript.

Notation

A	Coefficient matrix of the variational equation system	p	Load parameter
A	Work	p_σ	Multiplicity of the eigenvalue λ_σ
a_{ij}	Element of **A**	$p_i{}^0, q_i{}^0$	Generalized coordinates of the unperturbed motion
B	Characteristic matrix, damping matrix	$p_i{}^\mathfrak{S}, q_i{}^\mathfrak{S}$	Generalized coordinates of the perturbed motion
$D[x]$	Differential operator	$p_{ik}(t)$	Element of the time-dependent coefficient matrix of the variational equations
D	Angular momentum vector		
d_σ	Decrease in rank belonging to the eigenvalue λ_σ		
E	Unit matrix	Q	Force, shearing force
E	Modulus of elasticity, total energy	$q_{i,s}$	Coordinates of the stationary point
$f_i(t)$	Time-dependent part of the ith displacement eigenfunction	q	Line load
		$R[x]$	Boundary-value operator
g	Gravity acceleration	ds	Vector line element
$H, \mathfrak{H}$	Hamiltonian	T	Kinetic energy
H_i	ith Hurwitz determinant	t	Time
H	Hurwitz matrix	U	Potential energy
h	Position parameter	u	Displacement
$i = \sqrt{-1}$	Imaginary unit	u_{ij}	Sensitivity coefficient
J	Moment of inertia	V	Lyapunov function
K	Force	V_a, V_i	Potential energy of the external and internal forces, respectively
k_a	Density of the external forces		
		dV	Volume element
$L[x]$	Linear operator	**v**	Velocity vector
L	Lagrangian	v	Velocity
l	Length	W	Wronskian determinant
M	Torque vector	w	Displacement
m	Mass	$x_i{}^0, y_i{}^0$	Coordinates of the unperturbed motion
$N[x]$	Nonlinear operator		
n_k	kth component of the normal vector	$x_i{}^\mathfrak{S}, y_i{}^\mathfrak{S}$	Coordinates of the perturbed motion
P	Force	X_i	Volume force

α	Bending stiffness, parameter which specifies the ratio of the deflection angle between the compressive load and the tip tangent for a compression column
$\alpha_i{}^0,\ \alpha_i{}^{\ominus}$	Parameters values for unperturbed and perturbed motion, respectively
β_i	Change of the parameter due to a perturbation
γ	Characteristic number
$\gamma_x,\ \gamma_y,\ \gamma_z$	Direction cosines
δ_{ij}	Kronecker delta
ϵ	Measure for the perturbation of the state variable, small difference
η	Measure for perturbation of the initial conditions, reference load
Θ	Moment of inertia
κ	Characteristic number
λ^*	Lamé constant
λ_σ	σth eigenvalue
$\lambda_{\mathbf{A}}$	Eigenvalue of the matrix $\mathbf{A}$
μ^*	Lamé constant
μ	Parameter of the perturbation method, mass density, characteristic number
ξ	Reference eigenvalue
ξ_i	Variation of the ith state variable
$\xi_i^{(k)}$	ith component of the kth eigenvector
$\Pi_i,\ \Pi_a,\ \Pi_g$	Potential energy of the internal forces, external forces, and ground state, respectively
ρ	Characteristic multiplier, mass density
ρ_i	Fourier coefficient
σ_{ij}	Stress tensor element
τ	Time
φ	Angle
$\varphi_i(x)$	Position-dependent component of the ith displacement eigenfunction
ω	Characteristic exponent, circular frequency, time duration of an oscillation period

PART
1
___

Fundamentals

1.1. Basic Concepts and Definitions

There is no absolute definition of stability. It has continuously evolved during the course of history, and has always been adjusted to the special requirements of particular problems. The multilateral meanings of the stability concept must be studied in detail, as well as the various methods of determining stability, before stability theory proper in mechanics may be examined. Previous studies on the subject are either monographs devoted to questions of stability in a more or less restricted region of mechanics (*1*), or they comprise chapters of special books dealing with particular problems in mechanics (*2*), so that in this case the concept of stability is studied with respect to the particular problem in question. The majority of the cited references contain only *one* definition of stability as well as more or less extensive generalizations of it. If the different definitions included in these works are compared, the features they have in common are not always readily apparent.

An attempt will be made in this book to present the features that the various stability definitions have in common. We shall trace the bifurcation of the stability concept and the extent to which the various nuances of the concept have penetrated into the different branches of mechanics. In this way we may clarify the relations between the various points of view, as well as the manner in which they are interwoven, and may thus arrive at a mutual stimulation.

1.1.1. *Historical Development of the Stability Concept*

Based on an article by Magnus,[1] it may be established that stability studies have their first beginnings in history in the works of Aristotle and

---

[1] Magnus, K., Development of the stability concept in mechanics. *Naturwissenschaften* **46**, 590–595 (1959).

Archimedes. Two different points of view were expressed by these two men. The former author investigated the motion occurring after a perturbation, and he determined the stability of the unperturbed state from the *course of the motion*. In the latter case, the *purely geometric situation* occurring after the perturbation of a system was used to determine the stability of the unperturbed system. There are therefore two methods. The first is called the kinematic method. Beginning with Aristotle, it was used up until the period of Galileo. Except for astronomy, which used it continuously to investigate the stability of motion, it has again increased in importance with respect to the stability of motion of technical systems. The second method is called the geometric method. It is connected with the names of Archimedes, Torricelli, Baldi, and Lagrange and dominated the field of mechanics for an extended period of time, from about the 17th to the 19th century greatly overshadowing the first method. As Magnus pointed out, this was of little benefit to the general storehouse of knowledge. The energy method represents a third method. Energy criteria are used here to determine the stability of an equilibrium position. This method, which is of great importance in the theory of elastic bodies (the numerous results obtained by S. Timoshenko when using this method need only be recalled), also contributed to the development of a stability theory which is almost complete but has only limited regions of applicability. However, it must be pointed out that E. J. Routh made an attempt to generalize this method. He tried to extend the energy criteria to an investigation of the stability of states of motion. This was the first attempt to establish a unified stability theory for all branches of mechanics. However, it was not completely successful. Even today the various methods exist separately, and independent theories of stability have been formulated, depending on whether dynamics is being studied—in which the kinematic method is predominant—or whether the theory of elasticity and statics is being studied, where the geometric–static and energy methods are used in most cases. At the present time, the foundations of the latter are being subjected to a critical examination, which will be discussed at a later point.

Just as the methods have led to a division between the theories, the various stability definitions have contributed to the division. Even if the multiple meanings of the stability concept in everyday usage are disregarded (durability, constancy, steadiness, immobility, etc.), numerous possibilities of interpretation remain even if the precise definitions of technical physics are used. It must first be stressed that the concept is relative and depends on the reference system employed. Inexact formulations can lead to a great deal of confusion. It is also very important to determine which characteristics are used for determining the stability of a system. Some well-known examples of this are given by the definition of Lagrange, who states that the trajectory of a planet is stable if its major axis remains bounded in spite of perturbations. On the other hand, we have the definition of Poisson stipulating that a planet is stable when it repeatedly passes arbitrarily close to every point of its trajectory during the further course of its motion. These represent two different opinions regarding the same subject: stability of a planet! How many more different stability concepts will there be when different problems are considered! The stability concept of Lyapunov has found wide acceptance recently and will be discussed in detail later on. However, in spite of a desire for unification, this concept is not all inclusive, as it is frequently too restricted. Features these concepts have in common, which will be different for each case and must be justified in each individual case, must be sought below the surface. The manner in which this may be done will now be discussed.

1.1.2. *Common Characteristics of Modern Stability Definitions*

Let us first present the common characteristics of all stability definitions. They consist of the following stipulations and concepts: an unperturbed state, whose stability is being studied, is specified. A perturbation is then applied to the unperturbed state, so that it is transformed into a perturbed state. Certain characteristics are emphasized, which we shall call norms, which characterize the states at any desired time. The change in the norms during the transition from the unperturbed state to the perturbed state under the influence of the perturbation is determined. Based on this behavior, a conclusion may be reached regarding the stability of the unperturbed state or its instability. In order to define the change in the norms, a measure for the perturbation is defined, as well as a measure for the norm. The following definition is used: If the perturbation does not exceed this defined measure, the unperturbed state is called stable when the change in the norm caused by the perturbation does not exceed its established measure. Otherwise, the unperturbed state is unstable.

The manner in which a great number of different, special stability concepts emerge from this small number of simple stipulations will now be illustrated:

A certain reference system is established when the unperturbed state is selected. Thus, the relative nature of the stability concept becomes apparent, which was already mentioned above. The perturbed state, as well as what is commonly designated as stability, has a direct relationship to the unperturbed state selected and its reference system.

In addition, which characteristic is chosen as the norm is an important factor in reaching a conclusion regarding stability. There are again differences in the stability concept, depending on the choice of the norm. As a specific example of this, we would like to point to the opinions of Lagrange and Poisson regarding the stability of planet trajectories that do not agree, as was already mentioned in Section 1.1.1. As a more abstract example, we shall select the following as the norm: In a metric space $\mathfrak{R}$ with elements f, g and the metric d, let the point set $\mathfrak{M}\{f_k\}$, $k = 1, 2, 3,...,$ with elements f_k from $\mathfrak{R}$ be given. Let $\mathfrak{M}\{f_k\}$ characterize a state that we shall assume is an unperturbed state. A sphere $\mathfrak{R}_r(f_k) = \{f/d(f, g, f_k) < r\}$ corresponds to each f_k with the center at f_k and the radius r. The sphere $\mathfrak{R}_r$ is the norm, and the radius r is the measure with which the characteristic of the perturbed state—i.e., its distance from the unperturbed state—is determined in terms of the metric of the space. By combining all $\mathfrak{R}_r(f_k)$, we may define a subspace $\mathfrak{U} = \bigcup_k \mathfrak{R}_r(f_k) = \mathfrak{R}_r(f_1) \cup \mathfrak{R}_r(f_2) \cup \cdots$ of $\mathfrak{R}$. Let the perturbed state that is created due to the perturbation $\mathfrak{S}$ be specified by the point set $\mathfrak{N}_\mathfrak{S} = \mathfrak{N}_\mathfrak{S}\{g_k\}$, $k = 1, 2, 3,...,$ where the g_k are also elements of $\mathfrak{R}$. It may be stated that $\mathfrak{M}\{f_k\}$ is stable with respect to any perturbation $\mathfrak{S}$, if—in spite of the influence of $\mathfrak{S}$—the set $\mathfrak{N}_\mathfrak{S}$ always remains in the subspace $\mathfrak{U}$. In this way, a purely geometric–topological stability concept has been obtained.

For example, we may assume that $\mathfrak{R}$ is the three-dimensional Euclidean space R_3, d is the distance in the Euclidean sense, and f, g are points in R_3 with the coordinates $x_i{}^0$, $x_i{}^\mathfrak{S}$, $i = 1, 2, 3$. The unperturbed state of the ball B at the point $P(x_i{}^0)$ (see Fig. 1) is stable when the ball can only reach perturbed positions under the influence of perturbations of given magnitude, i.e., it can reach points $Q(x_i{}^\mathfrak{S})$, which are located inside the sphere $\mathfrak{R}_r(P)$ which have a finite radius r around the point P. Such is the

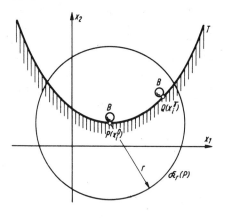

FIG. 1.   Stable position.

case, for example, when the ball is moved away from its equilibrium position P by a finite force along the funnel T, which is directed opposite to the force of gravity. If the disturbing force is not sufficient to displace the ball completely out of the funnel, then every perturbed position Q of the ball will lie within a sphere around P with a finite radius r, i.e., the position P is stable. In this way, we have become acquainted with the concept of equilibrium position stability (because P is the equilibrium position of the ball). This concept is of great importance in kinetics, as well as in statics, and was the primary subject of stability studies from the 17th to the 19th century. During this time, the geometric method prevailed.

Because of its importance, we shall mention another example of the geometric–topological point of view. Let $\Re$ be a $2n$-dimensional phase space R_{2n}, and let f, g be the points

$$P_k = P(q_i^0, p_i^0)_k\,, \qquad Q_k = Q(q_i^{\mathfrak{S}}, p_i^{\mathfrak{S}})_k\,, \qquad k = 1, 2, 3,...,$$

of this phase space with the generalized coordinates $(q_i^0, p_i^0)_k$, $(q_i^{\mathfrak{S}}, p_i^{\mathfrak{S}})_k$, $i = 1, 2,..., n$. Let the point set $\mathfrak{M} = \mathfrak{M}\{(q_i^0, p_i^0)_k\}$, $k = 1, 2, 3,...,$ be represented in phase space by the trajectory C_0, and let another point set $\mathfrak{N}_{\mathfrak{S}} = \mathfrak{N}_{\mathfrak{S}}\{(q_i^{\mathfrak{S}}, p_i^{\mathfrak{S}})_k\}$ be specified by the trajectory $C_{\mathfrak{S}}$ (Fig. 2). Let the

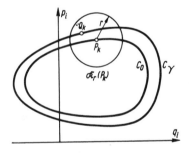

FIG. 2. Trajectories in phase space, an example of trajectory stability.

sequence of unperturbed states be specified by $\mathfrak{M}$ or C_0, respectively, and let the sequence of states created by the perturbation $\mathfrak{S}$ be described by $\mathfrak{N}_{\mathfrak{S}}$ or $C_{\mathfrak{S}}$. By combining the spheres $\Re_r(P_k)$, surrounding the points P_k of C_0, i.e., by means of $\bigcup_k \Re(P_k)$, we may specify the subspace $\mathfrak{U}$ that is in the shape of a "tube" with the radius r and which surrounds the trajectory C_0. It may be stated that the trajectory C_0 is stable when the trajectories $C_{\mathfrak{S}}$ resulting from the perturbation $\mathfrak{S}$ lie entirely within the tube $\mathfrak{U}$ that surrounds C_0, and this tube has a finite radius r when $\mathfrak{S}$ does not exceed a certain amount. In other words, the trajectory C_0 possesses orbital stability when the perturbed trajectories $C_{\mathfrak{S}}$ are suffi-ciently close to the unperturbed trajectory C_0. A simple picture of this orbital stability concept is obtained when the Euclidean space R_3 is again selected for $\mathfrak{R}$, which is a special case: it may be assumed that $q_i \equiv x_i$, $p_i \equiv 0$, $i = 1, 2, 3$, and that the trajectories are the orbits of point masses in motion. Thus, a rocket having an initial velocity which is very close to escape velocity of 11.3 km/sec with respect to an earth-fixed reference system has orbital instability, because a very small perturbation could make the rocket escape the gravitational field of the earth and continue to increase its distance from the earth. In this case, a hyperbolic trajectory would result that would deviate from the unperturbed, elliptical trajectory by an arbitrary amount. Therefore, the condition of orbital stability would be violated.

We can readily see the extent to which the norm selection influences the stability concept by abandoning the choice of the sphere $\mathfrak{R}_r(P_k)$ as the norm. We shall now assume that the points of the sets $\mathfrak{M}$ and $\mathfrak{N}_\mathfrak{S}$ (they are represented as trajectories C_0 and $C_\mathfrak{S}$ in phase space R_{2n} in Fig. 3) are identified by the value of a parameter t. Two image

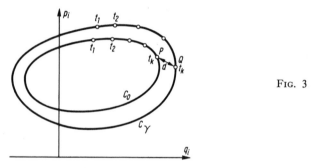

Fig. 3

points may be visualized, running along C_0 and $C_\mathfrak{S}$ when this parameter is continuously varied. The motion of these points indicates the change in the unperturbed and perturbed states. For example, if the parameter t is selected as the time, the changes in the states with time are obtained. This time we shall select the distance of the image points from each other as the norm, in the sense of the space metric, and this is the distance $d = d[P(t_k), Q(t_k)]$ of two points P and Q of the trajectories C_0 and $C_\mathfrak{S}$, which correspond to the same parameter value, i.e., the norm is the distance of two geometric points which are occupied by the image points at the "same time." For stability it is necessary that $d[P(t_k), Q(t_k)] < r$ for each value t_k of the parameter t, i.e., it is necessary that "at no time" can the image points on C_0 and $C_\mathfrak{S}$ be too far from each other—that is, no farther than the measure r. Two features of this norm selection become readily apparent. In the first place, we have passed from the geometric–topological concept of stability to the kinematic stability concept. The kinematic nature of this idea becomes very apparent from the motion of the image points in phase space. In the second place, the new stability requirement is stricter because it is possible that a process having orbital stability may no longer be stable according to the new definition. For example, let us consider two point masses m_0 and m_Ξ that move along concentric circles with a somewhat different period. If the differences in the circle radii are small enough, the geometric trajectories of the two points in motion will be close to each other and orbital stability is achieved. Owing to the difference in the period, which can be very small, the points will be out of phase repeatedly. In spite of the assumption that there be a small distance between the points at the beginning of motion, there will be an unacceptably large distance between them repeatedly, so that stability does not exist in the kinematic sense (Fig. 4).

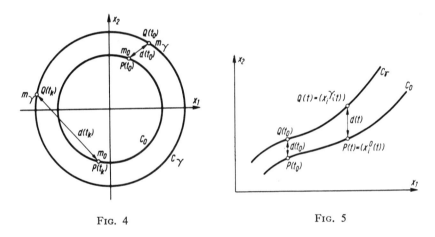

FIG. 4   FIG. 5

FIG. 4. Trajectories in phase space, an example of trajectory stability and instability according to Lyapunov.

FIG. 5. Trajectories in Euclidean space, an example of the stability definition of Lyapunov.

If this is to be illustrated graphically, we may substitute the Euclidean space R_3 for the abstract phase space R_{2n} (Fig. 5), which was done in the example given. The states are then motions, and the trajectories are the orbits. The image points moving in time coincide with the geometric points $P(t)$ (along the trajectory C_0 of the unperturbed motion) and $Q(t)$ (along the trajectory C_Ξ of the perturbed motion) which also change with time. Since we are dealing with Euclidean space, we may also regard P and Q as the totality of their coordinates, and we may set $P(t) = [x_i{}^0(t)]$, $Q(t) = [x_i{}^\mathfrak{S}(t)]$, $i = 1, 2, 3$. Instead of the stipulation $d[P(t), Q(t)] < r$, which is valid for any t, we may formulate the coordinate condition $|x_i{}^0(t) - x_i{}^\mathfrak{S}(t)| < r^*$ for any i, every t, and a suitable r^*, in the sense of the Euclidean metric. We therefore have a condition that is frequently encountered and is typical of the stability of motion.

We would like to draw particular attention to the case in which the trajectory C_0 representing the unperturbed state may be reduced to a single point P which coincides with one of the two image points at *every* time t. The point then represents an *equilibrium position* (Fig. 6).

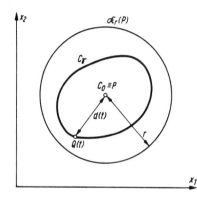

Fig. 6.   Stable equilibrium position.

If this equilibrium position is to be stable, all perturbed states represented by $C_\mathfrak{S}$ must satisfy the condition $d[P, Q(t)] < r$ at all times, i.e., the other image point describing the perturbed states must move along points $Q(t)$ in such a way that $C_\mathfrak{S}$ is always "in the vicinity" of P. However, not only the kinematic condition but also the geometric condition are simultaneously satisfied because all points of $C_\mathfrak{S}$ are located within the sphere $\mathfrak{R}_r(P)$. This indicates that, when an equilibrium condition is stable, it is stable in the kinematic and geometric sense. Consequently, we shall therefore regard an equilibrium position as a special case of motion, and shall apply the kinematic representation of stability, which is not an important restriction.

Further differences in the treatment of stability problems arise when various *types of perturbations* are considered. The perturbations can be instantaneous, periodic, and time dependent, and can differ in many ways in addition to their time dependence. This is most readily apparent if the mathematical representation is considered. Let us assume that a perturbed or unperturbed state can be described by a differential equation or a system of differential equations. It will customarily be assumed below that everything has been reduced to a system of first-order differential equations. The process is described in a multi-dimensional space R_n by means of the generalized coordinates q_i, $i = 1, 2,..., n$, so that the unperturbed state $q_i{}^0$ is a particular solution of

$$dq_i/dt = \dot{q}_i = F_i(q_1, q_2,..., q_n, \alpha_1{}^0, \alpha_2{}^0,..., \alpha_m{}^0, t) \equiv F_i{}^0, \qquad i = 1, 2,..., n,$$
(1.1.1)

and therefore satisfies the relationship

$$dq_i{}^0/dt \equiv \dot{q}_i{}^0 = F_i(q_1{}^0, q_2{}^0,..., q_n{}^0, \alpha_1{}^0, \alpha_2{}^0,..., \alpha_m{}^0, t), \qquad i = 1, 2,..., n.$$
(1.1.1')

The $\alpha_r{}^0$ and t are parameters (for example, let t be the time), and the particular solutions $q_i{}^0$ of (1.1.1) are coordinates of the point $P_k = (q_i{}^0)_k$, which represent the unperturbed set $\mathfrak{M}$ of the metric space R_n. Let the perturbed states by specified by the points $Q_k = (q_i{}^\mathfrak{S})_k$ of the set $\mathfrak{M}_\mathfrak{S}$. The coordinates $q_i{}^\mathfrak{S}$ of these points are either due to different initial conditions in solving (1.1.1) or due to changes from the parameters $\alpha_r{}^0$ to $\alpha_r{}^\mathfrak{S}$ of the functions F_i. They may also be due to the assumption of altered right sides in (1.1.1). The latter case—i.e., a change in the right sides of (1.1.1)—could, for example, be caused by the occurrence of perturbation functions $G_i(q_1, q_2,..., q_n, \alpha_1{}^\mathfrak{S}, \alpha_2{}^\mathfrak{S},..., \alpha_m{}^\mathfrak{S}, t)$.

We have thus presented three important types of perturbations: perturbations of the *initial conditions*, of the *parameters*, and of the *differential equations* (i.e., of the physical law that is described by these differential equations). We should consider parameter perturbations as a special case of perturbation of the differential equations. This case is of such great practical importance and has so many unique features that a separate study of it is justified. It frequently coincides with the case of perturbed differential equations when the parameter changes are so extensive that they lead to a qualitative change in the differential equations.

The most frequently studied case is that of *perturbation of the initial conditions*. Together with the kinematic representation of stability, this leads to the stability definition of Lyapunov, which was already mentioned. This definition is as follows: Let the unperturbed state be specified by $q_i{}^0(t)$, the perturbed state by $q_i{}^\mathfrak{S}(t)$, let $q_i{}^0(t_0)$, $q_i{}^\mathfrak{S}(t_0)$ be their initial conditions, and let ϵ, $\eta(\epsilon)$ be two arbitrarily small positive numbers. We assume that the perturbation of the initial conditions is small enough that

$$d[q_i{}^0(t_0), q_i{}^\mathfrak{S}(t_0)] \leqslant \eta$$

holds in the sense of the applicable metric. The sequence of unperturbed states is designated as stable when the condition

$$d[q_i{}^0(t), q_i{}^\mathfrak{S}(t)] < \epsilon \qquad \text{for} \quad t > t_0$$

is satisfied. The metric can be explicitly formulated for Euclidean space, and a special definition may therefore be employed: Stability as defined by Lyapunov occurs when, for

$$| q_i^0(t_0) - q_i^{\mathfrak{S}}(t_0)| \leqslant \eta(\epsilon)$$

$$| q_i^0(t) - q_i^{\mathfrak{S}}(t)| < \epsilon, \qquad t > t_0$$

(1.1.2)

always holds.

The case of *parameter perturbation* is of great importance. *Structure stability* is referred to when stability with respect to the variation of parameters is being considered. Parameter values making the perturbed states differ from the unperturbed states to a very great extent, and even qualitatively, are called *branch values* of the parameters. An entire theory for determining branch values has been developed that in essence is due to Poincaré.

Very frequently in practice it is necessary to consider perturbations caused by changes in initial conditions as well as parameter value changes. We shall discuss examples of this type at a later point.

The case of the continuous perturbation should also be mentioned, which mathematically corresponds to the addition of a *perturbation function* G_i to the right side of (1.1.1). Recently, the special case in which G_i is a statistical function has been studied in great detail. This leads to the concept of *statistical stability*, which is quite important.

In addition to the three types of perturbations mentioned, there are many others which influence and change the stability concept. Special conditions can also be applied to the perturbations, which then leads to the concept of *conditional stability*. We shall not discuss this subject in detail. We shall simply indicate the train of thought: The conditions imposed on the stability can, for example, imply restrictions regarding the *magnitude of the perturbation* and the continuous *dependence* of the state characteristics on the size of the perturbation. In the case of the Lyapunov stability definition, the latter statement would mean that η is a continuous function of ϵ. The continuous dependence condition was studied in particular by Klein.[1]

Let us consider Fig. 7 in which this is illustrated visually by considering the motion of a small ball. As can be seen from the figures, it is possible that the position P' in Fig. 7a is unstable with respect to small perturbations, because the ball passes immediately from the perturbed position Q' to the position P. Larger perturbations cause the ball to pass from the

---

[1] KLEIN, F., and SOMMERFELD, A., "The Theory of the Gyroscope," pp. 342–374. Johnson Reprint Corp. (Teubner), New York and Stuttgart, 1965.

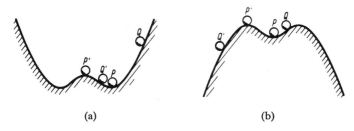

FIG. 7. (a) Stability in the large, instability in the small—"stability for practical purposes." (b) Stability in the small, instability in the large—"instability for practical purposes."

perturbed position Q to the stable position P. It may then be stated that the position P' is unstable *in the small*, and is stable *in the large*. The reverse of the situation is shown in Fig. 7b. There the position P is stable in the small, but is unstable in the large. In addition, it may be readily seen that in the case of Fig. 7a there is no continuous dependence of the perturbed state characteristics on the perturbation size. This is due to the fact that, no matter how small the perturbation may be, the ball will always jump from position Q' to position P. If the perturbation making the ball assume position Q is reduced and approaches zero, the ball will not return to the initial position P' in a continuous fashion but will always assume position P. Therefore, no stability exists in terms of Klein's definition. However, in the case of Fig. 7a, it is usually assumed that the position P' is stable for *practical purposes*, if P is sufficiently close to P'. The reverse is true in the case of Fig. 7b. In this case, position P is stable in terms of Klein's definition only for sufficiently small perturbations. It must be regarded as unstable for practical purposes if P' is close to P. The concepts of stability in the small and in the large may be clearly visualized when the corresponding limit cycles of an oscillator phase diagram are studied. This will be discussed in Section 1.4.2. Other definitions of stability for practical purposes lead to other differences. One of these is stability in a *finite time interval*. Even though stability over an infinite time interval is usually of interest, the requirement of an open time interval can frequently be disregarded. In order to surmount mathematical difficulties, stability for a finite time interval only may be considered. The Lyapunov definition modified in this way would result in the following stability condition:

$$| q_i^0(t) - q_i^{\mathfrak{S}}(t)| < k\eta$$

for

$$| q_i^0(t_0) - q_i^{\mathfrak{S}}(t_0)| \leqslant \eta \qquad \text{and} \qquad t_0 \leqslant t \leqslant t_0 + \tau.$$

The quantity k is a constant and τ is the time interval selected in such a way that stability, in terms of the selected definition, prevails over the entire interval τ during the changes in the corresponding states. By way of an example, we would like to mention the studies of de Bra[1] regarding the position control of satellites: The motion of satellites is considered to be stable when certain motion parameters remain within certain bounds during 20 revolutions of a satellite. This condition is sufficient to guarantee the desired stable performance over the predictable lifetime of the satellite.

Finally, we would like to mention the fact that a special stability concept results when the norm measure depends on time. In this case, reference is made to *asymptotic* stability, when the measure strives to zero as the time is increased. For example, for a purely geometric representation, let us assume that the condition $r \to 0$ for $t \to \infty$ holds for the sphere radius $\mathfrak{R}_r(P_k)$. Asymptotic orbital stability then prevails, because the trajectories $C_\mathfrak{S}$ of the perturbed states come as close as desired in time to the trajectory C_0 of the unperturbed states. For the kinematic representation, $d(t) \to 0$ for $t \to \infty$ must be satisfied for asymptotic stability, or in the case of the Lyapunov definition we must have

$$| q_i^0(t) - q_i^\mathfrak{S}(t)| \to 0 \qquad \text{for} \quad t \to \infty. \tag{1.1.3}$$

We have thus presented a general description of the common foundations and variations of the stability concept. We should like to mention the fact that we shall restrict ourselves below to initial value perturbations and parameter perturbations, because they are the most important cases in practice. In addition, we shall adopt the stability definition of Lyapunov, unless we indicate the opposite.

We shall now discuss the question of obtaining *criteria* which make it possible to estimate whether and when the stability conditions are satisfied. For this purpose, let us assume that the *unperturbed state* q_i^0 is specified as a particular solution of (1.1.1) by means of the condition

$$\dot{q}_i^0 = F_i(q_1^0, q_2^0, ..., q_n^0, \alpha_1^0, \alpha_2^0, ..., \alpha_m^0, t) \equiv F_i^0(q_k^0, \alpha_r^0, t).$$

The *perturbed state* is correspondingly specified by

$$\dot{q}_i^\mathfrak{S} = F_i(q_1^\mathfrak{S}, q_2^\mathfrak{S}, ..., q_n^\mathfrak{S}, \alpha_1^\mathfrak{S}, \alpha_2^\mathfrak{S}, ..., \alpha_m^\mathfrak{S}, t) \equiv F_i^\mathfrak{S}(q_k^\mathfrak{S}, \alpha_r^\mathfrak{S}, t). \tag{1.1.4}$$

[1] DE BRA, D. B., The large attitude motions and stability due to gravity of a satellite, etc. Stanford University, California, Dept. of Aeronautics and Astronautics, SUDAER No. 126 (1962).

We may set $q_i^{\circledS}(t) = q_i^0(t) + \xi_i(t)$ and designate $\xi_i(t)$ as the *variation* of $q_i^0(t)$. The following system of differential equations holds for the variations

$$\dot{\xi}_i(t) = F_i^{\circledS}(q_k^{\circledS}, \alpha_r^{\circledS}, t) - F_i^0(q_k^0, \alpha_r^0, t)$$

or

$$\dot{\xi}_i(t) = \Phi_i(\xi_1, \xi_2, ..., \xi_n, \beta_1, \beta_2, ..., \beta_m, t), \qquad (1.1.5)$$

where $\xi_i = q_i^{\circledS} - q_i^0$, $\beta_k = \alpha_k^{\circledS} - \alpha_k^0$. The *variational equations* (1.1.5) possess the trivial solution $\xi_1 = \xi_2 = \cdots = \xi_n = 0$ for $\beta_i = 0$, which represents the origin of the *n*-dimensional space. Since they simultaneously satisfy $\dot{\xi}_i(t) = 0$, the origin is a stationary solution, an equilibrium position. The stability definition of Lyapunov can be expressed as follows by employing the variations

$$|\xi_i(t)| < \epsilon \qquad \text{for} \quad t > t_0,$$
$$|\xi_i(t_0)| < \eta(\epsilon). \qquad (1.1.6)$$

Thus, everything has been reduced to investigating stability of the equilibrium position $\xi_i = 0$ for the case $\beta_i = 0$. In geometric terms this means that for stability the trajectory $C_{\circledS}$ described by $\xi_i(t)$ must remain in the vicinity of the origin, or even approaches the origin if the condition $|\xi_i(t)| \to 0$ for $t \to \infty$ is satisfied, i.e., in the case of asymptotic stability.

If a description of the *unperturbed state stability* is desired, the *time dependence of the variation* must be known or must be estimable. It is most natural to integrate the system (1.1.5) for the initial conditions $\xi_i(t_0) = q_i^{\circledS}(t_0) - q_i^0(t_0)$. The data necessary for determining the existence of stability are then immediately available.

In general, however, the integration of a system of differential equations will be avoided. The following possibilities then remain:

If the system of fundamental solutions is known, as is the case for linear differential equations, and if it is known which characteristic parameters influence the stability behavior of the solution, it is then possible to restrict the stability investigation to determining the course of these *characteristic parameters* and to obtain stability criteria in this way.

In other cases, *intermediate integrals* are used. A system of *n* differential equations of the first order for the $n + 1$ variables ξ_i, $i = 1, 2, ..., n$ and t possesses n independent first intermediate integrals

$$\Psi_i(\xi_k, t) = C_i, \qquad i, k = 1, 2, ..., n.$$

All n first integrals must be known in order to obtain the complete solution of the system of differential equations. It is frequently simpler to specify one or more intermediate integrals instead of completely solving the system, i.e., instead of integrating it. These intermediate integrals represent surfaces in the space R_{n+1} of the variables ξ_i, t. These surfaces represent geometric loci in phase space along which the proper solution curve (ξ_i, t) must lie. The behavior of the variation ξ_i and therefore the stability of the stationary, unperturbed solution follows from this more or less complete localization of the solution by means of the intermediate integral surfaces. For this purpose, the intermediate integral surfaces are often projected onto coordinate planes of the phase space, resulting in the well-known *phase curves* that—if they are multi-dimensional—can again be hypersurfaces. The use of energy integrals of conservative, dynamic systems represents a good example of employing intermediate integrals, from which the *energy criteria* of stability theory follow.

Example: Let the motion of a point mass be given by $\ddot{x} = -F(x)$. We set $x = \xi_1$, $\dot{x} = \xi_2$ and obtain the first-order system

$$\dot{\xi}_1 = \xi_2,$$
$$\dot{\xi}_2 = -F(\xi_1),$$

which can also be represented as the simultaneous system

$$\frac{d\xi_1}{\xi_2} = \frac{d\xi_2}{-F(\xi_1)} = \frac{dt}{1}.$$

A first integral is obtained from

$$\frac{d\xi_1}{\xi_2} = \frac{d\xi_2}{-F(\xi_1)},$$

because by integration it follows that

$$\tfrac{1}{2}\xi_2^2 + \int F(\xi_1)\, d\xi_1 = C_1.$$

Using the abbreviation

$$U(\xi_1) = \int F(\xi_1)\, d\xi_1,$$

we thus have

$$\Psi_1(\xi_1, \xi_2) = \tfrac{1}{2}\xi_2^2 + U(\xi_1) = C_1,$$

which represents a cylinder with its generator parallel to the t axis of the phase space (ξ_1, ξ_2, t). By projecting this cylinder onto the ξ_1, ξ_2 plane, one obtains the phase curve

$$\tfrac{1}{2}\xi_2{}^2 + U(\xi_1) = C_1 .$$

As can be readily seen, this is a geometric image of the energy integral, because $\tfrac{1}{2}\xi_2{}^2$ is the kinetic and $U(\xi_1)$ is the potential energy of the point mass.

If the phase curve is simply closed and surrounds the equilibrium position $\xi_1 = \xi_2 = 0$ then the *stability of the equilibrium position* is assured, because the phase curve surrounds the region G of the ξ_1, ξ_2 plane in which the total energy of motion is smaller than C_1. Any motion of the point mass with a total energy $C_2 < C_1$ will progress in such a way that its phase curve will be inside G. The magnitudes ξ_1, ξ_2 remain bounded for all times, so that stability prevails. Figure 8 shows

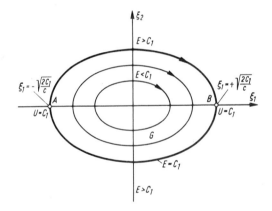

FIG. 8.   Elliptical phase curves

the relationships for $F(x) = cx$ for a point mass oscillating about the position $x = 0$. The phase curves are the ellipses $\tfrac{1}{2}(\xi_2{}^2 + c\xi_1{}^2) = C_1$, and we have $U(\xi_1) = c/2 \cdot \xi_1{}^2$.

If we set $\tfrac{1}{2}\xi_2{}^2 + U(\xi_1) = E$, where E is the total energy, then region G in which $E < C_1$ holds is surrounded by the ellipse $\tfrac{1}{2}(\xi_2{}^2 + c\xi_1{}^2) = C_1$, along which $E = C_1$ holds.

A somewhat different question is raised when the *stability of the geometric trajectory* is examined.

Example: Let us consider the motion of a point mass in the x, y plane under the influence of conservative forces. The equations of motion are

$$\ddot{x} = -\partial U(x, y)/\partial x, \qquad \ddot{y} = -\partial U(x, y)/\partial y.$$

By introducing new variables

$$\xi_1 = x, \qquad \xi_2 = y, \qquad \xi_3 = \dot{x}, \qquad \xi_4 = \dot{y},$$

we obtain the system of simultaneous differential equations

$$\frac{d\xi_1}{\xi_3} = \frac{d\xi_2}{\xi_4} = \frac{d\xi_3}{-L} = \frac{d\xi_4}{-M} = \frac{dt}{1},$$

with

$$L = \frac{\partial U}{\partial x} \equiv \frac{\partial U(\xi_1, \xi_2)}{\partial \xi_1} \qquad \text{and} \qquad M = \frac{\partial U}{\partial y} \equiv \frac{\partial U(\xi_1, \xi_2)}{\partial \xi_2}.$$

From

$$\frac{d\xi_1}{\xi_3} = \frac{d\xi_3}{-L}, \qquad \frac{d\xi_2}{\xi_4} = \frac{d\xi_4}{-M}$$

a quadrature results in

$$\tfrac{1}{2}\xi_3{}^2 + \int L \, d\xi_1 = C_1, \qquad \tfrac{1}{2}\xi_4{}^2 + \int M \, d\xi_2 = C_2.$$

The following result is obtained by addition and by taking into account the meaning of L and M:

$$\tfrac{1}{2}(\xi_3{}^2 + \xi_4{}^2) + \int \left[\frac{\partial U(\xi_1, \xi_2)}{\partial \xi_1} \, d\xi_1 + \frac{\partial U(\xi_1, \xi_2)}{\partial \xi_2} \, d\xi_2 \right] = C_1 + C_2.$$

If we set

$$\int \left[\frac{\partial U}{\partial \xi_1} \, d\xi_1 + \frac{\partial U}{\partial \xi_2} \, d\xi_2 \right] = \int dU = U(\xi_1, \xi_2)$$

and $C_1 + C_2 = C$, we have the intermediate integral

$$\tfrac{1}{2}(\xi_3{}^2 + \xi_4{}^2) + U(\xi_1, \xi_2) = C,$$

which is the energy theorem at the same time, because $\tfrac{1}{2}(\xi_3{}^2 + \xi_4{}^2)$ is the kinetic and $U(\xi_1, \xi_2)$ is the potential energy of motion.

Since $\tfrac{1}{2}(\xi_3{}^2 + \xi_4{}^2) \geqq 0$, the relationship $U \leqq C$ must always hold for motions having the total energy $E \leqq C$. Let us assume that the simple closed, continuous curve $U(\xi_1, \xi_2) = C$ delimits a bounded region $U \leqq C$ in the ξ_1, ξ_2 plane. Motions that originate in this region and whose total energies satisfy $E = C$ or $E < C$, also satisfy the relationship $U \leqq C$. Consequently, their orbits must remain entirely in this region, because they could only cross its boundaries for $U > C$, which is not possible. Therefore, these motions are stable. The use of this "energy criterion" for the determination of orbital stability is common practice in celestial mechanics.

If the criterion is applied to the oscillating point mass considered previously, we then have $U(\xi_1) = C_1$, from which it follows that $\xi_1 = \pm(2C_1/c)^{1/2}$. The points A and B along the ξ_1 axis in Fig. 8 are thus specified. The ξ_1, ξ_2 plane of the preceding example has now degenerated into the ξ_1 axis, and the region bounded by $U(\xi_1, \xi_2) = C$ has degenerated into the closed interval $[A, B]$ of the ξ_1 axis. In any case, it follows that the oscillator motion is stable about the origin, because for $E \leqslant C_1$ the oscillation amplitude will be smaller than $(2C_1/c)^{1/2}$.

The following holds for the intermediate integral $\Psi_i(\xi_k, t) = C_i{}^1$:

$$\frac{d\Psi_i}{dt} = \frac{\partial \Psi_i}{\partial t} + \frac{\partial \Psi_i}{\partial \xi_k}\frac{d\xi_k}{dt} = 0.$$

This means that the solution curves corresponding to the integration constants C_i lie along the surface $\Psi_i = C_i$. We may also express this relationship in a geometric manner as follows:

$$d\Psi_i/dt = \operatorname{grad} \Psi_i(ds/dt) = 0$$

using the vectors

$$\operatorname{grad} \Psi_i = (\partial \Psi_i/\partial t, \partial \Psi_i/\partial \xi_k)$$

and

$$ds = (dt, d\xi_k).$$

Since the solution curve remains on the integral surface Ψ_i, the vector ds indicating the advance of an image point along the solution curve is perpendicular to the normal to Ψ_i, which is parallel to the vector $\operatorname{grad} \Psi_i$. We therefore have $\operatorname{grad} \Psi_i \cdot ds = 0$.

The purpose of identifying the solution curve in R_{n+1} space can be generalized in the following way. Instead of finding intermediate integrals and therefore integral surfaces of the corresponding differential equation, we may select *arbitrary* but suitable hypersurfaces $V_i(\xi_k, t) = C_i$, which divide the R_{n+1} space in such a way that the solution curves intersect them when they pass from one region to another. The course of the solution curves may also be obtained by considering the penetration points. If this generalized procedure is employed, i.e., if the solution curves are made to penetrate arbitrary

---

[1] We shall employ the *Einstein summation convention* from this point on.

families of surfaces $V_i = C_i$, then $d\mathbf{s}$ will no longer be perpendicular to grad V_i, so that we have

$$dV_i/dt = \text{grad } V_i(d\mathbf{s}/dt) \neq 0.$$

The use of such *test surfaces* may be traced back to Lyapunov. Therefore, the functions $V_i(\xi_k, t) = C_i$ are called *Lyapunov functions*. The intermediate integrals $\Psi_i = C_i$ can be regarded as special cases of Lyapunov functions for which we have $dV_i/dt = 0$.

If the $V_i = C_i$ are closed surfaces around the origin of R_{n+1} which enclose themselves for increasing C_i and if

$$dV_i/dt = \text{grad } V_i(d\mathbf{s}/dt) \equiv dC_i/dt < 0$$

always holds, the conclusion must be reached that the solution curves intersect the family of surfaces in the direction of *decreasing* values of the parameter C_i. They always are closer to the origin as t increases. Asymptotic stability may then exist.

Example: Let us consider the damped motion of an oscillating point mass, for which the system of differential equations is $\dot{x} = v$, $\dot{v} = -cx + D(x, v)$. We shall again employ the notations $x = \xi_1$, $v = \xi_2$, and we obtain the system

$$\dot{\xi}_1 = \xi_2,$$
$$\dot{\xi}_2 = -c\xi_1 + D(\xi_1, \xi_2).$$

Using the ideas of Lyapunov, we neither integrate this system nor do we determine its intermediate integrals. Instead, we select as suitable functions $V_i(\xi_1, \xi_2) = C_i$, the energy integrals of the corresponding conservative system that is obtained for $D \equiv 0$ (Fig. 9). They may be

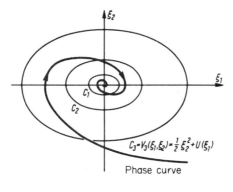

$$C_3 = V_3(\xi_1, \xi_2) = \tfrac{1}{2}\xi_2^2 + U(\xi_1)$$

Phase curve

Fig. 9. Phase curve of a damped oscillator.

written as follows:

$$V_i = \tfrac{1}{2}(\xi_2{}^2 + c\xi^{\,2}) = C_i$$

and represent closed curves (ellipses) in the ξ_1, ξ_2 plane that envelope each other. They satisfy the following equation:

$$dV_i/dt = c\xi_1\dot{\xi}_1 + \xi_2\dot{\xi}_2 = \xi_2 D(\xi_1, \xi_2),$$

as may readily be seen by using the right sides of the differential equation system. The condition $dV_i/dt < 0$, the course of the phase curve shown in Fig. 9, and therefore asymptotic stability of motion is assured if $D(\xi_1, \xi_2)$ and ξ_2 always have opposite sign. The damping D just satisfies this requirement, so that the well-known fact that damped oscillations of point masses are asymptotically stable has been verified.

The preceding statements have shown that there is a tendency toward obtaining stability criteria while trying to avoid the integration of differential equations as much as possible. The use of the Lyapunov functions has represented the biggest step in this direction. The method of Lyapunov functions is somewhat inconvenient and incomplete, as we shall see later. This is due to the fact that the stability criteria obtained with these functions are only sufficient, and there is no rigorous and general method of obtaining and formulating suitable Lyapunov functions.

1.2. SENSITIVITY EQUATIONS AND VARIATIONAL EQUATIONS

We assume an undisturbed state $q_i{}^0$, which satisfies the relationship

$$\dot{q}_i{}^0 = F_i(q_k{}^0, \alpha_r{}^0, t), \qquad i, k = 1, 2,..., n,$$
$$r = 1, 2,..., m, \tag{1.2.1}$$

in a general phase space R_n. The system of first-order differential equations $\dot{q}_i = F_i$ which correspond to a mechanical system and are basic to Eqs. (1.2.1), can, for example, be given by the set of canonical Hamilton differential equations. The $\alpha_r{}^0$ are parameters, and t is the time. Owing to perturbations, which as a special case should also consist of parameter changes, perturbed state $q_i{}^\mathbb{S}$ is obtained. If it is assumed that parameter changes cause $\alpha_r{}^0$ to pass into $\alpha_r{}^\mathbb{S} = \alpha_r{}^0 + \beta_r$, and that the structure of the differential Eqs. (1.2.1) is not changed, then the following holds if we restrict ourselves to terms which are linear in β_r:

$$q_i{}^\mathbb{S} = q_i{}^0(\alpha_r{}^0 + \beta_r, t) + \xi_i = q_i{}^0(\alpha_r{}^0, t) + (\partial q_i{}^0/\partial \alpha_r{}^0)\beta_r + \xi_i. \tag{1.2.2}$$

The sum $(\partial q_i{}^0 / \partial \alpha_r{}^0) \beta_r$ is the change of $q_i{}^0$ caused by parameter changes. For purposes of notation abbreviation, we shall set

$$u_{ir} = \partial q_i{}^0 / \partial \alpha_r{}^0. \tag{1.2.3}$$

It follows from (1.2.2) that

$$\dot{q}_i{}^{\mathbb{S}} = \dot{q}_i{}^0(\alpha_r{}^0, t) + \dot{u}_{ij}\beta_j + \dot{\xi}_i, \tag{1.2.4}$$

where we have used (1.2.3). From

$$\dot{q}_i{}^{\mathbb{S}} = F_i(q_k{}^{\mathbb{S}}, \alpha_r{}^0 + \beta_r, t)$$

together with (1.2.2), it follows that

$$\dot{q}_i{}^{\mathbb{S}} = F_i[q_k{}^0(\alpha_r{}^0, t) + u_{kl}\beta_l + \xi_k, \alpha_r{}^0 + \beta_r, t]. \tag{1.2.5}$$

If this expression is expanded in a Taylor series, the following expression is obtained if all terms containing products of ξ_k, β_r and containing the terms ξ_k, β_r themselves raised to higher powers are neglected:

$$\dot{q}_i{}^{\mathbb{S}} = F_i[q_k{}^0(\alpha_r{}^0, t), \alpha_r{}^0, t]$$
$$+ [(\partial F_i / \partial q_l)(q_k{}^0, \alpha_r{}^0, t)](u_{lj}\beta_j + \xi_l) + (\partial F_i / \partial \alpha_j{}^0)(q_k{}^0, \alpha_r{}^0, t)\beta_j. \tag{1.2.6}$$

Owing to (1.2.4) and (1.2.1), we have, on the other hand,

$$\dot{q}_i{}^{\mathbb{S}} = F_i[q_k{}^0(\alpha_r{}^0, t), \alpha_r{}^0, t] + \dot{u}_{ij}\beta_j + \dot{\xi}_i. \tag{1.2.7}$$

If (1.2.7) and (1.2.6) are compared, it can then be seen that the relationships

$$\dot{u}_{ij} = [(\partial F_i / \partial q_l)(q_k{}^0, \alpha_r{}^0, t)] u_{lj} + (\partial F_i / \partial \alpha_j{}^0)(q_k{}^0, \alpha_r{}^0, t) \tag{1.2.8}$$

and

$$\dot{\xi}_i = (\partial F_i / \partial q_l)(q_k{}^0, \alpha_r{}^0, t)\xi_l \tag{1.2.9}$$

must hold.

It follows from this that, in this degree of approximation, the influence of parameter changes can be treated independently of the changes in the initial conditions. The system of differential Eqs. (1.2.8) for the sensitivity coefficients u_{ir} can be used for this purpose. In general, the initial conditions $u_{ir}(0) = 0$ are appropriate for this system.

On the other hand, if the influence of changes in the initial conditions is of interest, the variations ξ_i must be employed, and the system of differential Eqs. (1.2.9) must be used. In both cases, it is assumed the $q_k{}^0$ are known, because at the beginning of the calculation a decision must be made as to which undisturbed state will be investigated.

System (1.2.9) is known as the system of variational equations. It was introduced in the literature by H. Poincaré. In general, it is nonlinear, because, if terms of higher order are considered in the expansion of (1.2.6) and if changes in the parameters α_k are not considered, the following system is obtained for the variation ξ_i :

$$\dot{\xi}_i = \frac{\partial F_i}{\partial q_l}(q_k{}^0, \alpha_r{}^0, t)\,\xi_l + \cdots + \frac{1}{n!}\frac{\partial^n F_i}{(\partial q_l)^n}\,\xi_l{}^n + \cdots. \qquad (1.2.10)$$

In the following, we shall first consider the linear Eqs. (1.2.9), and shall deal with the nonlinear Eqs. (1.2.10) later on. In this connection, it might be of interest to note that there is a certain connection between (1.2.8) and (1.2.9): as can be seen, (1.2.8) is composed of equation subgroups, which formally coincide with (1.2.9) with the exception of the additional terms $\partial F_i/\partial \alpha_j{}^0$.

1.3. LINEAR VARIATIONAL EQUATIONS

1.3.1. *Linear Variational Equations with Constant Coefficients*

We shall assume that the time t does not appear explicitly in the functions F_i in the right side of (1.2.1). In this case, the process described by $q_i{}^0(t)$ is autonomous. In addition, we select the stationary state as the undisturbed state, for which $\dot{q}_i{}^0 = 0$ holds. We obtain it by solving the algebraic or transcendental system of equations $F_i(q_k{}^0, \alpha_r{}^0) = 0$. It can have one or more solutions $q_{i,s}^0$, which do not depend on the time, but rather represent fixed points in phase space. If we formulate the system of variational equations with respect to this stationary solution, which represents the undisturbed state, according to (1.2.9), we then obtain

$$\dot{\xi}_i = (\partial F_i/\partial q_j)(q_{k,s}^0, \alpha_r{}^0)\xi_j .$$

As an abbreviation, we set

$$(\partial F_i/\partial q_k)(q_{j,s}^0, \alpha_r{}^0) = a_{ik,s}$$

or simply

$$(\partial F_i/\partial q_k)(q_{j,s}^0, \alpha_r{}^0) = \alpha_{ik}$$

because we carried out the calculation for a fixed s, so that the subscript s in the a_{ik} can be suppressed.

We then arrive at the system of first-order differential equations

$$\dot{\xi}_i = a_{ik}\xi_k, \qquad i, k = 1, 2,..., n, \tag{1.3.1}$$

which has constant coefficients a_{ik} .

The structure of the matrix $\mathbf{A} = (a_{ik})$ is decisive for the type of solutions. We shall discuss this, and shall assume that matrix theory and the theory of systems of first order differential equations with constant coefficients are known to the reader. The reader is referred to a book by Zurmühl (3) for a discussion of these topics.

By means of the so-called elementary transformations, the characteristic matrix $\mathbf{B} = (a_{ik} - \lambda \delta_{ik})$,[1] which corresponds to any nth-order matrix $\mathbf{A} = (a_{ik})$, can be converted into the Smith canonical form

$$N = \begin{pmatrix} E_1 & & 0 \\ & \ddots & \\ 0 & & E_n \end{pmatrix}.$$

Except for the sign, or a constant factor, we obtain the relationship

$$\det \mathbf{B} \cong E_1 E_2 \cdots E_n ,$$

where E_ν are the elementary polynomials of the matrix $\mathbf{B}$. Let $\mathbf{A}$ have the $m \leqslant n$ different eigenvalues λ_σ , $\sigma = 1, 2,..., m$, which are obtained as the roots of the characteristic equation

$$\det(a_{ik} - \delta_{ik}\lambda) = 0,$$

as is well known. The following representation of the elementary polynomials then holds:

$$E_\nu = (\lambda - \lambda_1)^{e_{\nu 1}}(\lambda - \lambda_2)^{e_{\nu 2}} \cdots (\lambda - \lambda_m)^{e_{\nu m}}.$$

The $(\lambda - \lambda_\sigma)^{e_{\nu\sigma}}$ are called the elementary divisors of the matrix $\mathbf{A}$. The relationship

$$\sum_{\nu=1}^{n} e_{\nu\sigma} = p_\sigma ,$$

holds for the elementary divisor exponents $e_{\nu\sigma}$ that correspond to an eigenvalue λ_σ , where p_σ is the multiplicity of the eigenvalue λ_σ .

If the Smith canonical form of $\mathbf{B}$ or the Jordan standard form of $\mathbf{A}$ are

---

[1] $\delta_{ik} = \begin{cases} 1 & \text{for } i = k \\ 0 & \text{for } i \neq k \end{cases}$ is the Kronecker delta.

known, it may then be used to determine the characteristics of $\mathbf{A}$, which can be represented in the following way:

	λ_1	λ_2	$\cdots$	λ_σ	$\cdots$	λ_m	
E_n	$e_{n,1}$	$e_{n,2}$	$\cdots$	$e_{n,\sigma}$	$\cdots$	$e_{n,m}$	
E_{n-1}	$e_{n-1,1}$	$e_{n-1,2}$	$\cdots$	$e_{n-1,\sigma}$	$\cdots$	$e_{n-1,m}$	
	$\cdots$	$\cdots$	$\cdots$	$\cdots$	$\cdots$	$\cdots$	
	p_1	p_2	$\cdots$	p_σ	$\cdots$	p_m	$\sum p_\sigma = n$
	d_1	d_2	$\cdots$	d_σ	$\cdots$	d_m	$\sum d_\sigma = \kappa$

In addition to known quantities, this table also contains d_σ, the rank decrease or defect, which the characteristic matrix $\mathbf{B} = (a_{ik} - \delta_{ik}\lambda)$ undergoes when the eigenvalue λ_σ is substituted for λ, and κ, which is the number of linearly independent eigenvectors of the matrix $\mathbf{A}$. The d_σ are equal to the number of elementary divisor exponents appearing in the column under λ_σ. The following relationship is of decisive importance for this discussion. If $d_\sigma = p_\sigma$, then all elementary divisor exponents $e_{\nu\sigma} = 1$ that correspond to λ_σ. It may be stated that the elementary divisors corresponding to λ_σ are linear. On the other hand, if $d_\sigma < p_\sigma$, then at least one of the elementary divisor exponents $e_{\nu\sigma} > 1$. These facts will be employed shortly.

As is well known, the general solution of (1.3.1) is comprised of particular solutions having the form

$$\xi_{i,\sigma} = KP(t)\, e^{\lambda_\sigma t}, \tag{1.3.2}$$

where K is a constant and $P(t)$ is a polynomial in t of order $e_{\nu\sigma} - 1$. As can be seen, the structure of these particular solutions is determined by λ_σ and $e_{\nu\sigma}$, which are quantities appearing in the characteristic of $\mathbf{A}$. *If the characteristic of $\mathbf{A}$ is known, all the possible particular solutions of* (1.3.1) *may be determined without having integrated the system at all.*

In order that the undisturbed solution q_i^0 be stable, we must require that $\xi_i(t)$ remain sufficiently bounded for all times $t > t_0$. As Eq. (1.3.2) shows, the bounded nature of the particular solution and thus the general solution is essentially determined by the exponential term $e^{\lambda_\sigma t}$, and therefore by the behavior of λ_σ. H. Poincaré called the quantities λ_σ the characteristic exponents because they are decisive for stability.

It is obvious that the possible particular solutions (1.3.2) remain bounded or even decrease monotonically with time, when the real parts of all characteristic exponents are negative. We therefore obtain the following.

THEOREM 1: *When all characteristic exponents λ_σ [roots of the characteristic equation of $\mathbf{A} = (a_{ik})$] have negative real parts, then the stationary solution is asymptotically stable.*

It may also be readily seen that, if the exponent λ_σ in only one of the particular solution (1.3.2) has a positive real part, the general solution of (1.3.1) will contain at least one term which constantly increases with time. In this case, ξ_i does not remain bounded, and we have the following.

THEOREM 2: *If only one of the characteristic exponents λ_σ has a positive real part, then the stationary solution q_i^0 is unstable.*

We have a special case when a characteristic exponent λ_σ has a zero real part. Let the multiplicity of λ_σ be p_σ. Everything now depends on the relationship between the decrease in rank d_σ and the multiplicity p_σ. If we first assume $d_\sigma = p_\sigma$, then all the $e_{\nu\sigma}$ are equal to 1, as was shown previously. Then the polynomial $P(t)$ that appears in (1.3.2), which is of order $e_{\nu\sigma} - 1$, degenerates to a constant, which can be absorbed in K, and the particular solutions corresponding to $\lambda_\sigma = {}^+_{(-)}i\beta$ are of the form $\xi_{i,\sigma} = Ke^{(\pm)i\beta t}$, which are therefore bounded. The stability has therefore not been impaired. If, however, we consider the second case $d_\sigma < p_\sigma$, then at least one of the $e_{\nu\sigma}$ will be larger than one, and $P(t)$ appearing in (1.3.2) is a true function of time t, which will be at least linear. For $\lambda_\sigma = {}^+_{(-)}i\beta$ there will be at least one particular solution in the form $\xi_{i,\sigma} = KP(t)\,e^{(\pm)i\beta t}$ that apparently does not remain bounded, so that the bounded nature of the general solution ξ_i is canceled. In this case, the stability disappears. This can be stated by means of the following.

THEOREM 3: *When one of the characteristic exponents λ_σ has a zero real part, then the stationary solution q_i^0 is stable for $d_\sigma = p_\sigma$ (however, no longer asymptotically) and is unstable for $d_\sigma < p_\sigma$.*

Based on the preceding conclusions, we may outline the following procedure for investigating stability:

1. The characteristic equation $\det(a_{ik} - \delta_{ik}\lambda) = 0$ is solved, and the characteristic exponents λ_σ are obtained as the roots of this equation, which have the multiplicity p_σ.

2. If the real parts of λ_σ are all negative, then stability prevails. No questions remain, and the investigation is concluded.

3. If at least one of the real parts of λ_σ is positive, then instability prevails. Again, no questions remain, and the investigation is concluded.

4. If there is a λ_σ with zero real part, and if $p_\sigma = 1$, then we also have $d_\sigma = 1$, which means that the relationship $p_\sigma = d_\sigma$ is satisfied. Since the stability is not influenced by this, it is not necessary to make an additional investigation. If $p_\sigma > 1$, the root $\lambda_\sigma = {}^+_{(-)}i\beta$ must be substituted for λ in the characteristic matrix $\mathbf{B} = (a_{ik} - \delta_{ik}\lambda)$, and the decrease in rank d_σ obtained in this way must be determined. If $d_\sigma = p_\sigma$, the stability is not influenced. On the other hand, if $d_\sigma < p_\sigma$, this characteristic exponent $\lambda_\sigma = {}^+_{(-)}i\beta$ will cause instability. In this way, everything has been clarified.

Example: Let

$$\dot{\xi}_1 = -3\xi_1,$$

$$\dot{\xi}_2 = -3\xi_2,$$

$$\dot{\xi}_3 = -3\xi_1 + \xi_4 - \xi_5,$$

$$\dot{\xi}_4 = 6\xi_2,$$

$$\dot{\xi}_5 = -\xi_5,$$

be the system of variational equations. Then the matrix $\mathbf{A}$ is of the form

$$\mathbf{A} = \begin{pmatrix} -3 & 0 & 0 & 0 & 0 \\ 0 & -3 & 0 & 0 & 0 \\ -3 & 0 & 0 & 1 & -1 \\ 0 & 6 & 0 & 0 & 0 \\ 0 & 0 & 0 & 0 & -1 \end{pmatrix},$$

and from its characteristic equation

$$\begin{vmatrix} -(3+\lambda) & 0 & 0 & 0 & 0 \\ 0 & -(3+\lambda) & 0 & 0 & 0 \\ -3 & 0 & -\lambda & 1 & -1 \\ 0 & 6 & 0 & -\lambda & 0 \\ 0 & 0 & 0 & 0 & -(1+\lambda) \end{vmatrix} = 0$$

it follows that $(3 + \lambda)^2\lambda^2(1 + \lambda) = 0$, from which the three characteristic exponents are obtained

$$\lambda_1 = -3, \qquad \lambda_2 = 0, \qquad \lambda_3 = -1$$

with the multiplicities

$$p_1 = 2, \qquad p_2 = 2, \qquad p_3 = 1.$$

We shall not expect the exponents λ_1, λ_3 to impair the stability, because they are negative and real. On the other hand, the zero exponent λ_2 must be investigated more closely. In order to do this, we substitute the value $\lambda_2 = 0$ for λ in

$$\mathbf{B} = \begin{pmatrix} -(3+\lambda) & 0 & 0 & 0 & 0 \\ 0 & -(3+\lambda) & 0 & 0 & 0 \\ -3 & 0 & -\lambda & 1 & -1 \\ 0 & 6 & 0 & -\lambda & 0 \\ 0 & 0 & 0 & 0 & -(1+\lambda) \end{pmatrix}.$$

We then obtain the matrix

$$\mathbf{B}_{(\lambda=0)} = \begin{pmatrix} -3 & 0 & 0 & 0 & 0 \\ 0 & -3 & 0 & 0 & 0 \\ -3 & 0 & 0 & 1 & -1 \\ 0 & 6 & 0 & 0 & 0 \\ 0 & 0 & 0 & 0 & -1 \end{pmatrix},$$

which in this special case must naturally coincide with $\mathbf{A}$. By repeatedly performing the elementary operation of adding a row multiplied by an arbitary number with another row, in the sense of the Gaussian algorithm, and by interchanging rows, we transform $\mathbf{B}_{(\lambda=0)}$ into

$$\left.\begin{pmatrix} -3 & 0 & 0 & 0 & 0 \\ 0 & -3 & 0 & 0 & 0 \\ 0 & 0 & 0 & 1 & -1 \\ 0 & 0 & 0 & 0 & -1 \\ \hline 0 & 0 & 0 & 0 & 0 \end{pmatrix}\right\}\begin{matrix} r \\ \\ \end{matrix}\ \ \} \ d$$

from which it can be seen that $\mathbf{B}_{(\lambda=0)}$ is of order $n = 5$, and of rank $r = 4$. Therefore, the decrease in rank is $d = 1$.

In this case we have $d_2 = 1 < p_2 = 2$ for $\lambda_2 = 0$, and therefore the zero root $\lambda_2 = 0$ destroys the stability. It may therefore be concluded that the undisturbed and stationary state corresponding to the variational equations is unstable.

As a matter of interest, we would also like to mention the fact that, by substituting $\lambda_1 = -3$ in $\mathbf{B}$, we obtain

$$\mathbf{B}_{(\lambda=-3)} = \left.\begin{pmatrix} 0 & 0 & 0 & 0 & 0 \\ 0 & 0 & 0 & 0 & 0 \\ -3 & 0 & +3 & +1 & -1 \\ 0 & 6 & 0 & 3 & 0 \\ 0 & 0 & 0 & 0 & +2 \end{pmatrix}\right\}\begin{matrix} d \\ \\ r \end{matrix}$$

from which it follows that $d_1 = 2 = p_1$. In this simple case, it is therefore possible to write the characteristic of the matrix $\mathbf{A}$. It is given by

	$\lambda_1 = -3$	$\lambda_2 = 0$	$\lambda_3 = -1$	
E_5	1	2	1	
E_4	1	0	0	
p	2	2	1	$\sum = 5$
d	2	1	1	$\sum = 4$

We therefore have $E_5 = (\lambda + 3)\lambda^2(\lambda + 1)$, $E_4 = \lambda + 3$, $E_3 = E_2 = E_1 = 1$, and in fact $\det \mathbf{B} = E_1 E_2 \cdots E_5 = (\lambda + 3)^2 \lambda^2 (\lambda + 1)$, as was already stated previously. In the other situations, which are not so clear, it is necessary to go a step further to determine the various elementary divisor exponents by making the transformation to the Jordan canonical form, from which the characteristic can be read off immediately.

It is always desirable to minimize the work entailed in treating the formulated differential equations when investigating stability. Since we can decisively settle the question of stability by means of the first theorem mentioned in this chapter, according to which all roots of the characteristic equation must have a negative real part for stability to prevail, an attempt will be made to establish it without first completely solving the characteristic equation. The criterion used to decide whether an algebraic equation of nth order in λ has only roots with negative real parts is then simultaneously the decisive stability criterion for the case of linear variational equations with constant coefficients postulated here. If it is satisfied, then stability is definitely present.

There are many such criteria, and the two best known of these criteria will now be derived.

FIRST CRITERION: This results in a graphical method, which dates back to Nyquist–Michailov (1932/1938), and is very extensively used in control theory.

The characteristic equation is obtained by solving

$$\det(a_{ik} - \lambda \delta_{ik}) = 0$$

in the form

$$\lambda^n + a_1 \lambda^{n-1} + a_2 \lambda^{n-2} + \cdots + a_{n-1}\lambda + a_n = 0,$$

where the coefficients a_m of this nth-order algebraic equation (Fig. 10) are certain functions of the elements a_{ik} of the matrix $\mathbf{A}$. The n roots $\lambda_\nu = x_\nu + iy_\nu$, $\nu = 1, 2,..., n$ of the polynomial

$$f(\lambda) = \lambda^n + a_1\lambda^{n-1} + \cdots + a_{n-1}\lambda + a_n$$

are assumed to satisfy the condition of only negative real parts.

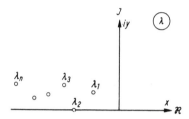

FIG. 10.   Position of the roots of the polynomial $f(\lambda) = 0$ in the λ plane.

In the following such a $f(\lambda)$ will be called a "Hurwitz polynomial." Under these conditions the roots represent points in a complex λ plane, which all lie to the left of the imaginary axis. We shall now regard $f(\lambda)$ as the image of the λ plane. Then the origin of the complex f plane will be the image of the n roots λ_ν, and the imaginary iy axis of the λ plane will become a curve $f(iy)$ in the f plane (Fig. 11). This curve is called the

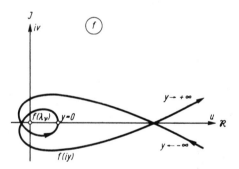

FIG. 11.   Locus curve.

characteristic curve or *locus curve* in control theory. If all roots λ_ν are on the left of the imaginary axis, then the origin in the f plane will always be on the left of the characteristic curve, when we proceed from $y = -\infty$ to $y = +\infty$.

The method of Nyquist–Michailov consists of deriving the variational equations for a stability problem, forming the matrix $\mathbf{A}$ and formulating the characteristic equation for $\mathbf{A}$. Thus the polynomial $f(\lambda)$ is obtained. The characteristic curve $f(iy)$ is then plotted in a f plane point by point. The manner in which this curve passes around the origin of the f plane may be observed. If $f = 0$ always remains on the left of the characteristic curve, when the latter is traversed in the direction of increasing y, it may then be concluded that $f(\lambda)$ is a Hurwitz polynomial. Stability then prevails.

There are a number of variations of the above method that we shall mention only briefly. Owing to symmetry of the characteristic curve with respect to the real axis, it is only necessary to plot one-half of the curve for $0 \leqslant y \leqslant +\infty$. There is another method of establishing the fact that the curve passes around the origin in such a way that stability occurs. The radius vector in the f plane drawn from the origin to the points of $f(iy)$ must rotate in the mathematically positive direction by the angle $(n/2)\pi$ when y increases from 0 to $+\infty$, and when the polynomial $f(\lambda)$ is of degree n. Finally, separate curves can be drawn for the real and imaginary parts of $f(iy)$, from which the Leonhardt and Cremer *criteria* follow. The reader is referred to the literature for more details.[1]

Up to the present we have discussed a graphic method. We shall now present an *algebraic method* advanced by Hurwitz (1895).

SECOND CRITERION: Let $f(\lambda) = \lambda^n + a_1\lambda^{n-1} + \cdots + a_n$ be a nth-degree *Hurwitz polynomial*, a polynomial whose n roots $\lambda_\nu = x_\nu + iy_\nu$ are all on the left of the imaginary axis iy. The polynomial

$$f*(\lambda) = \lambda^n - a_1\lambda^{n-1} + \cdots + (-1)^n a_n$$

is formed with the Hurwitz polynomial, which has the roots $-\bar{\lambda}_\nu$, which are the mirror images of λ_ν with respect to the imaginary axis. It may be postulated *that it is always possible to formulate a Hurwitz polynomial $F(\lambda)$ of degree $n + 1$ as well as another $\Phi(\lambda)$ of degree $n - 1$ by using $f(\lambda)$ and $f*(\lambda)$* under the assumption that the coefficient a_1 in $f(\lambda)$ is positive and nonzero.

For this purpose, we assume

$$F(\lambda) = (\lambda + 2c)f(\lambda) + \lambda f*(\lambda), \qquad c > 0$$

and

$$\Phi(\lambda) = (2a_1 - \lambda)f(\lambda) + \lambda f*(\lambda).$$

It is obvious that $F(\lambda)$ has degree $n + 1$. It is only necessary to determine whether $\Phi(\lambda)$ has degree $n - 1$. After multiplication and a simple calculation, we find

$$\Phi(\lambda) = 2[a_1{}^2\lambda^{n-1} + (a_1a_2 - a_3)\lambda^{n-2} \cdots],$$

[1] LEONHARDT, A., *Arch. Elektrotech.* **38**, 17 (1944); CREMER, L., *Z. Angew. Math. Mech.* **25/27**, 161–163 (1947).

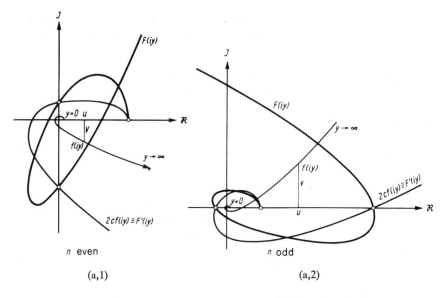

FIG. 12.   Graphic representations proving the Hurwitz criterion.

from which it follows that the degree of $\Phi(\lambda)$ is indeed $n - 1$ due to the assumption $a_1{}^2 \neq 0$.

We shall now show that $F(\lambda)$ and $\Phi(\lambda)$ are truly Hurwitz polynomials. We may do this by using the same ideas as those employed to establish the Nyquist–Michailov criterion. We shall attempt to find the course of the characteristic curves $F(iy)$ and $(\Phi(iy)$.

We have $F(iy) = 2cf(iy) + iy[f(iy) + f^*(iy)]$, and we must distinguish between two cases. Let $f(iy) = u + iv$ be a point in the complex f plane. Then $f^*(iy) = u - iv$ is its mirror image with respect to the real axis for even n. $f^*(iy) = -u + iv$ is its mirror image with respect to the imaginary axis for odd n. The following relationship then results:

If n is *even*, it is necessary to investigate the course of

$$F(iy) = 2cf(iy) + 2iyu$$

in the complex u, v plane. Figure 12a,1 shows the relationships for $0 \leqslant y \leqslant +\infty$. The curve $F(iy)$ is obtained by first elongating the given curve $f(iy)$ by the amount $2c$, which leads to the intermediate curve $F'(iy)$. The points of this curve are then displaced along the perpendicular vector $2iyu$ whose magnitude varies. Its magnitude equals zero when $f(iy)$ crosses the imaginary axis. The intersection points of $f(iy)$ with the imaginary axis correspond to the intersection points of $F'(iy)$ with this axis. Consequently, these points remain fixed when they are displaced in

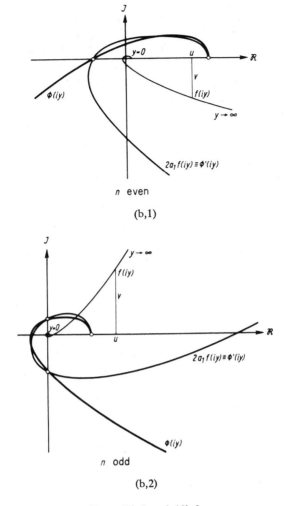

n even

(b,1)

n odd

(b,2)

FIGS. 12b,1 and 12b,2

the perpendicular direction by an amount $2iyu$, which is exactly zero for these points. The curve $F(iy)$ is thus obtained, which is a deformed image of $F'(iy)$. The intersection points of $F'(iy)$ and $F(iy)$ with the y axis coincide. Therefore, the fundamental condition that the origin of the complex plane always is on the left of $F(iy)$ is never violated when the curve is traversed in the direction of increasing y. It follows from this that $F(\lambda)$ is a Hurwitz polynomial.

For *odd n*, we have the characteristic curve $F(iy) = 2cf(iy) - 2yv$, which is shown in Fig. 12a,2 for $0 \leqslant y \leqslant +\infty$. It is obtained from the given curve $f(iy)$ similarly to the method described above. The curve $F'(iy) = 2cf(iy)$ obtained by elongation is displaced along the horizontal vector $2yv$. The magnitude of this vector changes in such a way that the points of $F'(iy)$ located along the real axis remain fixed. $F(iy)$ is again óbtained by deformation of $F'(iy)$ so that the origin always is to the left of $F(iy)$ when the curve is traversed in the direction of increasing y. The Hurwitz property of $F(\lambda)$ remains intact.

It may also be seen that in both cases $F(iy)$ passes through one more quadrant of the complex plane than does $f(iy)$. This is a sign that the degree of $F(\lambda)$ is one greater than the degree of $f(\lambda)$.

The corresponding studies must be made for

$$\Phi(iy) = 2a_1 f(iy) + iy[f^*(iy) - f(iy)].$$

For *even n*, the curve $\Phi(iy) = 2a_1 f(iy) + 2yv$ must be considered, and is shown in Fig. 12b,1. The quantity $\Phi'(iy) = 2a_1 f(iy)$ is first obtained by an elongation. The desired $\Phi(iy)$ is then obtained by a horizontal displacement by $2yv$. The points of $\Phi'(iy)$ along the real axis remain fixed. The quantity $\Phi(\lambda)$ is a Hurwitz polynomial due to the course of $\Phi(iy)$ around the origin. Finally, for *uneven n*, the curve $\Phi(iy) = 2a_1 f(iy) - 2iuy$ shown in Fig. 12b,2 is investigated. It is obtained by the vertical displacement by $-2iuy$ of the curve $\Phi'(iy)$, which was obtained by elongation. The points of $\Phi'(iy)$ remain fixed along the imaginary axis. In spite of the deformation, $\Phi(iy)$ will of necessity again pass around the origin on the left side. $\Phi(\lambda)$ remains a Hurwitz polynomial. Figures 12b,1 and 2 show that $\Phi(iy)$ pass through one quadrant less than does $f(iy)$ for even as well as odd n. This corresponds to the fact that the degree of $\Phi(\lambda)$ is one smaller than the degree of $f(\lambda)$.

We thus have proved the postulate that it is always possible to form two Hurwitz polynomials $F(\lambda)$ and $\Phi(\lambda)$ having the degree $n + 1$ and $n - 1$, respectively, from a given Hurwitz polynomial $f(\lambda)$.

After this preparation, two different procedures may be followed. In one case, the expression $\Phi(\lambda)$ is used, which is a Hurwitz polynomial if $f(\lambda)$ is a Hurwitz polynomial. In order to illustrate the Hurwitz nature of $f(\lambda)$, a progression of $\Phi(\lambda)$ polynomials is formed beginning with $f(\lambda)$ until a second-degree polynomial is obtained. We may readily determine whether it satisfies the condition of two roots with negative real parts.

If this is the case, the preceding third-degree polynomial must be of the Hurwitz type. This means that this is also true for the polynomial of the next higher degree, etc. It is possible to show that for $c = a_1$, the relationship between $f(\lambda)$ and $\Phi(\lambda)$ equals the relationship between $F(\lambda)$ and $f(\lambda)$, except for a constant factor. This means that $f(\lambda)$ is also a Hurwitz polynomial if $\Phi(\lambda)$ is a Hurwitz polynomial. In this way it is possible to show that $f(\lambda)$ is of the Hurwitz type by carrying out the proof in the reserve direction.

Example: Let us investigate $f(\lambda) \equiv \lambda^4 + 2\lambda^3 + 12\lambda^2 + 16\lambda + 16$. The following are formed:

$$\Phi_1(\lambda) = (4 - \lambda)(\lambda^4 + 2\lambda^3 + 12\lambda^2 + 16\lambda + 16)$$
$$+ \lambda(\lambda^4 - 2\lambda^3 + 12\lambda^2 - 16\lambda + 16),$$
$$\Phi_1(\lambda) = 8(\lambda^3 + 2\lambda^2 + 8\lambda + 8),$$
$$\Phi_2(\lambda) = 8(4 - \lambda)(\lambda^3 + 2\lambda^2 + 8\lambda + 8) + \lambda(\lambda^3 - 2\lambda^2 + 8\lambda - 8),$$
$$\Phi_2(\lambda) = 64(\lambda^2 + 2\lambda + 4) = 0.$$

The second-degree polynomial $\Phi_2(\lambda)$ has the roots $-1 \pm 3i$, and is therefore a Hurwitz polynomial. Thus, $\Phi_1(\lambda)$ and finally $f(\lambda)$ must also be Hurwitz polynomials.

In the other case, a *Hurwitz* theorem is used, which will now be described and proved.

THEOREM OF HURWITZ: *If the coefficients of the polynomial $f(\lambda)$ are used, we obtain the matrix*

$$\mathbf{H} = \begin{pmatrix} a_1 & 1 & 0 & 0 & 0 & 0 & \cdots \\ a_3 & a_2 & a_1 & 1 & 0 & 0 & \cdots \\ a_5 & a_4 & a_3 & a_2 & a_1 & 1 & \cdots \\ \cdot & \cdot & \cdot & \cdot & \cdot & \cdot & \cdot \cdot \cdot \end{pmatrix}.$$

The diagonal terms of this matrix contain the a_1, a_2, a_3,... in succession. In addition, all $a_k = 0$ for $k > n$. The following is a necessary and sufficient condition for the polynomial to be a Hurwitz polynomial and have only roots with negative real parts. The principal minors Δ_ν of the Hurwitz matrix $\mathbf{H}$, i.e., the

$$\Delta_1 = a_1, \qquad \Delta_2 = \begin{vmatrix} a_1 & 1 \\ a_3 & a_2 \end{vmatrix}, \qquad \Delta_3 = \begin{vmatrix} a_1 & 1 & 0 \\ a_3 & a_2 & a_1 \\ a_5 & a_4 & a_3 \end{vmatrix},..., \qquad \Delta_n = a_n \, \Delta_{n-1},$$

must all be positive.

Proof: Let $f(\lambda)$ be a Hurwitz polynomial. As was previously shown with the help of the characteristic curve $F(iy)$, this is a necessary and sufficient condition for $F(\lambda)$ to be a Hurwitz polynomial also. The Hurwitz matrix for the polynomial $\frac{1}{2}F(\lambda)$ is

$$\begin{pmatrix} c & 1 & 0 & 0 & \cdots \\ ca_2 & ca_1 + a_2 & c & 1 & \cdots \\ ca_4 & ca_3 + a_4 & ca_2 & ca_1 + a_2 & \cdots \end{pmatrix}.$$

The principal minors of this matrix are

$$D_1 = c, \quad D_2 = c^2 \varDelta_1, \quad D_3 = c^3 \varDelta_2, ..., D_n = c^n \varDelta_{n-1}, \quad D_{n+1} = c^{n+1} \varDelta_n.$$

We assume that the Hurwitz theorem holds for polynomials of degree n. Then $\varDelta_\nu > 0$ for $\nu = 1, 2,..., n$. We have $D_\nu > 0$ for $\nu = 1, 2,..., n + 1$ due to the relationship between D_ν and $\varDelta_\nu$, and because $c > 0$. Since $F(\lambda)$ is a Hurwitz polynomial, it follows that the Hurwitz condition of positive, principal minors is *necessary* for polynomials of degree $n + 1$.

This condition is also sufficient, because $\varDelta_\nu > 0$, $\nu = 1, 2,..., n$ follows from $D_\nu > 0$, $\nu = 1, 2,..., n + 1$. Since the Hurwitz conditions hold for polynomials having degree n, it follows that $f(\lambda)$ is a Hurwitz polynomial. If this is true for $f(\lambda)$, then $F(\lambda)$ must also be a Hurwitz polynomial due to its construction. Therefore, $D_\nu > 0$ was *sufficient* for this.

As we have just illustrated, if the Hurwitz condition holds for polynomials having degree n, then it holds very generally for polynomials having degree $(n + 1)$. This is because any arbitrary polynomial having degree $(n + 1)$ can be formed from one having degree n, just as $F(\lambda)$ was formed from $f(\lambda)$. The validity of the Hurwitz condition can be directly demonstrated for polynomials of the first and second degree. By mathematical induction from n to $n + 1$, it follows that the Hurwitz theorem holds for polynomials of arbitrary degree.

Example: Let us consider the polynomial

$$f(\lambda) = \lambda^4 + 2\lambda^3 + 12\lambda^2 + 16\lambda + 16$$

discussed previously. Its Hurwitz matrix is

$$\begin{pmatrix} 2 & 1 & 0 & 0 \\ 16 & 12 & 2 & 1 \\ 0 & 16 & 16 & 12 \\ 0 & 0 & 0 & 16 \end{pmatrix},$$

and the following principal minors are obtained:

$$\varDelta_1 = 2, \quad \varDelta_2 = \begin{vmatrix} 2 & 1 \\ 16 & 12 \end{vmatrix} = 8, \quad \varDelta_3 = \begin{vmatrix} 2 & 1 & 0 \\ 16 & 12 & 2 \\ 0 & 16 & 16 \end{vmatrix} = 64,$$

$$\varDelta_4 = \begin{vmatrix} 2 & 1 & 0 & 0 \\ 16 & 12 & 2 & 1 \\ 0 & 16 & 16 & 12 \\ 0 & 0 & 0 & 16 \end{vmatrix} = 1024.$$

They are all positive, and therefore $f(\lambda)$ is a Hurwitz polynomial.

The condition of positive, principal minors in the Hurwitz theorem is called the *Hurwitz stability criterion*. This is due to the fact that it is the condition for negative, real parts of the characteristic equation roots, and therefore the condition for stability.

The criterion of Routh (1875) is another algebraic criterion that we shall not discuss here. It is discussed in greater detail by Zurmühl (3).

We have seen that the position of the roots of the characteristic equation corresponding to a mechanical system in the complex plane determines whether the system is stable or unstable. If all the λ_ν are on the left of the imaginary axis, the system is asymptotically stable. If a few of the roots lie along the imaginary axis, it is stated that the system lies along the *stability limit*. In the case of an asymptotically stable system, the variation $\xi_i(t)$ of the unperturbed solution $q_i^0(t)$ will damp out in time. This damping is caused by the negative real parts of the roots λ_n included in the exponents of exponential functions. The smallest absolute magnitude h of the real parts of all λ_ν represents a certain measure of the damping rate, and therefore a measure of the stability, even though it is not completely conclusive. Geometrically, h is the distance from the imaginary axis to the closest root (Fig. 13). The

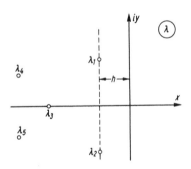

FIG. 13. h is the degree of stability.

quantity h is designated as the *degree of stability*. It is obtained by displacing the imaginary axis by h in the λ plane, so that the system, which was originally stable, now lies along the stability limit. The λ plane becomes the μ plane owing to this displacement, with the coordinates $\eta = y$ and $\xi = x + h$. Because $\lambda = x + iy$, $\mu = \xi + i\eta$, we have $\lambda = \mu - h$. From the original equation

$$f(\lambda) = \lambda^n + a_1\lambda^{n-1} + \cdots + a_n = 0,$$

we obtain the following new equation:

$$g(\mu) = \mu^n + A_1\mu^{n-1} + \cdots + A_n = 0$$

by substitution of $\lambda = \mu - h$. The coefficients A_ν are functions of h. We must set $h = h_0$ in such a way that at least one of the roots of the polynomial $g(\mu)$ has a vanishing real part but no root has a positive real part. The h_0 that satisfies this condition is the degree of stability.

Example: Let us consider $f(\lambda) = \lambda^2 + 2\lambda + 4 = 0$. Substituting $\lambda = \mu - h$, we obtain the equation

$$g(\mu) = \mu^2 + 2\mu(1 - h) + [(1 - h)^2 + 3] = 0,$$

whose roots are $\mu_{1,2} = (h - 1) \pm 3i$. For $h = 1$, $\mu_{1,2}$ have vanishing real parts; therefore this h indicates the degree of stability, as may also be seen in Fig. 14.

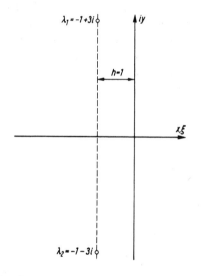

Fig. 14.   Example of calculating the degree of stability.

It is of great importance to know the manner in which the *parameters* of a system must be selected so that a certain stability limit is obtained. These parameters appear in the coefficients of the variational equations as the elements of their matrix **A**, and finally as coefficients of the characteristic equation. This question leads us to the topic of structural analysis and optimization of mechanical systems. The corresponding theory is well formulated in control theory, for example, and we therefore refer to the literature (2, 12).

We have already discussed the question of how certain parameters α_r can influence the stability of a mechanical system. We cannot ignore this question. For purposes of simplicity, we shall restrict ourselves to two parameters α_1 and α_2, because in this way everything can be represented in a *parameter plane*.

Example: Let the system of variational equations be specified by

$$\dot{\xi}_1 = (2 + \alpha_1)\xi_2 ,$$
$$\dot{\xi}_2 = -\xi_1 + \xi_2 - 3\xi_3 ,$$
$$\dot{\xi}_3 = (\alpha_2 - 1)\, \xi_1 + \xi_2 - 2\xi_3 .$$

Its matrix is

$$\mathbf{A} = \begin{pmatrix} 0 & 2 + \alpha_1 & 0 \\ -1 & 1 & -3 \\ \alpha_2 - 1 & 1 & -2 \end{pmatrix},$$

and the characteristic equation is

$$\begin{vmatrix} \lambda & -(2 + \alpha_1) & 0 \\ 1 & (\lambda - 1) & 3 \\ -(\alpha_2 - 1) & -1 & (2 + \lambda) \end{vmatrix} = \lambda^3 + \lambda^2 + (3 + \alpha_1)\,\lambda + 3(\alpha_2 - \tfrac{1}{3})(\alpha_1 + 2) = 0.$$

The coefficients of this equation are

$$a_1 = 1, \qquad a_2 = 3 + \alpha_1 , \qquad a_3 = 3(\alpha_2 - \tfrac{1}{3})(\alpha_1 + 2).$$

The Hurwitz matrix is

$$\mathbf{H} = \begin{pmatrix} a_1 & 1 & 0 \\ a_3 & a_2 & a_1 \\ a_5 & a_4 & a_3 \end{pmatrix}$$

and leads to the stability conditions

$$\Delta_1 = a_1 > 0, \qquad \Delta_2 = a_1 a_2 - a_3 > 0, \qquad \Delta_3 = a_3 \Delta_2 > 0.$$

These may also be written as follows:

$$a_1 > 0, \qquad a_2 > 0, \qquad a_3 > 0, \qquad \Delta_2 = a_1 a_2 - a_3 > 0,$$

which may be readily verified. Owing to the conditions imposed on the a_k, this is equivalent to the conditions

$$a_1 = 1 > 0$$

$$a_2 = (3 + \alpha_1) > 0 \qquad\qquad\qquad \text{i.e.,} \quad \alpha_1 > -3$$

$$a_3 = 3(\alpha_2 - \tfrac{1}{3})(\alpha_1 + 2) > 0 \qquad \text{i.e.,} \begin{cases} \alpha_2 > \tfrac{1}{3}, & \alpha_1 > -2 \\ \alpha_2 < \tfrac{1}{3}, & \alpha_1 < -2 \end{cases}$$

$$\Delta_2 = a_1 a_2 - a_3 = 3 + \alpha_1 - 3(\alpha_2 - \tfrac{1}{3})(\alpha_1 + 2) > 0, \quad \text{i.e.,} \quad \begin{aligned} -3\alpha_1\alpha_2 - 6\alpha_2 \\ + 2\alpha_1 + 5 > 0. \end{aligned}$$

The curves $\alpha_1 = -3$, $\alpha_2 = \frac{1}{3}$, $\alpha_1 = -2$, $3\alpha_1\alpha_2 + 6\alpha_2 - 2\alpha_1 - 5 = 0$ can be plotted in the α_1, α_2 plane. The boundaries of the region whose points correspond to the stability of the system are thus obtained (Fig. 15). For example, if the value of $\alpha_1 = 1$ is held constant and if we proceed

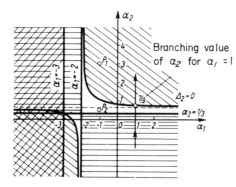

FIG. 15. Stability map in the parameter plane.

along a line parallel to the α_2 axis in a mathematically positive direction, we pass from an instability region to a stability region, and then again pass into an instability region. The *branching values* of the parameter α_2 are obtained when the stability limits are traversed. These values correspond to points at which the stability of the solution changes. For the example considered above, the branching value $\alpha_2 = 7/9$ is obtained at the limit $\varDelta_2 = 0$ when the vertical line $\alpha_1 = 1$ is traversed.

If *several* parameters occur, the corresponding studies may be performed in n-dimensional parameter space, or the study is performed in various parameter planes. The *D decomposition (2.5)* of J. I. Nejmark which is frequently used in control theory represents another possibility for investigating the influence of parameters on stability. Let us again advance the basic concepts of the method using two parameters α_1, α_2.

Let us assume that a characteristic equation $f(\lambda, \alpha_1, \alpha_2) = 0$ is given. This equation represents an implicit equation for mapping the roots λ_ν of $f = 0$ in the *parameter plane* α_1, α_2. If $\lambda = iy$ is substituted in $f = 0$ and if we then solve for α_1, α_2, the α_1, α_2 values (and therefore points of the parameter plane) that correspond to purely imaginary roots of $f = 0$ are obtained.

A *limit curve* $\varGamma$ is specified by $f(iy, \alpha_1, \alpha_2) = 0$. This curve divides the parameter plane into regions. Some of the regions contain points which correspond to the right half of the λ plane. The other regions contain points corresponding to roots in the left half of the λ plane. For purposes of stability, all roots of $f = 0$ must lie in the left λ half-plane. In the α_1, α_2 plane, we shall therefore determine the region delimited by the curve Γ corresponding to the mapping of points in the *left* λ plane. We thus obtain the stability region.

The method first calls for obtaining the curve Γ by using $f(iy, \alpha_1, \alpha_2) = 0$. At times it must be drawn point by point. The coordinates $\alpha_1{}^p$, $\alpha_2{}^p$ of points P are then substituted in $f(\lambda, \alpha_1, \alpha_2) = 0$. These points are selected from the various regions of the parameter plane obtained by plotting Γ. Finally, the location of the roots λ in the λ plane obtained from $f(\lambda, \alpha_1{}^p, \alpha_2{}^p) = 0$ is determined. If $f = 0$ is an algebraic equation in λ having degree n, then the stability regions are defined by points P_{st} having the following property. The equation $f = 0$ has exactly n roots λ with negative real parts using the coordinates $(\alpha_1, \alpha_2)_{P_{st}}$ of P_{st}. In this case, all n roots lie in the left half of the λ plane. Let us describe such a calculation.

Example: We assume the characteristic equation

$$f(\lambda, \alpha_1, \alpha_2) = \lambda^3 + \lambda^2 + (3 + \alpha_1)\lambda + 3(\alpha_2 - \tfrac{1}{3})(\alpha_1 + 2).$$

Substituting $\lambda = iy$, we obtain the following:

$$i[-y^3 + y(3 + \alpha_1)] - y^2 + 3(\alpha_2 - \tfrac{1}{3})(\alpha_1 + 2) = 0.$$

The imaginary and real parts must vanish separately, so we have

$$-y^3 + y(3 + \alpha_1) = 0, \qquad -y^2 + 3(\alpha_2 - \tfrac{1}{3})(\alpha_1 + 2) = 0.$$

These two conditions are satisfied by $y = 0$ and $3(\alpha_2 - \tfrac{1}{3})(\alpha_1 + 2) = 0$. This leads to the branches $\alpha_2 = \tfrac{1}{3}$, $\alpha_1 = -2$ of the limit curve Γ. They appear in Fig. 15 as a result of the condition $a_3 = 0$. On the other hand, if $y \neq 0$ holds, then we must have $y^2 = 3 + \alpha_1$ due to the left-hand equation. If this is substituted into the right-hand equation,

$$-(3 + \alpha_1) + 3(\alpha_2 - \tfrac{1}{3})(\alpha_1 + 2) = 0$$

is obtained, which corresponds to the equation

$$3\alpha_1\alpha_2 + 6\alpha_2 - 2\alpha_1 - 5 = 0$$

and therefore $\Delta_2 = 0$. This branch of the limit curve Γ is shown in Fig. 15.

In order to determine which region of the parameter plane divided by the limit curve Γ is the stability region, we substitute the coordinates of the points $P_1(\alpha_1 = -1, \alpha_2 = 3)$ and $P_2(\alpha_1 = -1, \alpha_2 = 0.5)$ into $f(\lambda, \alpha_1, \alpha_2) = 0$ (see Fig. 15).

The coordinates of P_1 result in the equation $\lambda^3 + \lambda^2 + 2\lambda + 8 = 0$ having the roots $\lambda_1 = -1.98$, $\lambda_{2/3} = 0.53 \pm 1.93i$. Since only one root of the cubic equation has a negative real part, the region of the α_1, α_2 plane containing the point P_1 cannot be a stability region.

The coordinates of P_2 result in the equation

$$\lambda^3 + \lambda^2 + 2\lambda + 0.5 = 0$$

having the roots $\lambda_1 = -0.278$, $\lambda_{2,3} = -0.361 \pm 1.29i$. This time all three roots have a negative real part. We therefore have $P_2 \equiv P_{st}$, which defines a stability region of the α_1, α_2 plane. It is located below the right branch of the hyperbola $\varDelta_2 = 0$ above the line $\alpha_2 = \frac{1}{3}$ and to the right of the vertical line $\alpha_1 = -2$. This result coincides with the previous calculation using the Hurwitz criterion.

It is interesting to note that the Nyquist–Michailov stability criterion may be regarded as a *special case of a D decomposition*. The limiting curve $f(iy) = \alpha$ which divides the α plane for the equation $f(\lambda) - \alpha = 0$ containing only one complex parameter α is identical to the characteristic curve of the Nyquist–Michailov stability criterion.

1.3.2. *Linear Variational Equations with Variable Coefficients*

The literature contains a great number of important general theorems (*4*) corresponding to a system of linear differential equations with variable coefficients. We assume that these theorems are known to the reader. We shall only consider the stability theory of such a system.

A system with variable coefficients is obtained if the right sides of the system of equations $\dot{q}_i^0 = F_i(q_k^0, \alpha_r^0, t)$ contains time t explicitly. The mechanical process is then called *nonautonomous*. The linear or linearized variational equations can be obtained by the method outlined in Sections 1.2 and 1.3.1 as follows:

$$\dot{\xi}_i = a_{ik}(t)\xi_k .   \tag{1.3.3}$$

The terms

$$a_{ik}(t) = (\partial F_i/\partial q_k)(q_j^0, \alpha_r^0, t)   \tag{1.3.4}$$

are variable coefficients. It is assumed that they are continuous and bounded.

If we have a nth-order system, then (1.3.3) has n linearly independent particular *solution vectors* $\xi_i^{(k)}$, $k = 1, 2,..., n$ representing a *fundamental system*. The *general solution* $\xi_i = C_k\xi_i^{(k)}$ is obtained from the fundamental system and the n integration constants C_k. The solution vectors $\xi_i^{(k)}$ (k_{fixed}) themselves are composed of n solution components $\xi_i^{(k)}(t)$, $i = 1, 2,..., n$ (k_{fixed}). Therefore, a total of n^2 solution components

$\xi_i^{(k)}(t)$, $i = 1, 2, ..., n$, $k = 1, 2, ..., n$ corresponds to (1.3.3). The *Wronskian determinant* $W(t)$ of the particular solutions at different times t_0 and t satisfies the relationship

$$W(t) = W(t_0) \exp\left(\int_{t_0}^{t} a_{ii}(t)\, dt\right).\qquad(1.3.5)$$

This relationship will be employed below.

An attempt will be made to estimate the time dependence of the general solution $\xi_i(t)$, which itself depends on the behavior of the particular solution $\xi_i^{(k)}(t)$. If the *magnitudes* of all the $\xi_i^{(k)}(t)$ strive to zero with time, or their magnitude remains below a certain *limit*, it may be assumed that the unperturbed state q_i^0 underlying the variational equations is *stable*.

In the case of the linear differential equations with constant coefficients, it was possible to estimate their solutions (1.3.2) using the real part of the characteristic exponents. This possibility no longer exists for solutions of equations with variable coefficients. An attempt will be made to estimate their growth pattern using similar methods. It is possible to use a method suggested by Lyapunov using the so-called *characteristic numbers* (abbreviation, CN).

Let us assume that the continuous function $x(t)$ defined for $t \geqslant 0$ is given. It is bounded if $|x(t)| < M$ is satisfied for $t > 0$ and $M > 0$, where M is finite. It is unbounded if it is possible to find values $t > 0$ for an arbitrarily large M for which $|x(t)| > M$ holds. The unbounded property of the function is defined by: $\overline{\lim} |x(t)| = \infty$ for $t \to \infty$.

Let us assume that two numbers λ and μ exist for $x(t)$ so that the product $x(t)e^{\lambda t}$ is unbounded and the product $x(t)e^{\mu t}$ tends to zero asymptotically in time. In this case there exists a *characteristic number* κ for which $\mu \leqslant \kappa \leqslant \lambda$ holds and which has the following properties: for arbitrarily small $\epsilon > 0$, the function $x(t)\, e^{(\kappa + \epsilon)t}$ is unbounded and $x(t)\, e^{(\kappa - \epsilon)t}$ tends to zero asymptotically with time. The number κ is therefore a measure for the growth behavior of $x(t)$. Let us assume $x(t) = t^2$, for example, and the characteristic number is $\kappa = 0$. This is because $xe^{\kappa t} = t^2 e^{\kappa t}$ equals $t^2 e^{0 + \epsilon t} = t^2 e^{\epsilon t}$ for $\kappa + \epsilon$, which is unbounded. For $\kappa - \epsilon$, it tends to zero in the same way as $t^2 e^{0 - \epsilon t} = t^2 e^{-\epsilon t}$.

Let us consider the solutions (1.3.2) of the equations with *constant* coefficients. We must form $KP(t)\, e^{x_\sigma t} e^{iy_\sigma t} e^{\kappa t}$, where $\lambda_\sigma = x_\sigma + iy_\sigma$ has been used. It may be readily seen that we must have $\kappa = -x_\sigma$ because $KP(t)\, e^{iy_\sigma t} e^{\epsilon t}$ is obtained for $\kappa + \epsilon$, which is unbounded. We obtain $KP(t)\, e^{iy_\sigma t} e^{-\epsilon t}$ for $\kappa - \epsilon$, which strives to zero in time. It is apparent that the *characteristic number equals the real part of the characteristic exponent* λ_σ *multiplied by* -1. This is not surprising, because in Section 1.3.1 we have already employed this real part of λ_σ in order to establish stability by estimating the growth pattern of the variations. Therefore, the theory of characteristic numbers presented here may be regarded as a logical generalization of the theory of characteristic exponents discussed in Section 1.3.1.

The following three *theorems* hold for the characteristic numbers.

THEOREM 1: *The characteristic number of the sum of two functions* x_1 *and* x_2 *equals the* smallest *characteristic number of these functions.*

Proof: Let $\kappa_1 < \kappa_2$ be the characteristic numbers of x_1, x_2, so that the products $x_1 e^{(\kappa_1+\epsilon)t}$, $x_2 e^{(\kappa_2+\epsilon)t}$ are unbounded. On the other hand, we have $x_1 e^{(\kappa_1-\epsilon)t} \to 0$ and $x_2 e^{(\kappa_2-\epsilon)t} \to 0$ for $t \to \infty$. The sum $x = x_1 + x_2$ with $\kappa = \mathrm{CN}(x)$ and $\epsilon > 0$ is assumed to satisfy the following conditions: $x e^{(\kappa+\epsilon)t}$ becomes unbounded, and $x e^{(\kappa-\epsilon)t} \to 0$ for $t \to \infty$. We now have

$$x e^{(\kappa+\epsilon)t} = x_1 e^{(\kappa+\epsilon)t} + x_2 e^{(\kappa+\epsilon)t}.$$

If we set $\kappa = \kappa_1 = \kappa_2 - \rho$, $\rho > 0$, we then have

$$x e^{(\kappa+\epsilon)t} = x_1 e^{(\kappa_1+\epsilon)t} + x_2 e^{(\kappa_2-\rho+\epsilon)t},$$

which will definitely become infinitely large in time owing to the same property of the first term, regardless of the behavior of the second term. On the other hand, we have

$$x e^{(\kappa-\epsilon)t} = x_1 e^{(\kappa_1-\epsilon)t} + x_2 e^{(\kappa_2-\rho-\epsilon)t},$$

which tends to zero with time because both terms continuously decrease. Therefore, the number $\kappa = \kappa_1$ satisfies all the requirements of a characteristic number for the sum $x = x_1 + x_2$. It is therefore the characteristic number of x.

THEOREM 2: *The characteristic number of the* product *of two functions* x_1 *and* x_2 *is equal to the* sum *of the characteristic numbers of* x_1 *and* x_2 *or is larger than this sum.*

Proof: Let us assume that $x = x_1 \cdot x_2$. We set $\mu = \kappa_1 + \kappa_2 - \epsilon$, where κ_1 and κ_2 are the characteristic numbers of x_1 and x_2 and $\epsilon > 0$, which is assumed to be arbitrarily small. According to the definition of μ, the condition $x e^{\mu t} \to 0$ is to be satisfied for $t \to \infty$. This is the case because according to our selection of μ, we have $x e^{\mu t} = x_1 x_2 e^{(\kappa_1+\kappa_2-\epsilon)t}$, which can be transformed into the expression $x e^{\mu t} = x_1 e^{(\kappa_1-\epsilon/2)t} x_2 e^{(\kappa_2-\epsilon/2)t}$. This expression indeed tends to zero, because there are two continuously decreasing factors on the right side due to the definitions of κ_1 and κ_2. Therefore, the relationship $\kappa_1 + \kappa_2 \leqslant \kappa$ holds for the characteristic number κ of the product x according to the definition of μ. This is what has been postulated. Using the functions $x_1 = e^{a \, t \cos t}$, $x_2 = e^{-a \, t \cos t}$, $a > 0$ as an example, we shall show that Theorem 2 is true not only for $\kappa = \kappa_1 + \kappa_2$ but also for $\kappa > \kappa_1 + \kappa_2$. It may be readily seen that $\kappa_1 = CN(x_1) = \kappa_2 = CN(x_2) = -a$. We have

$$x_{1,2} e^{-(a+\epsilon)t} = e^{[a(\pm \cos t - 1) - \epsilon]t} \to 0,$$

because we always have $a(\pm \cos t - 1) - \epsilon < 0$. On the other hand,

$$x_{1,2} e^{-(a-\epsilon)t} = e^{[a(\pm \cos t - 1) + \epsilon]t},$$

and we may always find t values such that $a(\pm \cos t - 1) + \epsilon > 0$. Therefore, $x_{1,2} e^{-(a-\epsilon)t}$ becomes infinitely large with time. Therefore, $-a$ is the common characteristic number for x_1, x_2. Finally, the product $x = x_1 x_2 = 1 = $ const, and it has the characteristic number zero. Hence the relationship $\kappa = CN(x) = 0 > \kappa_1 + \kappa_2 = -2a$ holds, which was postulated previously.

THEOREM 3: *If the characteristic number κ of a function x is positive, then $\lim x = 0$ for $t \to \infty$. If it is negative, then $\overline{\lim} |x| = \infty$ holds for $t \to \infty$.*

Proof: We have $xe^{(\kappa - \epsilon)t} \to 0$ for $t \to \infty$ according to the definition of κ. We assume that $\kappa > 0$ and ϵ is arbitrarily small. Therefore, ϵ can be selected so that $\kappa - \epsilon > 0$. Then the relationship $xe^{(\kappa - \epsilon)t} \to 0$ can only hold for $t \to \infty$ and positive κ, if $x \to 0$ for $t \to \infty$. This is due to the fact that the factor $e^{(\kappa - \epsilon)t}$ continuously increases because of $\kappa - \epsilon > 0$. On the other hand, $\kappa < 0$ and ϵ is arbitrarily small. According to the definition of κ, $xe^{(\kappa + \epsilon)t}$ will become unbounded. We have

$$xe^{(\kappa + \epsilon)t} = xe^{(\epsilon - |\kappa|)t},$$

and ϵ may be selected small enough that $\epsilon - |\kappa| < 0$. Because of this, $e^{(\epsilon - |\kappa|)t}$ will continuously decrease. In spite of this, the quantity $xe^{(\epsilon - |\kappa|)t}$ will become unbounded for $t \to \infty$. This is only possible if x becomes unbounded, i.e., $\overline{\lim} |x(t)| = \infty$ for $t = \infty$.

We shall now introduce the concept of the characteristic number κ_k of a particular solution vector $\xi^{(k)}$, $i = 1, 2, ..., n$ (k fixed). Let $\kappa_{j,k} = CN(\xi_j^{(k)})$ (j fixed, k fixed) be the characteristic numbers of the *components* of the vector $\xi^{(k)}$. We shall then assume that

$$\kappa_k = \min(\kappa_{j,k}), \quad j = 1, 2, ..., n \quad (k \text{ fixed})$$

holds, i.e., the characteristic number of the vector is assumed to be the *smallest* of the characteristic numbers of its components.

Using κ_k, we may now formulate two theorems that are completely analogous to those given in Section 1.3.1.

THEOREM 4:   *If* all *characteristic numbers* κ_k *of the solution vectors* $\xi_i^{(k)}$ *of the fundamental system of the system of differential equations* (1.3.3) *are* positive, *then the stationary solution* $q_i{}^0$ *is asymptotically stable.*

THEOREM 5:   *If only one of the characteristic numbers* κ_k *is negative, then the stationary solution* $q_i{}^0$ *is unstable.*

Proof: If $\kappa_k > 0$ holds for all k, then the $\kappa_{j,k}$ of the components $\xi_i^{(k)}$ of all solution vectors are also greater than zero for all j and k. Therefore, all these particular solution components strive asymptotically to zero in time. The general solution ξ_i of (1.3.3) is the sum of the particular solution components and therefore strives asymptotically to zero, which proves the stability of $q_i{}^0$.

On the other hand, if only one of the characteristic numbers κ_k is negative, then at least one of the characteristic numbers $\kappa_{j,k}$ of the components of the solution vectors is negative. There is at least one particular solution component that becomes infinitely large with time. Because of this component, the general solution ξ_i of the variational equations must also become infinitely large, which brings about the instability of $q_i{}^0$.

The characteristic numbers $\kappa_{j,k}$ of variational equations with *constant* coefficients are the real parts of the characteristic exponents λ_k multiplied by -1. If the two theorems given above are applied to variational equations with constant coefficients, it follows from $\kappa_{j,k} > 0$ for all j, k that all characteristic exponents λ_k have negative real parts. If at least one of the κ_k and therefore at least one of the $\kappa_{j,k}$ are negative, it follows that at least one of the characteristic exponents λ_k will have a positive real part. Asymptotic stability results in the first case and instability results in the second case. It can be seen that the corresponding theorems from Section 1.3.1 are only special cases of Theorems 4 and 5. Later on we will use the concept of a *regular system*. We will now introduce this concept. We first determine the *characteristic number of the Wronskian determinant* $W(t)$ of the solutions of the variational equations. On the one hand, we have

$$W(t) = \det(\xi_i^{(k)}(t)],$$

so that

$$W(t) = \sum (-1)^{J(\alpha,\beta,\gamma,\ldots,\nu)} \xi_\alpha^{(1)} \xi_\beta^{(2)} \xi_\gamma^{(3)} \cdots \xi_\nu^{(n)}.$$

The sum signs $\sum$ indicate that all permutations of the subscripts 1, 2, 3,..., n are to be used in the terms of the sum. $J(\alpha, \beta, \gamma,..., \nu)$ is the number of inversions of the corresponding permutation. On the other hand, we have the relationship

$$W(t) = W(t_0) \exp \left(\int_{t_0}^{t} a_{ii} \, dt \right),$$

where a_{ii} is the trace of the matrix $\mathbf{A} = [a_{ik}(t)]$ of the differential equations.

According to the theorem of the characteristic number of a product, we have

$$\mathrm{CN}(\xi_\alpha^{(1)} \xi_\beta^{(2)} \cdots \xi_\nu^{(n)}) \geqslant \mathrm{CN}(\xi_\alpha^{(1)}) + \mathrm{CN}(\xi_\beta^{(2)}) + \cdots + \mathrm{CN}(\xi_\nu^{(n)}).$$

The characteristic numbers of the solution components are always larger than the characteristic numbers κ_k of the corresponding solution vectors corresponding to them, or they are equal to them. We also therefore have the relationship

$$\mathrm{CN}(\xi_\alpha^{(1)} \xi_\beta^{(2)} \cdots \xi_\nu^{(n)}) \geqslant \sum_{k=1}^{n} \kappa_k \,.$$

The characteristic number γ of the determinant $W(t)$ is the characteristic number of a sum, the terms of which have the characteristic numbers $\mathrm{CN}(\xi_\alpha^{(1)} \xi_\beta^{(2)} \cdots \xi_\nu^{(n)})$. Consequently, γ is equal to the smallest of the characteristic numbers $\mathrm{CN}(\xi_\alpha^{(1)} \xi_\beta^{(2)} \cdots \xi_\nu^{(n)})$, but always larger than

$$\sum_{k=1}^{n} \kappa_k \,.$$

This is because all characteristic numbers $\mathrm{CN}(\xi_\alpha^{(1)} \cdots \xi_\nu^{(n)})$, even the smallest one, are larger than $\sum_{k=1}^{n} \kappa_k$. We therefore have

$$\gamma \geqslant \sum_{k=1}^{n} \kappa_k \,.$$

We now form the product of the functions $W(t)$ and

$$e^{-p(t)} \left(p(t) = \int_{t_0}^{t} a_{ii}(t) \, dt \right),$$

and we will assume that they have the characteristic numbers γ and μ. The product

$$W(t) \, e^{-p(t)} = W(t_0) \, e^{p(t)} \, e^{-p(t)} = W(t_0) = \mathrm{const}$$

has the characteristic number zero. According to the theorem of characteristic numbers of a product, we must therefore have

$$0 \geqslant \mu + \gamma \geqslant \mu + \sum_{k=1}^{n} \kappa_k \, .$$

Lyapunov called the system of variational equations *regular*, if

$$\mu + \sum_{k=1}^{n} \kappa_k = 0.$$

is satisfied. We will show that the variational equations *having constant coefficients are always regular.*

Assume such a system is given by $\dot{\xi}_i = a_{ik}\xi_k$. Its coefficient matrix is $\mathbf{A} = (a_{ik})$ with the constant elements a_{ik} . The particular solutions of the system have the form $K_{kj}P_{kj}e^{iy_k t}e^{x_k t}$ and therefore have the characteristic numbers $\kappa = -x_k$. The quantities x_k are the real parts of the roots λ_k of the characteristic equation

$$f(\lambda) = \lambda^n + a_1\lambda^{n-1} + \cdots + a_n = 0.$$

As is well known, the coefficients a_1 of this equation satisfy $a_1 = -\operatorname{tr} \mathbf{A}$, that is, $a_1 = -a_{ii}$. The real parts of its roots satisfy the relationship $\sum_k x_k = -a_1$, so that

$$\sum_k x_k = +a_{ii}$$

also holds. These relationships will be used.

The terms of $W(t)$ are equal to

$$(-1)^J K_{\alpha 1}P_{\alpha 1}(t)\, e^{\lambda_1 t}K_{\beta 2}P_{\beta 2}(t)\, e^{\lambda_2 t} \cdots K_{\nu n}P_{\nu n}(t)\, e^{\lambda_n t},$$

in this case, which leads to

$$(-1)^J \exp\!\left(\sum_k x_k t\right) \exp\!\left(i \sum_k y_k t\right) K_{\alpha 1}P_{\alpha 1}(t)\, K_{\beta 2}P_{\beta 2}(t) \cdots K_{\nu n}P_{\nu n}(t)$$

due to $\lambda_k = x_k + iy_k$. It is apparent that the characteristic number of such a sum is always equal to $-\sum_k x_k$, regardless of which term in the sum is under consideration. The characteristic number γ of $W(t)$ is the characteristic number of the sum of these terms. However, we have just seen that all the terms of the sum have the same characteristic number $-\sum_k x_k$. Consequently, the characteristic number γ of the sum is also equal to $-\sum_k x_k$. Therefore, the characteristic number γ of the

Wronskian determinant satisfies the relationship $\gamma = -\sum_k x_k \equiv -a_{ii}$. Since a_{ii} is constant, we have

$$p(t) = \int_{t_0}^t a_{ii}\, dt = a_{ii}(t - t_0),$$

Therefore, we have $e^{-p(t)} = \text{const } e^{-a_{ii}t}$, so that the characteristic number μ of $e^{-p(t)}$ must be equal to $+a_{ii}$.

If $\gamma = -a_{ii}$ and $\mu = +a_{ii}$, then we must have $\gamma + \mu = -a_{ii} + a_{ii} = 0$, which proves the regular property of systems of variational equations with constant coefficients.

1.3.3. *Linear Variational Equations with Periodic Coefficients*

An important spacial case of linear variational equations with variable coefficients occurs when the coefficient functions are periodic. Owing to their great practical importance in celestial mechanics or the theory of vibrations, for example, a special theory has been developed for systems of differential equations with periodic coefficients. We assume that this theory is known to the reader. Detailed descriptions of this theory are given in the literature, especially under such topics as *Hill* and *Mathieu differential equations*, which are the most important types of equations with periodic coefficients (5).

Assume that the system of variational equations is given by

$$\dot{\xi}_i = p_{ik}(t)\xi_k, \qquad i = 1, 2,..., n, \qquad k = 1, 2,..., n, \qquad (1.3.6)$$

in which the p_{ik} are real, bounded, and periodic functions of t with the common period ω. We shall assume that a fundamental system of solutions $\xi_i^{(k)}(t)$, $i = 1, 2,..., n$; $k = 1, 2,..., n$, is known, and we shall also assume that it satisfies the initial conditions $\xi_i^{(k)}(0) = \delta_{ik}$. Due to the periodic nature of the coefficients p_{ik}, the function $\xi_i^{(k)}(t + \omega)$ is also a solution system of (1.3.6). If the fundamental system $\xi_i^{(k)}(t)$ is substituted in (1.3.6), and if the transition from t to $t + \omega$ is made and the relationship $p_{ik}(t + \omega) = p_{ik}(t)$ is used, we have

$$\dot{\xi}_i^{(k)}(t + \omega) = p_{ir}(t + \omega)\, \xi_r^{(k)}(t + \omega) = p_{ir}(t)\, \xi_r^{(k)}(t + \omega)$$

which illustrates the fact that $\xi_i^{(k)}(t + \omega)$ also satisfies the system (1.3.6). If this is true, then $\xi_i^{(k)}(t + \omega)$ must be a linear combination of the solutions of the fundamental system, so that we have

$$\xi_i^{(k)}(t + \omega) = a_{rk}\xi_i^{(r)}(t), \qquad i, r, k = 1, 2,..., n, \qquad (1.3.7)$$

with the corresponding constants a_{rk}. These constants satisfy the relationship

$$\xi_i^{(k)}(\omega) = a_{rk}\xi_i^{(r)}(0) = a_{rk}\,\delta_{ir},$$

because of the initial conditions that are valid for the fundamental system. It is obtained by substituting $t = 0$ in (1.3.7).

The relationship

$$a_{ik} = \xi_i^{(k)}(\omega) \tag{1.3.8}$$

follows from this.

According to the theorem of *Floquet*, (1.3.6) has *multiplicative solutions* of the form

$$\xi_s(t + \omega) = \rho\xi_s(t), \qquad s = 1, 2, 3, ..., n, \tag{1.3.9}$$

and ρ is the *characteristic multiplier*. It must also be possible to obtain these solutions from the fundamental system solutions, and therefore the following relationships hold:

$$\xi_s(t) = b_i\xi_s^{(i)}(t), \tag{1.3.10}$$

$$\xi_s(t + \omega) = b_i\xi_s^{(i)}(t + \omega) \tag{1.3.11}$$

by introducing suitable constants b_i. By means of (1.3.7), we also obtain the following from (1.3.11):

$$\xi_s(t + \omega) = b_i a_{ri}\xi_s^{(r)}(t). \tag{1.3.12}$$

It is thus possible to transform (1.3.9) into the following by using (1.3.10) and (1.3.12):

$$b_i a_{ri}\xi_s^{(r)} = \rho b_r\xi_s^{(r)}. \tag{1.3.13}$$

The coefficients of the functions $\xi_s^{(r)}$ in the latter equation may be compared, leading to

$$(a_{ri} - \rho\,\delta_{ri})b_i = 0. \tag{1.3.14}$$

This is a system of determining equations for the constants b_i. Because of the fact that it is a linear, homogeneous, algebraic system of equations, the requirement that b_i have nontrivial solutions leads to the following:

$$\det(a_{ri} - \rho\,\delta_{ri}) = 0. \tag{1.3.15}$$

It may be seen from this that the *characteristic multipliers* ρ *are the eigenvalues of the characteristic matrix* $\mathbf{A} = (a_{ri})$. This is analogous to the case of linear variational equations with constant coefficients: As we shall see shortly, the structure of the characteristic matrix $\mathbf{A}$ is decisive in establishing the type of solutions for the system of differential equations (1.3.6).

Let us first assume the simplest case, in which all elementary divisors of $\mathbf{A}$ are *linear*. Equation (1.3.15) then has n simple roots ρ_k, $k = 1, 2,..., n$. The multiplicative solutions then comprise a fundamental system that satisfies

$$\xi_i^{(k)}(t + \omega) = \rho_k \xi_i^{(k)}(t), \qquad i, k = 1, 2,..., n. \tag{1.3.16}$$

This is a functional equation which is satisfied by

$$\xi_i^{(k)}(t) = e^{\lambda_k t} \varphi_i^{(k)}(t), \tag{1.3.17}$$

where

$$\lambda_k = (1/\omega) \ln \rho_k \tag{1.3.18}$$

is the *characteristic exponent* and $\varphi_i^{(k)}(t)$ is a function having the period ω. This may be readily demonstrated by substituting (1.3.17) and (1.3.18) in (1.3.16) because

$$\xi_i^{(k)}(t + \omega) = e^{\ln \rho_k} e^{(t/\omega)\ln \rho_k} \varphi_i^{(k)}(t + \omega)$$

$$= \rho_k e^{\lambda_k t} \varphi_i^{(k)}(t + \omega) = \rho_k e^{\lambda_k t} \varphi_i^{(k)}(t) = \rho_k \xi_i^{(k)}(t)$$

is identically satisfied only when (1.3.18) holds and when $\varphi_i^{(k)}(t)$ is periodic. It follows from this that (1.3.6) will have particular solutions of the form (1.3.17) if $\mathbf{A}$ only has linear elementary divisors.

On the other hand, if the elementary divisors are *not linear*, and if elementary divisor exponents of $\mathbf{A}$ exist which are larger than one, particular solutions are then obtained of (1.3.6) having the more complicated form

$$\xi_i^{(k)}(t) = P(t) e^{\lambda_k t}. \tag{1.3.19}$$

The polynomials $P(t)$ are polynomials in t with periodic coefficients. The degree of these polynomials is one less than the corresponding elementary divisor exponent. Since this proof is very similar to that given in Section 1.3.1 for equations with constant coefficients, we shall not elaborate upon this further nor present the proof. The reader is referred to the literature (*1.1*) for a detailed discussion of this subject. A knowledge of the fundamental form of the (1.3.6) solutions is sufficient for performing stability studies; this is provided by (1.3.17) or (1.3.19). It may again be seen that the properties of the characteristic exponent λ_k are decisive for stability. The following theorems can thus be formulated, corresponding to those presented in Section 1.3.1:

THEOREM 1: *When* all *the characteristic exponents* $\lambda_k = 1/\omega \cdot \ln \rho_k$ *have* negative *real parts*, i.e., *the magnitudes of the characteristic multipliers* ρ_k *are smaller than* 1, *the stationary solution basic to the variational equations is asymptotically* stable.

Proof: If the conditions presented in this theorem hold, all factors $e^{\lambda_k t}$, and therefore all particular solutions (1.3.17) or (1.3.19) and the general solution of (1.3.6), decrease continuously with time.

THEOREM 2: *If* only one *of the characteristic exponents* λ_k *has a* positive *real part—i.e., the magnitude of one of the characteristic multipliers* ρ_k *is larger than* 1—*the stationary solution is* unstable.

Proof: If the conditions presented in the theorem hold, at least one of these factors $e^{\lambda_k t}$, as well as one particular solution and the general solution of (1.3.6), increases continuously in time.

THEOREM 3: *When one of the characteristic exponents* λ_k *has zero real parts—i.e., the magnitude of the corresponding multiplier* ρ_k *is* 1—*the stationary solution is stable for* $d_k = p_k$ *(but no longer asymptotically stable), and is unstable for* $d_k < p_k$.

Proof: If the multiplicity p_k of the characteristic equation root ρ_k, whose magnitude is one, equals the decrease in rank d_k produced by it in the matrix $\mathbf{B} = (a_{ri} - \rho\,\delta_{ri})$, then the elementary divisors corresponding to ρ_k are linear and the corresponding polynomials $P(t)$ have degree zero with respect to t. Therefore, they degenerate to periodic functions. The solutions corresponding to this ρ_k therefore have the following form:

$$\xi_j^{(k)}(t) = e^{(\pm)i\arg\rho_k t/\omega}\varphi_j^{(k)}(t),$$

and are also periodic and do not disturb the stability. On the other hand, if the multiplicity p_k of ρ_k is larger than d_k, then elementary divisor exponents $e_{\nu k} > 1$ are possible. At least one of the polynomials $P(t)$ will have the degree $e_{\nu k} - 1 = m \geqslant 1$, so that a particular solution having the following form:

$$\xi_j^{(k)}(t) = e^{(\pm)i\arg\rho_k t/\omega}t^m\varphi_j^{(k)}(t)$$

exists. It represents a *secular term* of the (1.3.6) general solution and causes a continuous increase in time, thus destroying the stability.

The behavior of the roots ρ_k of the characteristic equation

$$\det(a_{ri} - \rho\,\delta_{ri}) = 0,$$

depending on the structure of the matrix $\mathbf{A} = (a_{ri})$, is decisive in establishing stability or instability. This equation may also be written in the following form:

$$\rho^n + a_1\rho^{n-1} + \cdots + a_{n-1}\rho + a_n = 0, \tag{1.3.20}$$

where the coefficients a_i are functions of the elements a_{ri} of $\mathbf{A}$. The unit circle of the complex ρ plane is transformed into the left half of the complex w plane by means of the transformation $\rho = 1 + w/1 - w$. The characteristic equation (1.3.20) becomes

$$\left(\frac{1+w}{1-w}\right)^n + a_1\left(\frac{1+w}{1-w}\right)^{n-1} + \cdots + a_{n-1}\left(\frac{1+w}{1-w}\right) + a_n = 0. \tag{1.3.21}$$

According to Theorem 1, the absolute magnitudes of the roots ρ_k of (1.3.20) must be smaller than one for stability to prevail. Points inside the unit circle of the ρ plane are therefore specified by the ρ_k. The ρ_k are transformed into w_k by means of the transformation $\rho_k = 1 + w_k/1 - w_k$, and the w_k are located in the left half of the w plane. It may therefore be assumed that stability is assured when all roots w_k of (1.3.21) have negative real parts. It is thus possible to apply the stability criteria already introduced in Section 1.3.1—for example, the Hurwitz criterion—to Eq. (1.3.21).

We shall now show that variational equations *with periodic coefficients are always regular*. This means we must postulate the following: If γ is the characteristic number of the Wronskian determinant and if μ is the characteristic number of the function $\exp(-\int_{t_0}^{t} p_{ii}\,dt)$ [where p_{ii} is the trace of the (1.3.6) coefficient matrix] we must then have $\gamma + \mu = 0$.

The following equation holds for the coefficients a_n of the characteristic Eq. (1.3.20):

$$(-1)^n a_n = \det\mathbf{A} = \det(a_{ik})$$

and therefore according to (1.3.8) we also have

$$(-1)^n a_n = \det[\xi_i^{(k)}(\omega)].$$

The following holds for the Wronskian determinant $W(t)$:

$$W(t) = W(t_0) \exp\left(\int_{t_0}^{t} p_{ii}\, dt\right) = \det[\xi_i^{(k)}(t)].$$

If we substitute $t_0 = 0$, $t = \omega$, it follows that

$$W(\omega) = W(0) \exp\left(\int_{0}^{\omega} p_{ii}\, dt\right) = \det[\xi_i^{(k)}(\omega)] = (-1)^n a_n\,.$$

However, we have $W(0) = \det[\xi_i^{(k)}(0)]$, and therefore, owing to $\xi_i^{(k)}(0) = \delta_{ik}$, we also have $W(0) = 1$, so that we finally obtain

$$(-1)^n a_n = \exp\left(\int_{0}^{\omega} p_{ii}\, dt\right). \tag{1.3.22}$$

Since $W(t) = \det[\xi_i^{(k)}(t)]$ and since in the most general case for $\xi_i^{(k)}(t)$, we have

$$\xi_i^{(k)}(t) = e^{\lambda_k t} P(t) = e^{(1/\omega)\ln|\rho_k| t} e^{i(1/\omega)\arg\rho_k t} P(t)$$

just as in Section 1.3.2, it follows from this that the characteristic number γ of $W(t)$ is given by

$$\gamma = -\frac{1}{\omega} \sum_k \ln|\rho_k| = -\frac{1}{\omega} \ln\left(\prod_k |\rho_k|\right).$$

According to the Vieta root theorems and based on (1.3.22), we have

$$\prod_k |\rho_k| = (-1)^n a_n = \exp\left(\int_{0}^{\omega} p_{ii}\, dt\right),$$

so that the result is

$$\gamma = -\frac{1}{\omega} \int_{0}^{\omega} p_{ii}\, dt. \tag{1.3.23}$$

Let us now attempt to find the characteristic number μ of

$$\exp\left(-\int_{t_0}^{t} p_{ii}\, dt\right) \equiv \exp\left(-\int_{0}^{t} p_{ii}\, dt\right).$$

We may find it from the requirement that $e^{\mu t} \exp(-\int_0^t p_{ii}\, dt)$ must remain bounded for $t \to \infty$. This the case if we set

$$\mu = (1/\omega) \int_{0}^{\omega} p_{ii}\, dt \tag{1.3.24}$$

because

$$\mu t - \int_0^t p_{ii}\, dt = (t/\omega) \int_0^\omega p_{ii}\, dt - \int_0^t p_{ii}\, dt$$

is a function having the period ω and is therefore bounded for any t. Therefore, $e^{\mu t} \exp(-\int_0 p_{ii}\, dt)$ always remains bounded. Consequently, $\mu = (1/\omega) \int_0^\omega p_{ii}\, dt$ is the characteristic number of $\exp(-\int_0^t p_{ii}\, dt)$.

With reference to (1.3.23) and (1.3.24), the regularity condition $\mu + \gamma = 0$ is satisfied.

Without discussing this further, we would like to mention that the variational equations with periodic coefficients may be *reduced in terms of Lyapunov's definition*, i.e., they may be transformed into equations with constant coefficients. We refer the reader to the literature (*1.1*) for a more detailed discussion.

As was already mentioned in Section 1.3.1, an attempt may be made to determine stability or instability by using the simplest method possible. The characteristic exponents λ_k may be used for this purpose in the case of variational equations with constant coefficients. These exponents are obtained as the roots of the characteristic equation. We wished to avoid solving the characteristic equation, or even the variational equations, by applying stability criteria—for example, the criteria of Nyquist–Michailov or Hurwitz.

The same situation apparently exists in the case of variational equations with periodic coefficients when λ_k or ρ_k or w_k are studied. The Hurwitz criterion could be applied to (1.3.21) in order to determine the behavior of w_k. Unfortunately, this is not true. According to (1.3.22), only the coefficient a_n of the characteristic equation is known. All other coefficients are functions of the elements a_{ik} of the characteristic matrix **A**. Equation (1.3.8) holds for the a_{ik}. Therefore, it would be necessary to already know particular solutions of the variational equations as well as specific values $\xi_i^{(k)}(\omega)$ in order to formulate the characteristic equation and to investigate the behavior of the characteristic multipliers ρ_k. These represent serious difficulties, and an attempt will be made to avoid them by applying approximation methods (perturbation method, series development method, infinite determinants method, etc.). This will now be discussed.

Problems encountered in practice often lead to differential equations with the following form:

$$\ddot{y} + [\lambda - hf(t)]\, y = 0,$$

in which λ and h are parameters, and $f(t)$ is a function having the period ω.

This is a *Hill differential equation.* By substituting $y = \xi_1, \dot{y} = \xi_2$, we obtain the following system from it:

$$\dot{\xi}_1 = \xi_2, \qquad \dot{\xi}_2 = (hf - \lambda)\xi_1,$$

i.e., we have

$$p_{11} = 0, \qquad p_{12} = 1, \qquad p_{21} = hf - \lambda, \qquad p_{22} = 0,$$

so that $p_{ii} = p_{11} + p_{22} = 0$, and the Wronskian determinant satisfies $W(t) = W(0) \exp(\int_0^\omega p_{ii}\, dt) = W(0) = \text{const}$. We have selected $\xi_i^{(k)}(t)$ as the fundamental system with the initial conditions $\xi_i^{(k)}(0) = \delta_{ik}$, $i, k = 1, 2$ so that $W(0) = \det[\xi_i^{(k)}(0)]$ holds. In this case, we have $W(t) = \text{const} = 1$ for any value of t and also for $t = \omega$, so that we obtain

$$W(\omega) = \xi_1^{(1)}(\omega)\xi_2^{(2)}(\omega) - \xi_1^{(2)}(\omega)\xi_2^{(1)}(\omega) = 1. \qquad (1.3.25)$$

The characteristic equation follows from

$$\begin{vmatrix} a_{11} - \rho & a_{12} \\ a_{21} & a_{22} - \rho \end{vmatrix} = 0$$

or from the following due to $a_{ik} = \xi_i^{(k)}(\omega)$

$$\begin{vmatrix} \xi_1^{(1)}(\omega) - \rho & \xi_1^{(2)}(\omega) \\ \xi_2^{(1)}(\omega) & \xi_2^{(2)}(\omega) - \rho \end{vmatrix} = 0$$

and is given by

$$\rho^2 - [\xi_1^{(1)}(\omega) + \xi_2^{(2)}(\omega)]\,\rho + [\xi_1^{(1)}(\omega)\,\xi_2^{(2)}(\omega) - \xi_1^{(2)}(\omega)\,\xi_2^{(1)}(\omega)] = 0.$$

We have $a_1 = \xi_1^{(1)}(\omega) + \xi_2^{(2)}(\omega)$ and, because of (1.3.25), we have $a_2 = \xi_1^{(1)}(\omega)\,\xi_2^{(2)}(\omega) - \xi_1^{(2)}(\omega)\,\xi_2^{(1)}(\omega) = 1$. It follows from $\rho^2 - a_1\rho + 1 = 0$ that

$$\rho_1 = \tfrac{1}{2}a_1 + \tfrac{1}{2}(a_1^2 - 4)^{1/2}, \qquad \rho_2 = \tfrac{1}{2}a_1 - \tfrac{1}{2}(a_1^2 - 4)^{1/2}.$$

When $a_1^2 < 4$, the ρ_1, ρ_2 are complex, and the magnitude of both is 1. This is due to the fact that we must have $|\rho_1| |\rho_2| = a_2 = 1$ according to the Vieta root theorems. Stability is thus guaranteed.

When $a_1^2 > 4$, the ρ_1, ρ_2 are real and have different magnitudes. Due to $|\rho_1| |\rho_2| = 1$, we must have $|\rho_1| > 1$ and $|\rho_2| < 1$. Instability consequently prevails. In the case $a_1^2 = 4$—that is, $a_1 = \pm 2$—we have $\rho_1 = \rho_2 = \rho = \tfrac{1}{2}a_1 = \pm 1$, i.e., we have $|\rho| = 1$, and this case corresponds to the stability limit.

· We have the following conditions:

$$\tfrac{1}{2} \mid a_1 \mid \, < 1 \qquad \text{for stability,}$$

$$\tfrac{1}{2} \mid a_1 \mid \, = 1 \qquad \text{for the stability limit.}$$

Using the definition of a_1, we may also write these as follows:

$$\tfrac{1}{2} \mid \xi_1^{(1)}(\omega) + \xi_2^{(2)}(\omega)\mid \, < 1 \qquad \text{for stability,}$$

$$\tfrac{1}{2} \mid \xi_1^{(1)}(\omega) + \xi_2^{(2)}(\omega)\mid \, = 1 \qquad \text{for the stability limit.}$$

It is now apparent that in order to evaluate these stability criteria, solutions $\xi_1^{(1)}(t)$, $\xi_2^{(2)}(t)$ must be obtained, which also satisfy the initial conditions $\xi_1^{(1)}(0) = 1$, $\xi_2^{(2)}(0) = 1$. Once these have been determined, they may be written in the following form:

$$\xi_1^{(1)} = \xi_1^{(1)}(\lambda, h, t), \qquad \xi_2^{(2)} = \xi_2^{(2)}(\lambda, h, t),$$

i.e., as functions of the parameters λ and h included in the differential equation. The stability criterion

$$\tfrac{1}{2} \mid \xi_1^{(1)}(\lambda, h, \omega) + \xi_2^{(2)}(\lambda, h, \omega)\mid \, = 1$$

may also be regarded as an implicit equation connecting λ and h. The solution of this equation results in the limiting curves of the stability regions in the λ, h plane, which is the parameter plane.

Example: If we set $f(t) = 2 \cos 2t$, we obtain the *Mathieu differential equation*. The *Strutt map* (Fig. 16) specifies the division of the λ, h plane. The small oscillations of a vertical, heavy pendulum having length l,

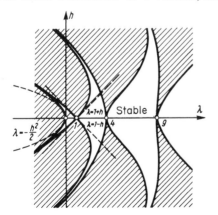

FIG. 16.   Stability map of Strutt.

whose point of suspension harmonically oscillates in the vertical direction according to $u = A \cos \omega t$, satisfy the differential equation

$$\varphi''(x) + (\lambda - 2h \cos 2x)\, \varphi(x) = 0,$$

where $x = \tfrac{1}{2}\omega t$, $\lambda = -4g/\omega^2 l$, $2h = -4A/l$. This is a Mathieu differential equation. The system

$$\dot{\xi}_1 = \xi_2, \qquad \dot{\xi}_2 = (2h \cos 2\tau - \lambda)\xi_1 \qquad (1.3.26)$$

is obtained by using the relationships $\varphi = \xi_1$, $\varphi' = \xi_2$, $x = \tau$. Differentiation with respect to the argument again is indicated by a point.

We have $\omega = \pi$, and the fundamental system

$$\xi_1^{(1)}(\tau), \qquad \xi_1^{(2)}(\tau),$$
$$\xi_2^{(1)}(\tau), \qquad \xi_2^{(2)}(\tau),$$

is employed.

Let us assume that τ is replaced by $-\tau$, and let $\xi_1^{(1)}(-\tau) = \xi_1^{(1)}(\tau)$ be a solution of the system. Substitution in (1.3.26) indicates that $\xi_2^{(1)}(-\tau) = -\xi_2^{(1)}(\tau)$ must hold. In the same way, we may assume that $\xi_1^{(2)}(-\tau) = -\xi_1^{(2)}(\tau)$ is a solution by transition from τ to $-\tau$. Substitution in (1.3.26) results in the relationship $\xi_2^{(2)}(-\tau) = \xi_2^{(2)}(\tau)$. We therefore have

$$\xi_1^{(1)}(\tau) = \xi_1^{(1)}(-\tau), \qquad \xi_1^{(2)}(\tau) = -\xi_1^{(2)}(-\tau),$$
$$\xi_2^{(1)}(\tau) = -\xi_2^{(1)}(-\tau), \qquad \xi_2^{(2)}(\tau) = \xi_2^{(2)}(-\tau). \qquad (1.3.27)$$

According to (1.3.7), we have

$$\xi_1^{(1)}(\tau + \pi) = a_{11}\xi_1^{(1)}(\tau) + a_{21}\xi_1^{(2)}(\tau),$$
$$\xi_1^{(2)}(\tau + \pi) = a_{12}\xi_1^{(1)}(\tau) + a_{22}\xi_1^{(2)}(\tau).$$

If we substitute $\tau = -\pi$, it follows that

$$\xi_1^{(1)}(0) = 1 = a_{11}\xi_1^{(1)}(-\pi) + a_{21}\xi_1^{(2)}(-\pi),$$
$$\xi_1^{(2)}(0) = 0 = a_{12}\xi_1^{(1)}(-\pi) + a_{22}\xi_1^{(2)}(-\pi).$$

According to (1.3.8) and (1.3.27), for $\omega = \pi$, we have

$$\xi_1^{(1)}(-\pi) = \xi_1^{(1)}(\pi) = a_{11},$$
$$\xi_1^{(2)}(-\pi) = -\xi_1^{(2)}(\pi) = -a_{12},$$

so that we finally obtain

$$1 = a_{11}^2 - a_{21}a_{12},$$

$$0 = a_{12}a_{11} - a_{22}a_{12}.$$

The relationship

$$1 = a_{11}a_{22} - a_{12}a_{21}$$

also follows from (1.3.8) and (1.3.25).

If the first equation is subtracted from this one, it follows that

$$a_{11}^2 - a_{11}a_{22} = 0.$$

Therefore, the relationships

$$a_{11}(a_{11} - a_{22}) = 0, \qquad a_{12}(a_{11} - a_{22}) = 0, \qquad a_{11}a_{22} - a_{12}a_{21} = 1$$

must hold simultaneously. There is no contradiction only for $a_{11} \neq 0$, $a_{12} \neq 0$, $a_{11} = a_{22}$. The following relationships therefore hold for the Mathieu differential equation:

$$a_{11} = a_{22} \qquad \text{or} \qquad \xi_1^{(1)'}(\pi) = \xi_2^{(2)'}(\pi).$$

Therefore,

$$\xi_1^{(1)'}(\pi) + \xi_2^{(2)'}(\pi) = 2\xi_1^{(1)'}(\pi),$$

and the stability conditions become $|\xi_1^{(1)'}(\pi)| < 1$ for stability, $|\xi_1^{(1)'}(\pi)| = 1$ for the stability limit.

We must now find the solution $\xi_1^{(1)}(\tau)$. This will be done according to a direction method formulated by Poincaré, which is known under the name of *perturbation method*. The trial solution

$$\xi_1^{(1)} = \xi_{10}^{(1)} + h\xi_{11}^{(1)} + h^2\xi_{12}^{(1)} + \cdots,$$

$$\xi_2^{(1)} = \xi_{20}^{(1)} + h\xi_{21}^{(1)} + h^2\xi_{22}^{(1)} + \cdots$$

(1.3.28)

is used, assuming that h is a sufficiently small parameter for the series to converge.

If (1.3.28) is substituted into the system (1.3.26), we obtain the following new systems, arranged according to powers of h:

$$\dot{\xi}_{10}^{(1)} = \xi_{20}^{(1)}, \dots, \qquad \dot{\xi}_{11}^{(1)} = \xi_{21}^{(1)}, \dots, \qquad \dot{\xi}_{12}^{(1)} = \xi_{22}^{(1)}, \dots,$$

$$\dot{\xi}_{20}^{(1)} = -\lambda\xi_{10}^{(1)}, \quad \dot{\xi}_{21}^{(1)} = -\lambda\xi_{11}^{(1)} + 2\cos 2\tau\,\xi_{10}^{(1)}, \quad \dot{\xi}_{22}^{(1)} = -\lambda\xi_{12}^{(1)} + 2\cos 2\tau\,\xi_{11}^{(1)}, \dots.$$

Instead of this we may also write

$$\ddot{\xi}_{10}^{(1)} = -\lambda\xi_{10}^{(1)}, \quad \ddot{\xi}_{11}^{(1)} = -\lambda\xi_{11}^{(1)} + 2\cos 2\tau\xi_{10}^{(1)},$$

$$\ddot{\xi}_{12}^{(1)} = -\lambda\xi_{12}^{(1)} + 2\cos 2\tau\xi_{11}^{(1)},$$

$$\vdots \tag{1.3.29}$$

$$\ddot{\xi}_{1k}^{(1)} = -\lambda\xi_{1k}^{(1)} + 2\cos 2\tau\xi_{1(k-1)}^{(1)},$$

as can easily be seen. The corresponding boundary conditions are $\xi_{10}^{(1)}(0) = 1$, $\xi_{1k}^{(1)}(0) = 0$ for $k \geqslant 1$, which follows from the fact that $\xi_{1}^{(1)}(0) = 1$ should hold for every h.

Taking into account the boundary conditions, the solutions of (1.3.29) are

$$\xi_{10}^{(1)} = \cos\lambda^{1/2}\tau,$$

$$\xi_{ik}^{(1)} = \frac{2}{\lambda^{1/2}}\int_0^{\tau} \xi_{1(k-1)}^{(1)}(x)\cos 2x\sin\lambda^{1/2}(\tau - x)\, dx, \qquad k \geqslant 1.$$

We thus obtain

$$\xi_1^{(1)}(\pi) = \sum_{k=0}^{\infty} h^k \xi_{1k}^{(1)}(\pi)$$

with

$$\xi_{10}^{(1)}(\pi) = \cos\lambda^{1/2}\pi, \qquad\qquad \xi_{11}^{(1)}(\pi) = 0,$$

$$\xi_{12}^{(1)}(\pi) = \frac{\pi\lambda^{1/2}\sin\pi\lambda^{1/2}}{4\lambda(\lambda - 1)}, \qquad \xi_{13}^{(1)}(\pi) = 0.$$

If this series is already truncated at the first term, the following condition is obtained for the stability limit

$$\xi_{10}^{(1)}(\pi) = \cos\lambda^{1/2}\pi = 1 \qquad \text{or} \qquad \lambda = n^2.$$

As can be seen, we have obtained the limiting points for $h = 0$ in the Strutt map along the λ axis. If the third term in the series is considered, then the following condition holds for the stability limit

$$\cos\lambda^{1/2}\pi + h^2\frac{\pi\lambda^{1/2}\sin\pi\lambda^{1/2}}{4\lambda(\lambda - 1)} = 1$$

However, if we postulate that not only h but also λ is small, we may stipulate the following:

$$\cos\lambda^{1/2}\pi \approx 1 - \frac{\lambda\pi^2}{2}, \qquad \frac{\pi\lambda^{1/2}\sin\pi\lambda^{1/2}}{4\lambda(\lambda - 1)} \approx -\frac{\pi^2}{4}.$$

The condition of the stability boundary becomes

$$1 - \frac{\lambda \pi^2}{2} - \frac{h^2 \pi^2}{4} = 1, \qquad \lambda = -\frac{h^2}{2}.$$

As can be seen from Fig. 16, we have indeed obtained the approximate course of the limiting curve of the Strutt map in the vicinity of the origin of the λ, h plane. The course of the remaining limiting curves of this map can be obtained to any degree of accuracy by applying other methods, for example, by a Mathieu[1] series development.

If stability is to prevail, we must require that

$$1 - \frac{\lambda \pi^2}{2} - \frac{h^2 \pi^2}{4} < 1, \qquad \lambda + \frac{h^2}{2} > 0.$$

For the vertical pendulum mentioned previously, this means that

$$4g/\omega^2 l < 2A^2/l^2,$$

so that its point of suspension must be displaced with a frequency $\omega^2 > 2gl/A^2$ in order that the vertical position of the pendulum be stable.

1.4. INVESTIGATIONS IN PHASE SPACE

Let us assume that a mechanical process, similar to that we have already discussed, is described by the system of differential equations $\dot{q}_i = F_i(q, \alpha^0)$. We will restrict ourselves to autonomous systems. In order to investigate the stability of a particular solution $\dot{q}_i^0 = F_i(q^0, \alpha^0)$ we will attempt to obtain our answer as easily as possible, as described in previous chapters. In other words, we will attempt to reach conclusions regarding stability *without the complete integration* of the system of differential equations whenever possible. In addition to the possibilities discussed in Section 1.3, we may use the *intermediate integrals* for this purpose, which geometrically represent *phase surfaces* and which have already been mentioned. The course of the solutions lying along the phase surfaces can be obtained from the properties of the phase surfaces. We may at least determine whether *orbital stability* exists or not.

The simplest and most easily understandable condition is obtained for a system with two variables, which is described by

$$\dot{q}_1 = F_1(q_1, q_2, \alpha^0), \qquad \dot{q}_2 = F_2(q_1, q_2, \alpha^0). \tag{1.4.1}$$

[1] See, for example, ROTHE, R., and SZABÓ, I., "Higher Mathematics," Vol. VI, p. 195. Teubner, Stuttgart, 1965.

The phase space then becomes the *phase plane,* and the phase surfaces become *phase curves.* This is shown in Fig. 17 for the case of orbital stability: the undisturbed solution q_i^0 corresponds to the phase curve C_0, and the adjacent, disturbed solutions correspond to the phase curves $C_{\mathfrak{S},1} C_{\mathfrak{S},2}$, which remain entirely in the "tube" (strip, in this case) $\bigcup_k \mathfrak{R}_r(P_k)$ (Section 1.1.2). Therefore, we have orbital stability.

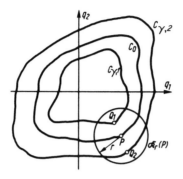

Fig. 17.   Phase curves with orbital stability.

We should mention that the representation of an entire family of phase curves is called a *phase diagram.* With a phase diagram it is possible to obtain the stability behavior of not only one, but an arbitrary number of solutions.

We have restricted ourselves to *autonomous* processes for purposes of simplicity. Naturally, it is possible to treat *nonautonomous* systems in this same way. In this case it is no longer possible to use the q_1, q_2 *plane,* but it is necessary to employ the q_1, q_2, *t space.* The phase curves become curves in space which often have a very complicated course, which makes the problem difficult to follow. In this case the method of phase curves cannot be recommended in most cases, and will therefore not be treated here. On the other hand, the use of a *phase cylinder* is to be recommended when the phase curves of autonomous processes are periodic in at least one of the variables, for example in q_1 (Fig. 18). In this case, there is

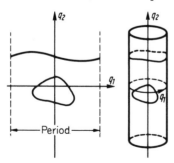

Fig. 18.   Phase plane (development of the phase cylinder), phase cylinder.

only an apparent spatial representation. In reality, the phase curve is cut off along the length of a period of q_1 , and this piece is rolled up into a folder.

On the other hand, it is not difficult to study *nonlinear* processes if it is possible to find intermediate integrals for them. This is especially true for *conservative* processes where the energy theorem represents such an intermediate integral. This method is primarily used when dealing with the theory of nonlinear oscillations of a system having one degree of freedom.

Example: Let us assume that the vibration process is given by $\ddot{x} + f(x) = 0$, where $f(x)$ is a nonlinear restoring force. By means of the substitutions $x = q_1$, $\dot{x} = q_2$ we obtain the system of differential equations

$$\dot{q}_1 = q_2 , \qquad \dot{q}_2 = -f(q_1).$$

As can easily be seen, the intermediate integral is

$$\tfrac{1}{2}q_2{}^2 + V(q_1) = h \qquad (1.4.2)$$

with $V(q_1) = \int_0^{q_1} f(q_1)\, dq_1$ and the parameter h. Since it is possible to interpret $\tfrac{1}{2}q_2{}^2$ as the kinetic energy, $V(q_1)$ as the potential energy, and h as an energy constant, we have indeed found the energy integral of the process.

The relationship (1.4.2) can be represented as a phase curve in the q_1 , q_2 plane for a certain value of the parameter h, (Fig. 19). A complete family of phase curves $C_{k,i}$, $i = 1, 2,...,$[1] is obtained—that is, the phase diagram—when the contour curves of the energy surface $h = \tfrac{1}{2}q_2{}^2 + V(q_1) = H(q_1 , q_2)$ resulting from the selection of the parameter h_i , $i = 1, 2,...$, are projected onto the q_1 , q_2 plane.

In the most general case, the phase curves corresponding to a particular solution of interest are first plotted. Perturbations (we first only consider perturbations in the initial conditions) are then carried out. The phase curves corresponding to the perturbed solutions are then plotted in the phase diagram. Their course with respect to the curve for the unperturbed solution makes it possible to conclude whether or not orbital stability exists. In most cases, the stability of two types of solutions—that is, of *stationary* and periodic solutions—will be of most interest. Stationary solutions are characterized by $\dot{q}_i = 0$. They represent *equilibrium positions* and are represented by points $q_{i,s}$ in phase space. Periodic

---

[1] k is the number of the intermediate integral.

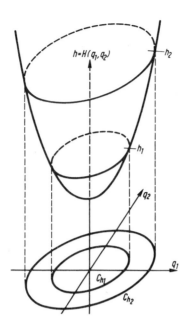

FIG. 19. Energy surface and phase curves.

solutions, i.e., solutions that are periodic with respect to the parameter t (which usually represents time), lead to closed curves in phase space. These curves are called *limit cycles* if the process corresponding to them is the limiting state of *self-excited vibrations*.

1.4.1. Stationary Points

The course of the phase curves in the vicinity of stationary points is obtained as follows.

We set $q_i \equiv q_{i,s}$, $\dot{q}_i \equiv \dot{q}_{i,s} = 0$ in the system of differential equations $\dot{q}_i = F_i(q, \alpha^0)$. A system of determining equations for the stationary points $q_{i,s}$, is thus obtained with $F_i(q_s, \alpha^0) = 0$. Then $q_i = q_{i,s} + \xi_i$ is substituted into the original system of differential equations, and one obtains the following system of differential equations for the variations ξ_i, considering that $\dot{q}_{i,s} = 0, F_i(q_s, \alpha^0) = 0$ holds:

$$\dot{\xi}_i = (\partial F_i/\partial q_k)_s \xi_k + \Phi_i[(\partial^n F_i/\partial q_k{}^n)_s, \xi_k]. \tag{1.4.3}$$

The terms of the Taylor series development are represented by $\Phi_i[\partial^n F_i/\partial q_k{}^n)_s, \xi_k]$. They contain the higher derivatives of F_i and the ξ_k raised to higher powers.

If the coefficients $(\partial F_i/\partial q_k)_s$ are not zero, the point is a *simple* stationary point. On the other hand, if they are all equal to zero for a given i, then we have $\dot{\xi}_i = \Phi_i(\xi_k)$ for this i. This point is then a *nonsimple* stationary point.

The course of the phase surfaces in the vicinity of the stationary point is described by Eq. (1.4.3). If the point is a simple stationary point and if we restrict ourselves to its immediate vicinity, then only the linear terms will be important. We may then use the *abbreviated, linear* variational equations

$$\dot{\xi}_i = (\partial F_i/\partial q_k)_s \xi_k = a_{ik}\xi_k .$$

The results presented in Section 1.3 may be used when discussing the solutions of these equations.

For purposes of simplicity, we shall again assume below that we are dealing with variational equations and two variables. The abbreviated variational equations are then the following:

$$\begin{aligned}
\dot{\xi}_1 &= a_{11}\xi_1 + a_{12}\xi_2 , \\
\dot{\xi}_2 &= a_{21}\xi_1 + a_{22}\xi_2 .
\end{aligned} \tag{1.4.4}$$

The form of the solutions depends on the roots $\lambda_{1,2}$ of the characteristic equation $\det(a_{ik} - \delta_{ik}\lambda) = 0$. If particular solutions of (1.4.4) have been based on certain initial conditions and on the $\lambda_{1,2}$, the phase curves may be plotted in the ξ_1, ξ_2 plane in the vicinity of the stationary point. The stationary point itself is transformed to the origin of the ξ_1, ξ_2 plane by variable transformation. Once the phase diagram surrounding it is known, conclusions may be reached regarding its stability in the small.

For example, if the roots $\lambda_{1,2} = \alpha \pm i\beta$ are conjugate complex roots, a stable focus is obtained for $\alpha < 0$, and an unstable focus is obtained for $\alpha > 0$ (Fig. 20). In the first case, we have (1.1.6), i.e.,

$$|\xi_i(t)| < \epsilon \quad \text{for} \quad |\xi_i(t_0)| < \eta(\epsilon) \quad \text{and} \quad t > t_0 , \quad i = 1, 2,...,$$

are satisfied. Orbital stability as well as stability satisfying the stricter requirement of Lyapunov then prevail. In the second case, instability is produced in exactly the corresponding way.

We shall now discuss the *types of stationary points* in general. The roots of the characteristic equation satisfy

$$\lambda_{1,2} = \tfrac{1}{2}\{(a_{11} + a_{22}) \pm [(a_{11} + a_{22})^2 + 4(a_{12}a_{21} - a_{11}a_{22})]^{1/2}\}.$$

If both roots are real and have the same sign, a *node* is obtained. It is stable for $a_{11} + a_{22} < 0$ and unstable for $a_{11} + a_{22} > 0$ (Fig. 21).

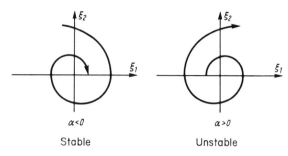

Fig. 20.   Unstable and stable focus.

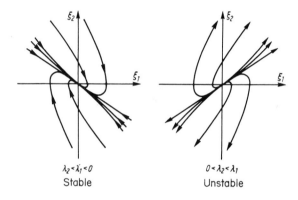

Fig. 21.   Stable and unstable node.

If both roots are real and have opposite sign, a *saddle* (Fig. 22) is obtained corresponding to instability.

If both roots are purely imaginary, a *center* (Fig. 23) is obtained which is stable.

If the roots are conjugate, complex roots, a *focus* is obtained, which is stable for $\alpha = a_{11} + a_{22} < 0$ and unstable for $a_{11} + a_{22} > 0$ (Fig. 20).

Let us again assume only two variables q_1, q_2 or ξ_1, ξ_2, respectively, and we shall restrict ourselves to the immediate vicinity of a *nonsimple*, stationary point. For a given i, we have $(\partial F_i / \partial q_k)_s = 0$. The abbreviated, nonlinear, variational equation $\dot{\xi}_i = \frac{1}{2}(\partial^2 F_i / \partial q_j \, \partial q_k)_s \xi_j \xi_k$, j, $k = 1$, 2,... is obtained from

$$\xi_i = \Phi_i[(\partial^n F_i / \partial q_k{}^n), \xi_k]. \tag{1.4.5}$$

This is due to the fact that second-order terms primarily determine the

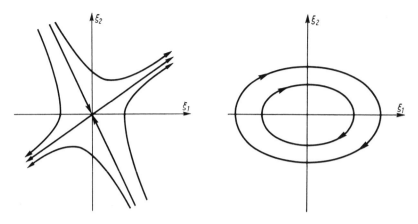

FIG. 22. Saddle.   FIG. 23. Center.

course of the phase curves in the ξ_1, ξ_2 plane[1] in the vicinity of the stationary point. A system of nonlinear, differential equations is obtained. An attempt will first be made to find an intermediate integral, so that approximate phase curves can be plotted in the vicinity of the stationary point. These curves may then be employed to determine stability. If this is not possible, the course of the phase curves must be determined indirectly from (1.4.5) using the *isocline method*, which is a graphic method.

Example: Let us consider the following system:

$$\dot{q}_1 = F_1(q_1, q_2) = q_2, \qquad \dot{q}_2 = F_2(q_1, q_2) = aq_1 + bq_1{}^2 + cq_1{}^3. \quad (1.4.6)$$

We shall assume that $F_2(q_s) \equiv aq_{1s} + bq_{1s}^2 + cq_{1s}^3 = 0$ and

$$(\partial F_2/\partial q_1)_s = a + 2bq_{1s} + 3cq_{1s}^2 = 0$$

are satisfied for the coordinates $q_{1s} = \alpha$, $q_{2s} = 0$ of the stationary point. (These two conditions are satisfied by $b = -2a/\alpha$, $c = a/\alpha^2$.)

According to (1.4.3), (1.4.6) becomes

$$\dot{\xi}_1 = (\partial F_1/\partial q_1)_s \xi_1 + (\partial F_1/\partial q_2)_s \xi_2 + \Phi_1,$$
$$\dot{\xi}_2 = (\partial F_2/\partial q_1)_s \xi_1 + (\partial F_2/\partial q_2)_s \xi_2 + \Phi_2. \qquad (1.4.7)$$

---

[1] If the second derivatives of F_i are also zero, third-order terms with respect to ξ_k must be considered, etc.

We have

$$\partial F_1/\partial q_1 \equiv 0, \qquad \partial F_1/\partial q_2 \equiv 1, \qquad \partial F_2/\partial q_2 \equiv 0,$$

and, according to our assumption, $(\partial F_2/\partial q_1)_s = 0$. The following linear equation results if Φ_1 is neglected in the first line of (1.4.7):

$$\dot{\xi}_1 = \xi_2 .$$

All the linear terms vanish in the second line of (1.4.7). The quadratic terms in Φ_2 must then be considered. The equation

$$\dot{\xi}_2 = \frac{1}{2}\left[\left(\frac{\partial^2 F_2}{\partial q_1^2}\right)_s \xi_1^2 + 2\left(\frac{\partial^2 F_2}{\partial q_1\,\partial q_2}\right)_s \xi_1\xi_2 + \left(\frac{\partial^2 F_2}{\partial q_2^2}\right)_s \xi_2^2 \right]$$

is approximately satisfied. The equations

$$\frac{\partial^2 F_2}{\partial q_1\,\partial q_2} \equiv 0, \qquad \frac{\partial^2 F_2}{\partial q_2^2} \equiv 0, \qquad \left(\frac{\partial^2 F_2}{\partial q_1^2}\right)_s = 2b + 6cq_{1s} = \frac{2a}{\alpha}$$

must also be taken into account. The special values of $(\partial^2 F_2/\partial q_1^2)_s$ for q_{1s}, b, and c have been substituted. The following nonlinear system of variational equations is finally obtained for the nonsimple stationary point $(\alpha, 0)$:

$$\dot{\xi}_1 = \xi_2 , \qquad \dot{\xi}_2 = (a/\alpha)\xi_1^2.$$

The following relationship may be derived from the variational equations:

$$\frac{d\xi_2}{d\xi_1} = \frac{a}{\alpha}\frac{\xi_1^2}{\xi_2} ,$$

so that along the isoclines

$$\xi_1 = 0, \qquad \xi_2 = 0, \qquad \xi_2 = \pm(a/\alpha)\xi_1^2$$

the tangent inclinations of the phase curves of the ξ_1, ξ_2 plane are

$$d\xi_2/d\xi_1 = 0, \qquad d\xi_2/d\xi_1 = \infty, \qquad d\xi_2/d\xi_1 = \pm 1.$$

This small amount of information is sufficient to plot the phase curves in the vicinity of the stationary point. This is found to be an unstable point (Fig. 24).

1.4.2. *Limit Cycles*

Periodic solutions of particular solutions are of as great interest as the stationary solutions. They correspond to closed phase curves. If a vibration problem is given by the system

$$\dot{q}_1 = q_2 , \qquad \dot{q}_2 = -\omega^2 q_1 ,$$

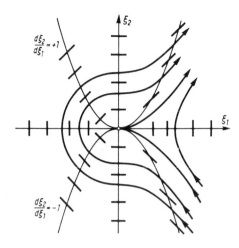

FIG. 24. Nonsimple
stationary point.

the intermediate integral

$$(q_2{}^2/2h) + (\omega^2/2h)\, q_1{}^2 = 1$$

is usually obtained. It leads to a family of ellipses in the phase plane
(q_1, q_2) whose semiaxes are $(1/\omega)(2h)^{1/2}$ and $(2h)^{1/2}$ depending on the
parameter h (Fig. 25). The origin becomes a center. All solutions are
periodic and pass around the stationary point (origin) and are closed
curves. This represents a simple illustration of the fact that closed phase
curves correspond to the periodic solutions of the vibration equation.

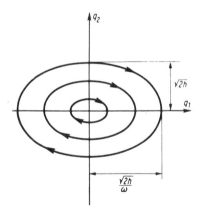

FIG. 25. Elliptical phase curves.

Isolated, periodic solutions may represent a special case in the general
case of a nonlinear system of differential equations. The phase diagram
then reveals isolated, closed phase curves. The course of the phase curves
in the vicinity of these curves indicates whether there are stable

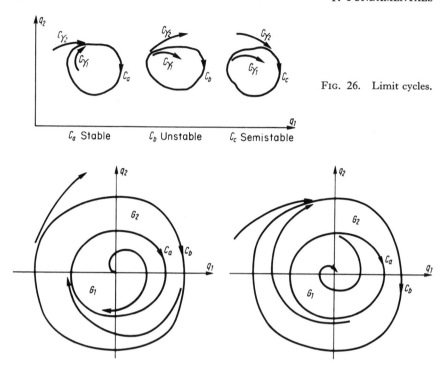

FIG. 26.   Limit cycles.

C_a Stable C_b Unstable C_c Semistable

FIG. 27.   Limit cycles.   FIG. 28.   Limit cycles.

or unstable periodic solutions (Fig. 26) in the sense of orbital stability. For example, if the trajectories of the perturbed solution (for example, $C_{\in 1}$, $C_{\in 2}$) come arbitarily close to the closed trajectory C_a for $t \to \infty$, as is the case for C_a, the corresponding periodic solution is stable. If the adjacent trajectories continue to move away for $t \to \infty$, as is the case for C_b, then the periodic solution is unstable. If both cases occur simultaneously, the periodic solution is semistable. The term C_c represents an example of this. A semistable solution is unstable for practical purposes.

Isolated periodic solutions are of special interest in the fields of self-sustained and self-excited vibrations. There are many mechanical and electrical examples, for example, clock movements, electric oscillation circuits with grid tubes, etc. Several closed curves may prevail in the phase diagram (Fig. 27). The phase curve C_a corresponds to a stable, periodic solution with *soft self-excitation*, because its *region of self-oscillation* G_1 surrounds the origin of the phase plane which is an unstable focus. At the same time, instability in the small with respect to the origin may occur concurrently with stability in the large, due to C_a. The phase curve C_b is unstable in a way such that trajectories within its *damping region* G_2 asymptotically approach the stable curve C_a. It may be readily seen that the closed phase curves may be regarded as asymptotic limiting curves of the adjacent trajectories. It is for this reason that they are also designated as *limit cycles*.

Figure 28 illustrates the opposite conditions. The quantity C_a is unstable, and C_b is stable. Stability in the small and instability in the large prevail. G_1 is the damping region, and G_2 is the self-excitation region. This is the case of *hard self-excitation*, because a perturbation of sufficient magnitude to make the perturbed trajectory enter the region of self-excitation G_2 is necessary to produce an asymptotic transition to the vibration represented by C_b.

There are several theorems predicting closed phase curves, particularly those advanced by Poincaré and Bendixson.

Let us first consider the criterion of Poincaré stipulating the requisite condition for closed phase curves. We shall restrict ourselves to a system of two variables. The gradient field in the phase plane is specified by

$$\dot{q}_1 = F_1(q_1, q_2), \qquad \dot{q}_2 = F_2(q_1, q_2).$$

The inclination of every line element of the field with respect to the q_1 axis of the phase plane is given by the angle $\Theta = \arctan F_2/F_1$. If a closed curve K not passing through any stationary point is traversed in the mathematically positive direction, line elements on K will be rotated by an angle $2\pi J$ if the image point on the curve K rotates by an angle 2π. According to Poincaré, the rational number J is called the *index* of the curve. It is calculated from

$$J = \frac{1}{2\pi} \int_K d\left(\arctan \frac{F_2}{F_1}\right) = \frac{1}{2\pi} \int_K \frac{F_1\, dF_2 - F_2\, dF_1}{F_1^2 + F_2^2}.$$

CRITERION OF POINCARÉ: *The index J of a curve K is zero if K encloses only regular, but no stationary points. It equals $+1$ if K encloses a focus, center, or node. It equals -1 if K encloses a saddle. These values may also be regarded as the indices of the points surrounded by K. If the region of the phase plane delimited by K contains the n stationary points $S_1,...,S_n$, we have*

$$J(K) = \sum_{k=1}^{n} J(S_k). \tag{1.4.8}$$

The index of a closed phase curve is $+1$. Since the index of a regular point is zero, it follows that a closed phase curve must enclose at least one stationary point. *In addition, it follows from* (1.4.8) *that* a closed phase curve can never enclose only one saddle S, *but must enclose other stationary points, for example, two centers W. Only in this way do we have $J(S) + 2J(W) = -1 + 2 = 1$, so that the index of the closed phase curve equals $+1$, as must be the case* (Fig. 29).

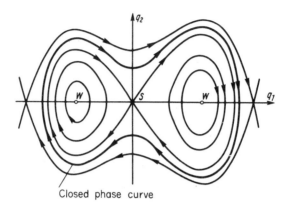

Closed phase curve

FIG. 29.   Closed phase curve contains a saddle S and two centers W.

The stipulation that the index of a closed phase curve equal $+1$ is necessary, but is not sufficient in itself. If $J = +1$, a limit cycle does not definitely exist. It is therefore advantageous to employ other theorems in addition. We shall first mention the theorem advanced by Bendixson. This theorem may be employed to demonstrate the existence of a limit cycle in a region B of the phase plane based on a sufficient condition.

THEOREM OF BENDIXSON: *If the sign of the expression $\partial F_1/\partial q_1 + \partial F_2/\partial q_2$ does not change in B, no closed phase curves can exist in B, provided this expression is not identically zero.*

Proof: If a closed phase curve C exists in B, the following equation must hold for the line integral taken along C:

$$\oint_C (F_1 \, dq_2 - F_2 \, dq_1) = 0.$$

According to Green's theorem, we have

$$\oint_C (F_1 \, dq_2 - F_2 \, dq_1) = \iint_{B^*} \left(\frac{\partial F_1}{\partial q_1} + \frac{\partial F_2}{\partial q_2} \right) dq_1 \, dq_2$$

for $B^* \in B$, and according to our assumption $\partial F_1/\partial q_1 + \partial F_2/\partial q_2$ must have the same sign everywhere in B and must always differ from zero. Consequently, the area integral and also the line integral cannot be zero—

i.e., the closed phase curve C cannot exist. This proof may also be applied to arbitrary partial regions B^* of B until B is completely covered. It has thus been demonstrated that there are no closed phase curves in B.

A third, important theorem is that proposed by Poincaré–Bendixson presenting the necessary and sufficient conditions for closed phase curves.

THEOREM OF POINCARÉ–BENDIXSON: *If a trajectory C remains within a finite region B of the phase plane without approaching any stationary points, C is itself a limit cycle or approaches one.*

The study by LaSalle[1] is an example of the application of this theorem in determining a limit cycle in the phase diagram of the Van der Pol equation. The various trajectories of the phase diagram correspond to different initial conditions. If one of these trajectories is called an "unperturbed" solution, the other trajectories will be "perturbed" solutions with respect to this trajectory and will correspond to different initial conditions. It may be concluded that they correspond to *"perturbations of the initial conditions"* in terms of the discussion presented above.

Let us now assume that a parameter α (or several of them) is included in the differential equations. A *perturbation of the parameter* may then also occur which can influence the phase diagram. The *branching values* α_{crit} of α, which are of particular interest, are characterized by the fact that the stationary points and limit cycles of the phase picture change qualitatively when the variable α passes through the value α_{crit}. For example, a stable point may become unstable, and at the same time a stable limit cycle may appear that justifies the name of "branching value." It may be assumed that the stable stationary solution has "branched" into an unstable stationary solution and a stable periodic solution. More frequently, when α changes, two limit cycles—a stable cycle and an unstable cycle—approach each other, coinciding as a semistable limit cycle for α_{crit}. The latter disappears completely when α has passed through the branching value. The calculation of the branching values is a very important problem in stability theory, because this determines whether stability or instability prevails.

Example: Let us assume that a load P is applied to two elastic rods which are supported and connected by pin joints, as is shown in Fig. 30. As long as the load P is small enough, there is a stable equilibrium position A that is maintained by the system. As the load P increases, a value P_{crit} is reached for which the framework *"snaps through"* into the new equilibrium position B. The load P plays the role of the

---

[1] LASALLE, J., *Quart. Appl. Math.* **7**, 1–19 (1949).

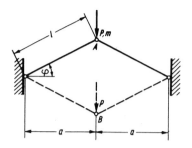

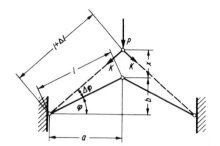

FIG. 30. Snap-through problem.   FIG. 31. Snap-through oscillator.

parameter α, and P_{crit} is the parameter value for which the nature of the initial, stable equilibrium position A changes in such a way that stability no longer exists. It may be shown that this value P_{crit} may be found by considering the phase diagram.

For this purpose, we shall assume that a single mass m is tied to the load P as the two rods of the framework are assumed to be massless, elastic "springs." We shall now assume that the mass m performs small oscillations around the equilibrium position A. If the deflections from position A are denoted by x, the small oscillations are described by the differential equation $m\ddot{x} + P + 2K \sin(\varphi + \varDelta\varphi) = 0$ (Fig. 31).

We have

$$K = EF\,\varDelta l/l = c\,\varDelta l,$$

$$\varDelta l = [a^2 + (x + b)^2 - l]^{1/2},$$

$$\sin(\varphi + \varDelta\varphi) = (x + b)/[a^2 + (x + b)^2]^{1/2},$$

and the differential equation becomes

$$m\ddot{x} + P + 2c(x + b)\{1 - l/[a^2 + (x + b)^2]^{1/2}\} = 0.$$

Introducing the new notation,

$$(x + b)/a = \xi_1, \qquad \dot{\xi}_1 = \xi_2, \qquad l/a = \alpha,$$

and performing simple transformations, we obtain the system of differential equations

$$\dot{\xi}_1 = \xi_2,$$

$$\dot{\xi}_2 = -\left[\frac{P}{am} + \frac{2c}{m}\xi_1 - \frac{2c\alpha}{m}\frac{\xi_1}{(1 + \xi_1{}^2)^{1/2}}\right] = -f(\xi_1).$$

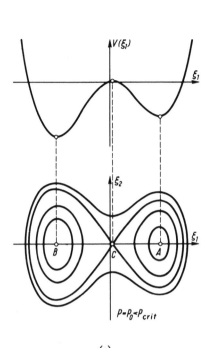

(a)

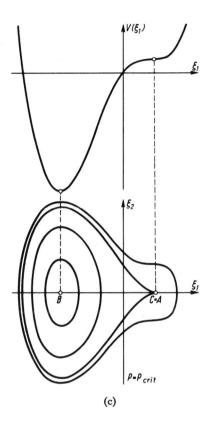

(c)

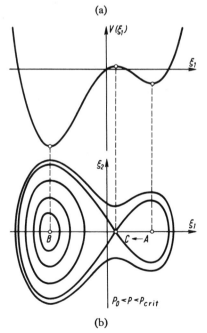

(b)

FIG. 32. Connection between potential function and phase diagram.

73

One intermediate integral, the energy integral, is given by

$$\tfrac{1}{2}\xi_2{}^2 + \int_0^{\xi_1} f(\xi_1)\,d\xi_1 = E^h.$$

The phase curves are then

$$\xi_2 = \pm 2\,[E^h - V(\xi_1)]^{1/2},$$

where

$$V(\xi_1) = \frac{P}{am}\,\xi_1 + \frac{c}{m}\,\xi_1{}^2 - \frac{2c\alpha}{m}\,[(1 + \xi_1{}^2)^{1/2} - 1].$$

As Fig. 32a illustrates, the phase diagram has the two stable equilibrium positions A and B as well as the unstable equilibrium position C for $P = P_0 < P_{\text{crit}}$.

If the load P is increased, the stable equilibrium position A approaches the unstable equilibrium position C (Fig. 32b). A and C coincide when the parameter P attains the "branching value" P_{crit}. Only *one* stable equilibrium position B remains, and the framework snaps through into this position for P_{crit} (Fig. 32c). If P exceeds P_{crit}, C and A vanish, and the stable position B remains. The framework remains in this position after snapping through.

In addition, a change in the phase diagram may be determined from the change in $V(\xi_1)$, which results from varying P. The condition for which A and C coincide is given by $dV/d\xi_1 = d^2V/d\xi_1{}^2 = 0$, from which we may calculate

$$\xi_1 = [(1/a)^{2/3} - 1]^{1/2} \quad \text{and} \quad P_{\text{crit}} = 2EF[1 - (a/1)^{2/3}]^{3/2}.$$

P_{crit} may be called the "branching value." If we pass through the values of P in reverse order, the phase curve is closed up to P_{crit}, where a separating curve occurs. In addition to B, the equilibrium positions A and C also occur.

1.5. The Direct Method of Lyapunov

1.5.1. *The Stability Definition of Lyapunov*

Let us consider a problem of mechanics defined by the system of differential equations

$$\dot{q}_i = F_i(t, q_1, q_2, ..., q_n), \qquad i = 1, 2, ..., n, \tag{1.5.1}$$

and let us investigate the stability of a particular solution $q_i^0(t)$. For purposes of simplicity, we have omitted the influence of any parameters. The following variational equations are obtained by the well-known transformation (1.2) $q_i = q_i^0 + \xi_i$:

$$\dot{\xi}_i = (\partial F_i/\partial q_k)(q_j^0, t)\,\xi_k + \Phi_i(t, \xi_1, ..., \xi_n), \qquad i, k = 1, ..., n. \quad (1.5.2)$$

According to Lyapunov, $q_i^0(t)$ is stable when the variations satisfy $|\xi_i(t)| < \epsilon$ for $t > t_0$ if $|\xi_i(t_0)| < \eta(\epsilon)$ holds at the initial time t_0. The quantity ϵ is an arbitrarily small positive number, just like η, which is a function of ϵ. This is equivalent to the postulate that the stationary solution $\xi_i = 0$, $i = 1, ..., n$ of the variations is stable if the conditions mentioned previously are satisfied. We shall demonstrate this in the following discussion.

As was already mentioned in Section 1.1.2, the Lyapunov stability definition pertains to *pertubations of the initial conditions*. This book will be restricted to this case. The stability concept may be extended, so that perturbations in the parameters or in the differential equation are taken into consideration. The reader is referred to the literature for a discussion of this subject, for example, Malkin (1.1).

The purpose of the direct *Lyapunov* method is to determine stability *without performing integration*. This goes one step further than previous methods, because solutions to the linear variational equations were known. We indirectly assume that the integration of the differential equations has been carried out. The characteristics of the solutions (for example, characteristic exponents) were used to determine stability. When the phase space method was used, intermediate integrals were found in order to establish phase surfaces along which the solutions of (1.5.2) were found. The direct method of *Lyapunov* uses *test surfaces* $V(t, \xi_1, ..., \xi_n) = $ const, so-called *Lyapunov functions*. These functions make it possible to determine the course of the solution curves of (1.5.2) in $t, \xi_1, ..., \xi_n$ space and to determine the stability of the stationary solution $\xi_i = 0$, $i = 1, ..., n$. For this purpose, let us first substitute the variable $t \equiv \xi_0$. We will use the relationships

$$\Omega_0 \equiv 1, \qquad \Omega_i = (\partial F_i/\partial q_k)(q_j^0, t)\,\xi_k + \Phi_i, \qquad i, k = 1, 2, ..., n.$$

Equation (1.5.2) can then be written in the form

$$\dot{\xi}_i = \Omega_i(\xi_0, ..., \xi_n), \qquad i = 0, 1, ..., n. \quad (1.5.3)$$

Let us assume that the path element $d\xi_i$, $i = 0, 1, ..., n$ describes a portion of the (1.5.3) solution curve. The vector grad $V = (\partial V/\partial \xi_i)$,

$i = 0,..., n$, is perpendicular to the test surface V in t, ξ_1,..., ξ_n space. Let us form the scalar product $(\partial V/\partial \xi_i)d\xi_i$. The course of the solution curve of (1.5.2) in ξ_1,..., ξ_n space can be inferred from the time derivative of the scalar product

$$Z = (\partial V/\partial \xi_i)\,\dot{\xi}_i = (\partial V/\partial \xi_i)\Omega_i, \qquad i = 0, 1,..., n,$$

which can also be written as

$$Z \equiv \frac{dV}{dt} = \frac{\partial V}{\partial t} + \frac{\partial V}{\partial \xi_i}\,\Omega_i, \qquad i = 1, 2,..., n. \qquad (1.5.4)$$

The decision regarding stability can be made on this basis. In order to do this, we must now deal with the special properties of Lyapunov functions.

1.5.2. *Lyapunov Functions*

Often it is necessary to select Lyapunov functions which are *definite* and under certain circumstances *have an arbitrarily small upper limit*.

Let us first consider the definite property. For this purpose, let us consider two functions $V(t, \xi_1,..., \xi_n)$ and $W(\xi_1,..., \xi_n)$, which vanish for $\xi_1 = \cdots = \xi_n = 0$. It is also assumed that they are single valued within the region in which they are defined and that they have continuous partial derivatives in this region. The function V is called definite if its sign is prescribed. $V(t, \xi_k)$ is *positive definite* or *negative definite* when the relationship

$$V(t, \xi_1,..., \xi_n) \geqslant W(\xi_1,..., \xi_n) \qquad \text{or} \qquad V(t, \xi_1,..., \xi_n) \leqslant -W(\xi_1,..., \xi_n)$$

holds for $t \geqslant t_0 > 0$ and $|\xi_k| \leqslant h$ (where t_0 and h are sufficiently large or small constants, respectively). We assume that $W(\xi_k)$ is a positive definite function which is independent of t.

A function $W(\xi_k)$ is called *positive definite* if it only takes on positive values for $|\xi_k| \leqslant h$ and vanishes for $\xi_1 = \cdots = \xi_n = 0$.

Example: The functions $W = \xi_1^2 + \xi_2^2 + \xi_3^4$ and

$$W = \xi_1^2 + \xi_2^2 + \xi_3^2 - \xi_3^3$$

(for sufficiently small h) are positive definite functions W for $n = 3$. On the other hand, the functions $W = \xi_1^2 + \xi_3^2$ and $W = \xi_1^2 + \xi_2^2 - \xi_3^4$ are not positive definite. In the first case, the function W can certainly be zero if ξ_1 and ξ_3 are both equal to zero, but at the same time ξ_2 can have an arbitrary value that can be different from zero. In the second case, the function W can change sign. A number of criteria for the property of being definite can be found in the work by Malkin (*1.1*).

The semidefinite property must also be mentioned in this connection. A function is called *semidefinite* if it has only one sign, but can become zero without all of its arguments being zero.

The function $W = \xi_1^2 + \xi_3^2$ mentioned above is therefore semidefinite.

Example: The function

$$V = t(\xi_1^2 + \xi_2^2) - 2\xi_1\xi_2 \cos t,$$

is a positive definite Lyapunov function for $n = 2$ and $t > 2$, which depends on t. If we set $W = \xi_1^2 + \xi_2^2$, we have

$$V - W = (t-1)(\xi_1^2 + \xi_2^2) - 2\xi_1\xi_2 \cos t > 0,$$

for $t > 2$, which must hold for a positive definite V. On the other hand, the function $V = e^{-t}(\xi_1^2 + \cdots + \xi_n^2)$ is not positive definite because we have $V \to 0$ for $t \to \infty$, which means that no positive definite W can exist that will always satisfy the inequality $V \geqslant W$.

The function $V(t, \xi_1, \xi_2, ..., \xi_n)$ has an infinitely small upper limit or is, in other words, *uniformly small*, if the inequality

$$|V(t, \xi_1, ..., \xi_n)| \leqslant l \qquad \text{for} \quad t \geqslant t_0, \quad |\xi_k| \leqslant \mu,$$

is satisfied for the arbitrarily small positive numbers l and μ.

Example: This condition is satisfied by $V = \sin t(\xi_1 + \cdots + \xi_n)$ but not by $V = \sin[t(\xi_1 + \cdots + \xi_n)]$. In the latter case, V is bounded but there is no infinitely small upper limit.

1.5.3. *The Stability and Instability Theorems of Lyapunov*

We assume that the variational equations are given by

$$\dot{\xi}_i = \Omega_i(t, \xi_1, ..., \xi_n), \qquad i = 1, ..., n,$$

where the Ω_i are continuous for $t \geqslant t_0$, $|\xi_k| \leqslant h$ in such a way that the variational equations have unique solutions for prescribed initial conditions.

The following theorem holds for stability.

THEOREM 1: *If a definite function $V(t, \xi_k)$ can be prescribed for the variational equations for which the expression $dV/dt = \partial V/\partial t + (\partial V/\partial \xi_i)\,\Omega_i$ is semidefinite with the opposite sign of V or is identically zero, the unperturbed solution is stable.*

Proof: We assume that V is positive definite, so that

$$V \geqslant W \geqslant 0$$

for $t \geqslant t_0$ and $\xi = \max(|\xi_1|,..., |\xi_n|) \leqslant h$. According to Theorem 1, $dV/dt \leqslant 0$. Owing to the positive definite property, $W = 0$ can only hold for $\xi_1 = \cdots = \xi_n = 0$.

Let $\xi = \epsilon$ where ϵ is arbitrarily small, and assume the condition $0 < \epsilon < h$ is satisfied. The continuous function W takes on its lower limit $l > 0$ under the assumptions $\xi = \epsilon$, $|\xi_k| \leqslant \epsilon$. The condition $W \geqslant l$ is just satisfied for $\xi = \epsilon$. Therefore, $\xi < \epsilon$ must be satisfied for $W < l$ and $|\xi_{k0}| < \xi$, where the ξ_{k0} are the initial values of ξ_k.

$V(t_0, \xi_k)$ is independent of t for $t = t_0$. If we select $|\xi_k| \leqslant \eta < \epsilon$, it is possible to have $V(t_0, \xi_k) < l$. We also have $V(t_0, \xi_{k0}) = V_0 < l$ for the initial conditions $|\xi_{k0}| \leqslant \eta$.

According to the assumptions made for dV/dt, we have

$$V - V_0 = \int_{t_0}^{t} (dV/dt)\, dt \leqslant 0,$$

so that $V \leqslant V_0$ results. This means $V \leqslant V_0 < l$ according to the statements made above. Owing to the positive definite property, we also have $W \leqslant V$. Both inequalities together result in $W \leqslant V \leqslant V_0 < l$, or $W < l$. This means that $\xi < \epsilon$ or $|\xi_k| < \epsilon$, respectively, are just satisfied for $k = 1,..., n$ and $t \geqslant t_0$. This is true for arbitrarily small ϵ. This means that the stability condition is satisfied for the stationary solution $\xi_i = 0$, $i = 1,..., n$.

The following represents another stability theorem.

THEOREM 2: *If the assumptions of Theorem 1 are satisfied and if dV/dt is definite and if V has an infinitely small upper limit, the unperturbed solution is asymptotically stable.*

It is possible to give a *geometric representation* for Theorems 1 and 2 if Ω_i does not depend on t and if $n \leqslant 3$. In this case, the positive definite function $V(\xi_1, \xi_2, \xi_3) = C$ specifies a family of surfaces that surround the coordinate origin. In addition, the surface $V = C_1$ lies entirely within the surface $V = C_2$ if $C_1 < C_2$.

For Theorem 1, consider the case where $dV/dt = (\partial V/\partial \xi_i)\,\dot{\xi}_i$ equals zero. This means that the corresponding perturbed solution is completely located on $V = C_k$. If the appropriate initial conditions were selected for the solution, then its "distance" $d[\xi_i, \xi_i^0 = 0]$ from the stationary solution $\xi_i^0 = 0$ will remain sufficiently small. Consequently $\xi_i^0 = 0$ is stable. As may be readily deduced, a definite intermediate integral of the variational equations can be used as a Lyapanov function, because the solutions of the variational equations must lie on the family of surfaces specified by the intermediate integral.

For Theorem 2, let us consider the case where dV/dt is always negative. The condition of an infinitely small upper bound for V means that the family of surfaces specified by $V = C$ is contracted to the origin. The relationship $dV/dt = (\partial V/\partial \xi_i)\,\dot{\xi}_i < 0$ means that an image point corresponding to the perturbed solution moving along the space curve will always penetrate the family of surfaces $V = C$ from the outside to the inside. Since, in addition, $V = C$ passes arbitrarily close to the origin for $C \to 0$, the image point must finally reach $\xi_i = 0$ for $t \to \infty$. This means that the stationary solution is asymptotically stable (Fig. 33).

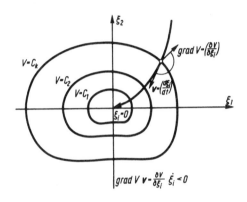

FIG. 33. Representation of the second stability theorem of Lyapunov.

Without presenting the proof, let us mention the instability theorems of A. M. Lyapunov and N. B. Chetayev. The proofs can be found in the book by Malkin (*1.1*).

THEOREM 3:   *Let us assume that a function $V(t, \xi_1 ,..., \xi_n)$ exists which has an infinitely small upper bound and that dV/dt is definite. Also, let us assume that V can have the same sign as dV/dt for arbitrarily large t when the ξ_k are arbitrarily small. The unperturbed solution is then unstable.*

THEOREM 4:   *Let us assume that a region $V > 0$ exists in any arbitrarily small vicinity of the origin or arbitrarily large values of t, in which V is bounded and dV/dt is positive. In this case the unperturbed solution is unstable for autonomous problems.*

Example   (for Theorem 1):

$$\dot{\xi}_1 = -(\xi_1 - 3\xi_2)(1 - 2\xi_1^2 - 4\xi_2^2), \qquad \dot{\xi}_2 = -(\xi_1 + \xi_2)(1 - 2\xi_1^2 - 4\xi_2^2),$$

where $V = \xi_1{}^2 + 3\xi_2{}^2$. We have $dV/dt = (\partial V/\partial \xi_1)\,\dot{\xi}_1 + (\partial V/\partial \xi_2)\dot{\xi}_2$, which results in

$$dV/dt = [2\xi_1(3\xi_2 - \xi_1) - 6\xi_2(\xi_1 + \xi_2)](1 - 2\xi_1{}^2 - 4\xi_2{}^2)$$
$$= -2(\xi_1{}^2 + 3\xi_2{}^2)(1 - 2\xi_1{}^2 - 4\xi_2{}^2).$$

V is positive definite and $dV/dt \leqslant 0$ for sufficiently small values of ξ_1, ξ_2, which means it is semidefinite with the opposite sign to V. Therefore, the unperturbed solution $\xi_1 = \xi_2 = 0$ is stable.

Example (for Theorem 2):

$$\dot{\xi}_1 = -\xi_1{}^3 - 3\xi_2, \qquad \dot{\xi}_2 = 3\xi_1 - 5\xi_2{}^3,$$

with the positive definite function $V = \xi_1{}^2 + \xi_2{}^2$. We have therefore

$$dV/dt = -2\xi_1(\xi_1{}^3 + 3\xi_2) + 2\xi_2(3\xi_1 - 5\xi_2{}^3) = -2(\xi_1{}^4 + 5\xi_2{}^4).$$

$dV/dt < 0$ is negative definite. Because V also has an infinitely small upper limit (the family of curves $V = C$ contracts to the origin for $C \to 0$), the unperturbed solution, $\xi_1 = \xi_2 = 0$ is asymptotically stable.

Example (for Theorem 4):

$$\dot{\xi}_1 = \xi_1{}^2 + \xi_2, \qquad \dot{\xi}_2 = \xi_1 + \xi_2{}^2,$$

where

$$V = (\xi_1{}^3/3) + 2\xi_1\xi_2 + (\xi_2{}^3/3).$$

Apparently a region surrounding the origin of the ξ_1, ξ_2 plane can be specified in which $V > 0$. In addition, V is bounded in it and

$$dV/dt = (\xi_1{}^2 + 2\xi_2)(\xi_1{}^2 + \xi_2) + (2\xi_1 + \xi_2{}^2)(\xi_1 + \xi_2{}^2) > 0.$$

The conditions of Theorem 4 are satisfied, and the unperturbed solution $\xi_1 = \xi_2 = 0$ is unstable.

As an important *example* from mechanics, let us derive the theorem of Lagrange–Dirichlet regarding the *stability of an equilibrium position*, which is as follows.

When the potential energy U has a minimum for an equilibrium position, the equilibrium position is stable.

Example: Let us go from the Cartesian coordinates x_i to the generalized Lagrangian coordinates q_i by means of a coordinate transformation in such a way that the kinetic energy T is a positive definite function of the velocities $\dot{q}_i$, and the equilibrium position $q_{i,s}$ corresponds to the origin. The second condition can always be satisfied. The first condition can be satisfied when the coordinate transformation is independent of time, which means that the system is time independent. In addition, we assume that the potential energy U is only a function of the coordinates.

Let us now consider the problem in $2n$-dimensional phase space q_i , $\dot{q}_i$. The function U should have a minimum equal to zero. Since U is a homogeneous function of the q_i , it takes on the value zero for $q_{i,s} = 0$. U is therefore positive for sufficiently small q_i , which means that U is a positive definite function. T is known to be positive definite, and therefore the Hamilton function $H = T + U$ is positive definite. The energy of the mechanical system is specified by H, and H represents an intermediate integral of the canonical system of differential equations. H can therefore be used as a Lyapunov function which also satisfies $dH/dt = 0$. Therefore, all the conditions of Theorem 1 are satisfied for the "Lyapunov function" H. Therefore, the equilibrium position is stable.

We started with the variational Eqs. (1.5.2). Up to the present, we have considered linear as well as nonlinear terms in the ξ_k appearing on the right side of the system. One very important question is whether or not stability can be determined if the nonlinear terms Φ_i are dropped. We would like to establish a criterion stipulating when so-called *stability in the first approximation* can be analyzed to a sufficient accuracy using the abbreviated, *linear equations of the first approximation*

$$\dot{\xi}_i = a_{ik}\xi_k , \qquad a_{ik} = (\partial F_i/\partial q_k)(q_j{}^0, t). \qquad (1.5.5)$$

In this case, the methods described in Section 1.3 would be sufficient.

The answer to this question is given in the following theorem of Lyapunov.

THEOREM 5: *If a system of the first approximation is regular and if all its characteristic numbers are positive, the unperturbed solution is stable. However, if only one of the characteristic numbers is negative, then the unperturbed solution is unstable. Whether stability or instability prevails under these assumptions can be decided without considering the nonlinear terms in* (1.5.2).

Two important special cases are contained in this theorem. Assume all the a_{ik} are *constant*. Equation (1.5.5) represents a system with constant coefficients which is regular. The characteristic numbers are all positive if the real parts of the characteristic exponents are all negative. If only one of the real parts of the characteristic exponents is positive, there will be one negative characteristic number. In this case, where the coefficients are constant, it is only necessary to analyze the real parts of the characteristic exponents.

Assume the a_{ik} are *periodic* functions ($a_{ik} \equiv p_{ik}$). System (1.5.5) is again a regular system. If the magnitudes of the characteristic multipliers are all smaller than one, all the characteristic numbers are positive. If only one characteristic multiplier is present with a magnitude larger than one, a negative characteristic number will exist. In this case, it is possible to analyze the situation with only the properties of the characteristic multipliers.

In Theorem 5 it was also assumed that the Φ_i are analytic, which means that they can be represented as a power series in the ξ_k that begins with terms of the second or higher order. This assumption can also be somewhat relaxed by requiring only that the Φ_i are continuous and satisfy the inequality

$$| \Phi_i(t, \xi_1, ..., \xi_n)| \leqslant M(| \xi_1 |^{1+\alpha} + \cdots + | \xi_n |^{1+\alpha}).$$

M and α are positive constants. The book by Petrovski contains an example of Theorem 5 under these assumptions for constant a_{ik}, and when the characteristic exponents all have negative real parts.[1]

Let us consider one case in more detail and make the following assumptions. In the equations

$$\dot{\xi}_i = a_{ik}\xi_k + \Phi_i, \qquad i, k = 1, 2, ..., n, \tag{1.5.6}$$

the a_{ik} are constant coefficients. The abbreviated equations (1.5.5), i.e. $\dot{\xi}_i = a_{ik}\xi_k$, have only characteristic exponents with negative real parts. The Φ_i are analytic, independent of t, and can be developed in power series. In this case the characteristic equation $\det(a_{ik} - \lambda \delta_{ik}) = 0$ corresponding to (1.5.5) has only roots λ_σ with negative real parts, and the coefficients of the characteristic equation satisfy the Hurwitz criterion.

Assuming asymptotic stability, let us construct a positive definite Lyapunov function V for the abbreviated Eqs. (1.5.5). It can be obtained from the condition that dV/dt is a negative definite form U. This is because dV/dt must be always negative for positive definite V if the assumption of asymptotic stability is to be satisfied. We therefore have

$$\frac{dV}{dt} = \frac{\partial V}{\partial \xi_i} \dot{\xi}_i = \frac{\partial V}{\partial \xi_i} a_{ik}\xi_k = U. \tag{1.5.7}$$

---

[1] PETROVSKI, J. G., "Lectures on the Theory of Ordinary Differential Equations," p. 162. Teubner, Leipzig, 1954.

Let A_k , $k = 1, 2,...$, and B_j , $j = 1, 2,...$, be the coefficients of the positive or negative, definite forms V and U. If the forms are expanded in (1.5.7) and if the coefficients of equal powers of ξ_i are compared, one obtains a system of algebraic equations that determine the A_k if the B_j are specified:

$$\alpha_{ik}A_k = B_i .\qquad(1.5.8)$$

The desired positive definite form V can indeed be produced in the form of a Lyapunov function if $\det(\alpha_{ik})$ is not zero. The coefficients a_{ik} of (1.5.5) are contained in the elements α_{ik} of this determinant. These elements have special properties, because the characteristic equation formed with them must satisfy the Hurwitz criterion. Based on this, it can be shown that $\det(\alpha_{ik}) \neq 0$ is always satisfied and therefore (1.5.8) always has a solution. Consequently, a form V can always be obtained based on (1.5.8) if the form U is specified.

Example: Let us show this for quadratic forms. Equations (1.5.5) then become

$$\begin{aligned}\dot{\xi}_1 &= a_{11}\xi_1 + a_{12}\xi_2 ,\\ \dot{\xi}_2 &= a_{21}\xi_1 + a_{22}\xi_2 .\end{aligned}\qquad(1.5.9)$$

The characteristic equation is

$$\lambda^2 - (a_{11} + a_{22})\lambda + a_{11}a_{22} - a_{12}a_{21} = 0,$$

and the Hurwitz criterion requires that

$$\begin{aligned}a_{11} + a_{22} &< 0,\\ a_{11}a_{22} - a_{12}a_{21} &> 0.\end{aligned}\qquad(1.5.10)$$

Let the quadratic forms be

$$V = A_1\xi_1{}^2 + A_2\xi_1\xi_2 + A_3\xi_2{}^2, \qquad U = B_1\xi_1{}^2 + B_2\xi_1\xi_2 + B_3\xi_2{}^2,$$

so that (1.5.7), making use of (1.5.9), becomes

$$\begin{aligned}(2A_1\xi_1 + A_2\xi_2)(a_{11}\xi_1 + a_{12}\xi_2) &+ (2A_3\xi_2 + A_2\xi_1)(a_{21}\xi_1 + a_{22}\xi_2)\\ &= B_1\xi_1{}^2 + B_2\xi_1\xi_2 + B_3\xi_2{}^2.\end{aligned}$$

From (1.5.8) it follows that

$$\begin{aligned}2A_1a_{11} + A_2a_{21} &= B_1 ,\\ 2A_1a_{12} + A_2(a_{11} + a_{22}) + 2A_3a_{21} &= B_2 ,\\ A_2a_{12} + 2A_3a_{22} &= B_3 .\end{aligned}\qquad(1.5.11)$$

Therefore, the determinant must satisfy

$$D \equiv \begin{vmatrix} 2a_{11} & a_{21} & 0 \\ 2a_{12} & a_{11} + a_{12} & 2a_{21} \\ 0 & a_{12} & 2a_{22} \end{vmatrix} = 4\{a_{11}[(a_{11}a_{22} - a_{21}a_{12}) + a_{22}^2] - a_{12}a_{21}a_{22}\} \neq 0.$$

Let us make the opposite assumption that $D = 0$. Then we would have to have

$$(a_{11} + a_{22})(a_{11}a_{22} - a_{12}a_{21}) = 0, \qquad (1.5.12)$$

as can easily be calculated. This cannot be the case because of (1.5.10), as the product on the left side of Eq. (1.5.12) would have to become smaller than zero. The equality sign in (1.5.12) is therefore incorrect: The determinant D must be different from zero, (1.5.11) always has a solution, and the form V can indeed be found if U is prescribed.

It may be readily seen that V must be positive definite if U is assumed to be negative definite, because if V were negative, it would agree with the sign of $dV/dt = U$ and (1.5.5) could not have an asymptotic stable stationary solution under these conditions, which disagrees with our assumption. Furthermore, V cannot be positive semidefinite because if dV/dt is always negative and if V became zero only at one point, it would have to become negative, which is not compatible with the assumption that it is positive semidefinite. The only possibility that remains is that V is positive definite, as was postulated.

Let us now use the positive definite form V constructed in this way as a Lyapunov function for the *unabbreviated* Eqs. (1.5.6). For this purpose let us form

$$dV/dt = (\partial V/\partial \xi_i)\dot{\xi}_i.$$

We obtain the following according to (1.5.7):

$$\frac{dV}{dt} = \frac{\partial V}{\partial \xi_i} a_{ik}\xi_k + \frac{\partial V}{\partial \xi_i} \Phi_i = U + \frac{\partial V}{\partial \xi_i} \Phi_i. \qquad (1.5.13)$$

Since the development of $(\partial V/\partial \xi_i)\Phi_i$ in (1.5.13) starts with terms of the third or higher degree in the ξ_i, and because U is negative definite according to our assumption, dV/dt is negative definite in this case also for $|\xi_i| < h$, regardless of the properties of Φ_i. This means that the assumptions for asymptotic stability of $\xi_i = 0$, $i = 1, 2, ..., n$, are satisfied according to Theorem 2, because V is assumed to be positive definite. Apparently the nonlinear terms Φ_i have no influence on the stability. As Theorem 5 states, under these conditions stability can be predicted using only the abbreviated Eqs. (1.5.5), that is, using the equations of the first approximation.

A special case occurs when there are no characteristic exponents with positive real parts (no characteristic multipliers with magnitudes greater than one) for the equations of the first approximation, but when a few of the characteristic exponents have *zero real part* (or if one of the characteristic multipliers has the magnitude one). Lyapunov showed that a so-called *critical case* exists in this case, in which the equations of the first approximation do not indicate whether stability or instability exists. In this case, the analysis must be performed using the unabbreviated Eqs. (1.5.2) and taking the nonlinear terms into account.

We will not consider the critical cases any further. They are treated in the book by Malkin mentioned previously. It must be stated, however, that the critical cases are by no means exceptions, but do indeed occur in problems of practical interest.

1.6. MATHEMATICAL APPROXIMATION METHODS

Very often the differential equations of mechanical systems whose stability is being investigated are linear but have varying coefficients. Often they are even nonlinear, or they can be partial differential equations. This can also be true for the variational equations. This leads to complicated problems. Since the central problem of stability theory is to estimate the solutions of the variational equations and to determine their behavior when parameters are changed, it will be necessary to use approximation methods in such difficult cases. These approximations have been developed in applied mathematics, mathematical physics (especially astronomy) or in control theory. We will now consider a few of these approximation methods.

1.6.1. *Isocline Method*

This method is primarily used to construct *phase curves* in the phase plane in the case of variational equations

$$\dot{\xi}_1 = f_1(\xi_1, \xi_2), \qquad \dot{\xi}_2 = f_2(\xi_1, \xi_2)$$

with nonlinear, time-independent functions. The expression

$$d\xi_2/d\xi_1 = f_2(\xi_1, \xi_2)/f_1(\xi_1, \xi_2) = J(\xi_1, \xi_2)$$

is formed and the curves $J(\xi_1, \xi_2) = $ const are constructed in the ξ_1, ξ_2 plane, the phase plane. These curves are called *isoclines* because they correspond to directional elements having a constant inclination.

If a family of isoclines has been constructed in the phase plane and if the corresponding directional elements have been plotted, this leads to an entire group of directional elements with which families of phase curves can be graphically constructed. An example of this procedure, as well as an example of determining stability from the phase curves, was already given in Section 1.4.1 and Fig. 24.

The book by Kauderer (*2.10*) contains similar procedures leading to the construction of phase curves from directional elements. These procedures were advanced by Liénard and Schäfer.

1.6.2. *Method of Small Oscillations*

In this case we assume that a first-order set of differential equations is given for the mechanical system. For example, it can be the Hamiltonian system

$$\dot{q}_r = \partial H/\partial p_r, \qquad \dot{p}_r = -\partial H/\partial q_r,$$

$$H = H(q_1, ..., q_f, p_1, ..., p_f, t), \qquad r = 1, 2, ..., f.$$

Introducing the new variables $q_1, ..., q_f = y_1, ..., y_f$ and $p_1, ..., p_f = y_{f+1}, ..., y_{2f}$, we obtain

$$\dot{y}_i = f_i(y_1, ..., y_{2f}, t), \qquad i = 1, 2, ..., 2f. \tag{1.6.1}$$

Let us investigate the stability of an "unperturbed" solution y_i^0. Just as was done in Section 1.2, the solutions y_i adjacent to y_i^0 are assumed to have the form

$$y_i(t) = y_i^0(t) + \xi_i(t).$$

By expanding the f_i into a Taylor series and neglecting terms of higher order, the following linear variational equations are obtained from (1.6.1):

$$\dot{\xi}_i = (\partial f_i/\partial y_k)_{y_r^0} \xi_k, \qquad i, k = 1, 2, ..., 2f. \tag{1.6.2}$$

This method is very often used, but can only be applied if the functions f_i are analytic and can be developed into a Taylor series. In addition, the linearized Eqs. (1.6.2) only determine the stability of the unperturbed solutions y_i^0 if the conditions of Lyapunov's Theorem 5 given in 1.5.3 are satisfied.

1.6.3. *Variation of Constants*

This is a method derived from the theory of linear differential equations. Let us assume that the system of variational equations is given by

$$\dot{\xi}_i = a_{ik}\xi_k + \Phi_i. \tag{1.6.3}$$

It consists of the linear system

$$\dot{\xi}_i = a_{ik}\xi_k \tag{1.6.4}$$

and the nonlinear perturbation terms of Φ_i. In the simpler case of linear differential equations, the Φ_i would only be functions of the independent variables, and the abbreviated system (1.6.4) becomes inhomogeneous because of them. Additional difficulties arise bacause, in the case assumed here, the Φ_i are nonlinear functions of the independent as well as of the dependent variables.

Let us assume that the fundamental system ξ_i^ν $(i, \nu = 1, ..., n)$ corresponds to (1.6.4). The trial solution $\xi_i = c_\nu \xi_i^\nu$ of (1.6.3) is assumed, and the c_ν are variables instead of constants. We therefore have

$$\dot{\xi}_i = \dot{c}_\nu \xi_i^\nu + c_\nu \dot{\xi}_i^\nu. \tag{1.6.5}$$

If this is substituted in (1.6.3), it follows that

$$\dot{c}_\nu \xi_i^\nu + c_\nu \dot{\xi}_i^\nu = a_{ik} c_\nu \xi_k^\nu + \Phi_i. \tag{1.6.6}$$

Because of (1.6.4) we have

$$\dot{\xi}_i^\nu = a_{ik}\xi_k^\nu, \tag{1.6.7}$$

which results in the following if this result is substituted into (1.6.6):

$$\dot{c}_\nu \xi_i^\nu = \Phi_i(c_1, ..., c_n, \xi_1^1, ..., \xi_1^n, \xi_2^1, ..., \xi_2^n, ..., \xi_n^n, t). \tag{1.6.8}$$

A nonlinear system of differential equations has thus been found for the c_k which can be solved for the c_k. The final result is

$$\dot{c}_k = \Psi_k(c_1, ..., c_n, \xi_1^1, ..., \xi_1^n, \xi_2^1, ..., \xi_2^n, ..., \xi_n^n, t), \tag{1.6.9}$$

where Ψ_k is a nonlinear function that results from the solution of (1.6.8).

Equation (1.6.9) does not appear to be a simplification, because we departed from the nonlinear system (1.6.3) in the ξ_k and obtained a nonlinear system (1.6.9) in the c_k, which could be possibly just as complicated.

Let us carry out the following transformation that is different in each case:

$$c_k = H_k(K_1, ..., K_n), \qquad k = 1, ..., n. \tag{1.6.10}$$

Equation (1.6.9) becomes

$$(\partial H_k/\partial K_i)\,\dot{K}_i = \Psi_k[H_1(K_1, ..., K_n), ..., H_n(K_1, ..., K_n), \xi_1^1, ..., \xi_n^n, t]. \tag{1.6.11}$$

If (1.6.11) is solved for the $\dot{K}_k$, we obtain the following equation in the place of (1.6.9):

$$\dot{K}_k = G_k(K_1, ..., K_n, \xi_1{}^1, ..., \xi_n{}^n, t). \tag{1.6.12}$$

It should be possible to find certain *average values* for the $\dot{K}_k$ over a time time interval T

$$[\dot{K}_k]_M = (1/T) \int_0^T G_k \, dt, \tag{1.6.13}$$

where the quantities $K_1, ..., K_n$ appearing in G_k can be held constant. In this way an approximation for the functions K_k

$$K_k = [\dot{K}_k]_M t + c_k{}^* \tag{1.6.14}$$

is found. This approximation is justified when the K_k change only slightly in the corresponding time interval T, as has been assumed. The $c_k{}^*$ represent constants of integration.

If this method is applied to nonlinear vibration problems, it is called the *method of slowly changing amplitude* and the time interval T is a period.

Based on (1.6.14), (1.6.10), and the trial solution $\xi_i = c_v \xi_i{}^v$, the following expression is found for the desired approximate solution ξ_i:

$$\xi_i = H_v([\dot{K}_1]_M t + c_1{}^*, ..., [\dot{K}_n]_M t + c_n{}^*) \xi_i{}^v.$$

This is the end of the calculation. The trick is to find a suitable transformation (1.6.10), so that the average value (1.6.13) can be formed in such a way that the K_k can be assumed to be constant over the time T.

Example: Let us consider the nonlinear differential equation $\ddot{x} + \omega_0^2 x + hx^3 = 0$ with the initial conditions $x(0) = A_0$, $\dot{x}(0) = 0$. With $\xi_1 = x$, $\xi_2 = \dot{x}$, it is transformed into

$$\dot{\xi}_1 = \xi_2,$$
$$\dot{\xi}_2 = -\omega_0^2 \xi_1 - h\xi_1^3,$$

in which $\Phi_1 = 0$, $\Phi_2 = -h\xi^3$. The linearized system has the fundamental solutions

$$\xi_1{}^1 = \cos \omega_0 t, \quad \xi_1{}^2 = -\sin \omega_0 t, \quad \xi_2{}^1 = -\omega_0 \sin \omega_0 t, \quad \xi_2{}^2 = -\omega_0 \cos \omega_0 t,$$

so that

$$\xi_1 = c_1 \cos \omega_0 t - c_2 \sin \omega_0 t,$$
$$\xi_2 = -\omega_0 c_1 \sin \omega_0 t - \omega_0 c_2 \cos \omega_0 t \tag{1.6.15}$$

is obtained.

By means of the transformation

$$c_1 = A \cos \Theta, \qquad c_2 = A \sin \Theta, \qquad (1.6.16)$$

the new quantities A and Θ are introduced to replace c_1 and c_2. From (1.6.8) it follows that

$$\dot{c}_1 \cos \omega_0 t - \dot{c}_2 \sin \omega_0 t = 0,$$

$$-\omega_0 \dot{c}_1 \sin \omega_0 t - \omega_0 \dot{c}_2 \cos \omega_0 t = -h(c_1 \xi_1^1 + c_2 \xi_1^2)^3,$$

which, together with (1.6.16), become

$$(\dot{A} \cos \Theta - A \sin \Theta \dot{\Theta}) \cos \omega_0 t - (\dot{A} \sin \Theta + A \cos \Theta \dot{\Theta}) \sin \omega_0 t = 0,$$

$$(\dot{A} \cos \Theta - A \sin \Theta \dot{\Theta}) \sin \omega_0 t + (\dot{A} \sin \Theta + A \cos \Theta \dot{\Theta}) \cos \omega_0 t$$

$$= (hA^3/\omega_0)(\cos \Theta \cdot \cos \omega_0 t - \sin \Theta \cdot \sin \omega_0 t)^3.$$

If trigonometric transformations are applied to this expression and the abbreviation $\varphi = \omega_0 t + \Theta$ is introduced, this system can be transformed into

$$\dot{A} \cos \varphi - \dot{\Theta} A \sin \varphi = 0,$$

$$\dot{A} \sin \varphi - \dot{\Theta} A \cos \varphi = (hA^3/\omega_0) \cos^3 \varphi.$$

Solving with respect to $\dot{A}$ and $\dot{\Theta}$, we obtain,

$$\dot{A} = (hA^3/\omega_0) \cos^3 \varphi \sin \varphi, \qquad \dot{\Theta} = (hA^2/\omega_0) \cos^4 \varphi,$$

which corresponds to Eqs. (1.6.12). Let us obtain the average values of the slowly varying quantities $\dot{A}$ and $\dot{\Theta}$. We find

$$[\dot{A}]_M = \frac{1}{2\pi} \int_0^{2\pi} \frac{hA^3}{\omega_0} \cos^3 \varphi \sin \varphi \, d\varphi = 0,$$

$$[\dot{\Theta}]_M = \frac{1}{2\pi} \int_0^{2\pi} \frac{hA^2}{\omega_0} \cos^4 \varphi \, d\varphi = \frac{3hA^2}{8\omega_0},$$

so that the following expression is obtained for (1.6.14):

$$A = c_1^*, \qquad \Theta = (3hA^2/8\omega_0)t + c_2^*. \qquad (1.6.17)$$

If (1.6.17) is substituted in (1.6.16) and the result substituted in (1.6.15), it follows that

$$\xi_1 = c_1^* \cos\left(\frac{3hA^2}{8\omega_0} t + c_2^*\right) \cos \omega_0 t - c_1^* \sin\left(\frac{3hA^2}{8\omega_0} t + c_2^*\right) \sin \omega_0 t,$$

$$\xi_2 = -\omega_0 \left[c_1^* \cos\left(\frac{3hA^2}{8\omega_0} t + c_2^*\right) \sin \omega_0 t + c_1^* \sin\left(\frac{3hA^2}{8\omega_0} t + c_2^*\right) \cos \omega_0 t \right].$$

These equations are transformed into the following expression by means of trigonometric transformations:

$$\xi_1 = c_1{}^* \cos[(\omega_0 + 3hA^2/8\omega_0)t + c_2{}^*],$$

$$\xi_2 = -\omega_0 c_1{}^* \sin[(\omega_0 + 3hA^2/8\omega_0)\,t + c_2{}^*].$$

The integration constants $c_k{}^*$ can now be determined using the initial conditions $\xi_1(0) = A_0$, $\xi_2(0) = 0$. This leads to $c_2{}^* = 0$, $c_1{}^* = A_0$. The following approximate solution, well known in the theory of vibrations, is thus obtained:

$$\xi_1 \equiv x = A_0 \cos[(\omega_0 + 3hA_0{}^2/8\omega_0)t].$$

1.6.4. Perturbation Method

This approximation method was initially resolved in astronomy but has been widely applied in other branches of physics. The fundamental concepts of this method were probably advanced by Poisson.

Let us assume that the motion of a body in the solar system is given by

$$\dot{q}_i = F_{i0} + \mu^k F_{ik}; \qquad i = 1, 2, ..., n; \qquad k = 1, 2, 3, ..., \qquad (1.6.18)$$

where the F_{i0} are due to the influence of the sun and therefore represent the most important terms of the system of differential equations. The terms $\mu^k F_{ik}$ represent the much smaller influences of the other planets and can therefore be considered as *perturbations* of the motion which is primarily caused by the dominant attraction of the sun. Therefore, we are justified in assuming the following trial solution for (1.6.18):

$$q_i = \mu^k q_{ik}, \qquad k = 0, 1, 2, ... \qquad (1.6.19)$$

If (1.6.19) is substituted into (1.6.18) and if *terms corresponding to equal powers of μ are set equal,* a calculation method is obtained which will now be described in more detail.

Let us consider a more general physical problem for which the canonical system

$$\dot{p}_i = -\partial H/\partial q_i, \qquad \dot{q}_i = \partial H/\partial p_i$$

holds. Assume that the Hamiltion function H is given by

$$H = H_0 + \mu^k H_k, \qquad k = 1, 2, ...$$

Here H_0 is again the dominant term and the $\mu^k H_k$, $k = 1, 2,...$ represent the perturbations. The obvious trial solution $q_i = \mu^k q_{ik}$, $p_i = \mu^k p_{ik}$

will be used. From this point of view, the assumption that the terms $\mu^k F_{ik}$ or $\mu^k H_k$, respectively, are indeed small perturbations is essential. This is certainly the case if μ is a small parameter. The perturbation method is therefore also often called the *method of the small parameter*. However, it has been extended far beyond its original application, and under certain circumstances it can also be used if μ is no longer a small parameter or if such a parameter does not exist and must be artificially introduced into the calculation. We will show this in examples. The perturbation method was developed especially by Poincaré, who used it mostly to find periodic solutions and to investigate their stability. We will not treat this subject further because it belongs in the field of nonlinear vibrations.

The description of the method is as follows.

A system of variational equations

$$\dot{\xi}_i = a_{ik}\xi_k + \mu\Phi_i(\xi_1, \xi_2, ..., \xi_n, t), \qquad i, k = 1, 2, ..., n,$$

is given where Φ_i represents the nonlinear terms. The trial solution for ξ_i is assumed in the form of a series $\xi_i = \xi_{i0} + \mu^r\xi_{ir}$, $r = 1, 2, ...$ This series converges if the parameter μ is sufficiently small. If it is substituted into the system of variational equations, one obtains

$$\dot{\xi}_{i0} + \mu^r\dot{\xi}_{ir} = a_{ik}(\xi_{k0} + \mu^r\xi_{kr}) + \mu\Phi_i(\xi_{s0} + \mu^r\xi_{sr}, t), \qquad s = 1, 2, ..., n.$$

By setting terms corresponding to equal powers of μ equal to each other, the following new system is obtained:

$$\dot{\xi}_{i0} = a_{ik}\xi_{k0},$$

$$\dot{\xi}_{ir} = a_{ik}\xi_{kr} + \Psi_i(\xi_{s0}, ..., \xi_{sm}, t), \qquad i, s, r = 1, 2, ..., n, m < r.$$

The system of the *generating solution* ξ_{i0} is shown on the first line. This solution can easily be found from the system because it is linear. The system of differential equations for the other terms ξ_{ir} of the trial solution is shown on the second line. It is also linear, which means that it is indeed possible to calculate the ξ_{ir}, because Ψ_i only contains known functions $\xi_{s0}, ..., \xi_{sm}$ of t that were determined in previous calculations. This is because $m < r$. Therefore, Ψ_i is a known function of t so that the system of the ξ_{ir} is inhomogeneous but no longer nonlinear.

Example: The differential equation

$$\dot{q} = -aq - \mu bq^2$$

was selected for the calculation. a and b are constants, and μ is the small parameter. The initial condition is assumed to be $q = K$ for $t = 0$. By means of the trial solution

$$q(t) = q_0(t) + \mu q_1(t) + \mu^2 q_2(t) + \cdots,$$

one obtains

$$\dot{q}_0 + \mu \dot{q}_1 + \mu^2 \dot{q}_2 + \cdots = -aq_0 - \mu aq_1 - \mu^2 aq_2 - \cdots - \mu bq_0{}^2 - 2\mu^2 bq_0 q_1 - \cdots.$$

The generating solution q_0 satisfies the equation

$$\dot{q}_0 = -aq_0.$$

The other terms of the trial solution satisfy the equations

$$\dot{q}_1 = -bq_0{}^2 - aq_1, \qquad \dot{q}_2 = -2bq_0 q_1 - aq_2, \dots.$$

The equation $\dot{q}_0 = -aq_0$ together with the initial condition $q_0 = K$ for $t = 0$ results in

$$q_0 = Ke^{-at}.$$

The equations $\dot{q}_1 = -bq_0{}^2 - aq_1 = -bK^2 e^{-2at} - aq_1$, together with the initial condition $q_1 = 0$ for $t = 0$, result in

$$q_1 = (bK^2/a)(e^{-2at} - e^{-at}),$$

from

$$\dot{q}_2 = -2bq_0 q_1 - aq_2 = -(2b^2 K^3/a)(e^{-3at} - e^{-2at}) - aq_2$$

and, together with $q_2 = 0$ for $t = 0$, we obtain

$$q_2 = (b^2 K^3/a^2)(e^{-3at} - 2e^{-2at} + e^{-at}),\dots,$$

etc.

The terms $q_0, q_1, q_2, \dots$ obtained step by step finally lead to the following expression using the trial solution:

$$q = Ke^{-at}[1 + (\mu bK/a)(e^{-at} - 1) + (\mu bK/a)^2(e^{-at} - 1)^2 + \cdots].$$

The *convergence proof* for solutions of this type has not yet been given. We may use the methods of Poincaré for this purpose. For simplicity, we will present the proof for a system of differential equations given by

$$\dot{q}_1 = F_1(q_1, q_2, \mu, t) = \sum c_{1\alpha\beta\gamma} q_1{}^\alpha q_2{}^\beta \mu^\gamma,$$

$$\dot{q}_2 = F_2(q_1, q_2, \mu, t) = \sum c_{2\alpha\beta\gamma} q_1{}^\alpha q_2{}^\beta \mu^\gamma,$$

$$(1.6.20)$$

with two independent variables q_1, q_2. This is the representation of N. Minorsky. It is assumed that the F_i are analytic and can therefore be represented as power series of q_1, q_2, and μ. The coefficients $c_{i\alpha\beta\gamma}$ are continuous functions of t in the interval $J = [t_0, t_1]$ and it is assumed that the series converge for $|q_i|$, $|\mu| < r$.

In addition, the series are assumed not to contain constant terms, so that the system (1.6.20) has the particular solution $q_i = 0$ for $\mu = 0$.

In order to obtain the solution, the following trial solution is used:

$$q_i = \mu^k q_{ik}(t), \qquad i = 1, 2, \qquad k = 1, 2,..., \qquad (1.6.21)$$

where $q_{i1}(t_0) = 0$ is assumed to hold. From (1.6.20) and with (1.6.21) it follows that

$$\dot{q}_{11} = c_{1100}q_{11}(t) + c_{1010}q_{21}(t) + c_{1001},$$
$$\dot{q}_{21} = c_{2100}q_{11}(t) + c_{2010}q_{21}(t) + c_{2001}. \qquad (1.6.22)$$

The $q_{11}(t)$, $q_{21}(t)$ can be determined with the initial conditions $q_{11}(t_0) = q_{21}(t_0) = 0$. The following differential equations are found for the other functions of the trial solution

$$\dot{q}_{1n} = c_{1100}q_{1n}(t) + c_{1010}q_{2n}(t) + \Psi_{1n},$$
$$\dot{q}_{2n} = c_{2100}q_{1n}(t) + c_{2010}q_{2n}(t) + \Psi_{2n}, \qquad (1.6.23)$$

in which the Ψ_{in} are certain polynomials that depend on $c_{i\alpha\beta\gamma}$ and q_{im} if $m < n$. All the q_{in} can be determined step by step from Eqs. (1.6.23). The q_{in} and their derivatives are continuous in the interval J.

In order to show that the series (1.6.21) converge, let us use the auxiliary system

$$\dot{Q}_i = F_i^*(Q_1, Q_2, \mu t) = \sum c_{i\alpha\beta\gamma}^* Q_1^\alpha Q_2^\beta \mu^\gamma, \qquad i = 1, 2, \qquad (1.6.24)$$

where the $c_{i\alpha\beta\gamma}^*$ are majorants of the coefficients $c_{i\alpha\beta\gamma}$ in the interval J.

The trial solution

$$Q_i = \mu^k Q_{ik}(t), \qquad Q_{i1}(t_0) = 0 \qquad (1.6.25)$$

is again assumed for (1.6.24). This first leads to the system

$$\dot{Q}_{11} = c_{1100}^* Q_{11} + c_{1010}^* Q_{21} + c_{1001}^*,$$
$$\dot{Q}_{21} = c_{2100}^* Q_{11} + c_{2010}^* Q_{21} + c_{2001}^*. \qquad (1.6.26)$$

The majorants Q_{i1} to q_{i1} and $\dot{Q}_{i1}$ to $\dot{q}_{i1}$ are obtained for the initial conditions $Q_i(t_0) = 0$. The majorants Q_{in} to q_{in} as well as $\dot{Q}_{in}$ to $\dot{q}_{in}$ are obtained in a completely analogous way. It is sufficient to show that the series $Q_i = \mu^k Q_{ik}(t)$ and their derivatives are uniformly convergent in the interval J if the majorant $c_{i\alpha\beta\gamma}^*$ is appropriately choosen and if μ is sufficiently small.

Let M be the upper limit of the F_i in J assuming that the necessary condition $|q_i|, |\mu| < r$ for the convergence of the F_i is satisfied. The quantity

$$\frac{M(q_1 + q_2 + \mu)}{1 - (q_1 + q_2 + \mu)/r} \tag{1.6.27}$$

is then a majorant for the F_i. The fact that the power series F_i do not have a constant term has been taken into account. The books by Petrovski[1] and Smirnov[2] show why (1.6.27) is a majorant. The system of differential equations

$$\dot{Q}_1 = \dot{Q}_2 = \frac{M(Q_1 + Q_2 + \mu)\left(1 + \dfrac{Q_1 + Q_2 + \mu}{r}\right)}{1 - \dfrac{Q_1 + Q_2 + \mu}{r}} \tag{1.6.28}$$

can be used for the majorants. It is only necessary to show that this system has a solution that vanishes for $t = t_0$ and can be developed in powers in μ in the interval J, provided that $|\mu|$ is sufficiently small.

If we set

$$Q_1 + Q_2 + \mu = r\tau,$$

(1.6.28) is transformed into the differential equation

$$\dot{\tau} = 2M\tau(1 + \tau)/(1 - \tau),$$

where τ is real and the initial conditions are $\tau = \mu/r$ for $t = t_0$. The equation

$$[(1 - \tau)/\tau(1 + \tau)]\,d\tau = 2M\,dt$$

can be integrated and yields the result $c[\tau/(1 + \tau)^2] = e^{2Mt}$. The initial condition determines the constant of integration c as follows:

$$c = [(r + \mu)^2/\mu r]\,e^{2Mt_0}.$$

The final result is

$$\tau/(1 + \tau)^2 = [\mu r/(r + \mu)^2]\,e^{2M(t-t_0)}.$$

[1] PETROVSKI, J. G., "Lectures on Partial Differential Equations," pp. 27–29. Teubner, Leipzig, 1955.

[2] SMIRNOV, V. J., "Higher Mathematics," Vol. IV, pp. 322, 323. Springer, Berlin, 1961.

It is seen that τ is given by $\tau = \alpha(1 + \tau)^2$, where $\alpha = \mu r e^{2M(t-t_0)}/(r+\mu)^2$ or it is given by

$$\tau^2 + [(2\alpha - 1)/\alpha]\, \tau + 1 = 0.$$

This means that τ can only be an analytic function of α if $[(2\alpha - 1)/2\alpha]^2 > 1$ or if $\alpha < \frac{1}{4}$, respectively. Owing to the properties of α, the following condition is obtained:

$$\mu r e^{2M(t-t_0)}/(r+\mu)^2 < \tfrac{1}{4},$$

which can be satisfied, provided that $|\,\mu\,|$ can be selected sufficiently small. This guarantees the convergence of the series (1.6.25) and their derivatives. These series really represent solutions, because they satisfy the differential equations and also are convergent. Consequently, the series (1.6.21) are also convergent solutions of (1.6.20), provided that $|\,\mu\,|$ is sufficiently small.

Let us now consider a case where the perturbation method is applied and *a parameter is introduced artificially.* Also, the trial solution, which is an expansion with respect to powers of the parameter, converges for arbitrary values of the parameter, not only small ones.

The system of variational equations

$$\dot{\xi}_i = a_{ik}(t)\xi_k; \qquad i, k = 1, 2, ..., n, \tag{1.6.29}$$

is cast into the new form

$$\dot{\xi}_i = \mu a_{ik}(t)\xi_k, \tag{1.6.30}$$

making use of the arbitrary parameter μ. The original form is obtained for $\mu = 1$. The following trial solution for (1.6.30) is selected

$$\xi_i = \xi_{i0} + \mu^r \xi_{ir}, \qquad i = 1, 2, ..., n, \qquad r = 1, 2, ... \tag{1.6.31}$$

Substituting (1.6.31) into (1.6.30), the following result is obtained:

$$\dot{\xi}_{i0} + \mu^r \dot{\xi}_{ir} = \mu a_{ik}(t)[\xi_{k0} + \mu^s \xi_{ks}].$$

From this it follows that

$$\dot{\xi}_{i0} = 0, \qquad \xi_{i0} = \text{const} = c_{i0},$$

$$\dot{\xi}_{ir} = a_{ik}(t)\,\xi_{k,r-1}, \tag{1.6.32}$$

$$\dot{\xi}_{ir} = \int_0^t a_{ik}(t)\xi_{k,r-1}\, dt, \qquad i, k = 1, 2, ..., \quad n, r = 1, 2, ...$$

For $t \geqslant 0$ let us assume that $|a_{ik}(t)| \leqslant a$, $i, k = 1, 2,..., n$. Therefore

$$| \xi_{i1} | = \left| \int_0^t a_{ik} \xi_{k0} \, dt \right| \leqslant \left| \sum_{k=1}^n c_{k0} \right| \int_0^t a \, dt$$

and

$$| \xi_{i2} | = \left| \int_0^t a_{ik} \xi_{k1} \, dt \right| \leqslant \left| \sum_{k=1}^n c_{k0} \right| n \int_0^t a \left\{ \int_0^t a \, dt \right\} dt.$$

Partial integration of the result results in

$$\int_0^t a \left\{ \int_0^t a \, dt \right\} dt = \frac{1}{2} \left[\int_0^t a \, dt \right]^2.$$

Therefore, we obtain the result

$$| \xi_{i2} | \leqslant (n/2) \left| \sum_{k=1}^n c_{k_0} \right| \left[\int_0^t a \, dt \right]^2,$$

and the general result is

$$| \xi_{is} | \leqslant \frac{n^{s-1}}{s!} \left| \sum_{k=1}^n c_{k_0} \right| \left[\int_0^t a \, dt \right]^s, \qquad s = 1, 2, 3,... . \qquad (1.6.33)$$

Substitution of (1.6.32) and (1.6.33) into (1.6.31) results in

$$| \xi_i | \leqslant | \xi_{i0} | + | \mu^r \xi_{ir} | \leqslant | c_{i0} | + \frac{\left| \sum_{k=1}^n c_{k_0} \right|}{n} \sum_{r=1}^\infty \frac{n^r}{r!} | \mu^r | \left[\int_0^t a \, dt \right]^r$$

or

$$| \xi_i | \leqslant | c_{i0} | + \frac{\left| \sum_{k=1}^n c_{k_0} \right|}{n} (e^\kappa - 1), \qquad \kappa = n | \mu | \int_0^t a \, dt, \qquad (1.6.34)$$

respectively.

It follows from (1.6.34) that the trial solution (1.6.31) is absolutely convergent for any value of the parameter μ, and therefore also for $\mu = 1$. The solution for the original problem (1.6.21) can be obtained from (1.6.31) by making the parameter μ equal to one in the series development (1.6.31) obtained by the perturbation method theory.

The way in which the perturbation method can be used for *stability investigations* will now be shown.

Example: Consider the differential equation

$$\ddot{y} + p(t) y = 0,$$

where $p(t)$ is a *periodic* function. The notation $\xi_1 = y$, $\xi_2 = \dot{y}$ is introduced which results in the system

$$\dot{\xi}_1 = \xi_2,$$

$$\dot{\xi}_2 = -p\xi_1.$$

It can be written in the form

$$\dot{\xi}_i = p_{ik}\xi_k, \qquad i, k = 1, 2, \tag{1.6.35}$$

where $p_{11} = 0$, $p_{12} = 1$, $p_{21} = -p, p_{22} = 0$. The case considered previously is obtained from

$$\dot{\xi}_i = \mu p_{ik}\xi_k. \tag{1.6.36}$$

The only difference is that the periodic coefficients $p_{ik}(t)$ have taken the place of the general coefficients $a_{ik}(t)$. Equation (1.6.36) becomes (1.6.35) for $\mu = 1$.

Let us consider the results given in Section 1.3.3. It was shown there that the characteristic equation

$$\det(a_{ri} - \rho\,\delta_{ri}) = \begin{vmatrix} a_{11} - \rho & a_{12} \\ a_{21} & a_{22} - \rho \end{vmatrix} = 0$$

must be used for the investigation of stability. We have the equations $a_{ik} = \xi_i^{(k)}(\omega)$. The quantity ω is the common period of the coefficient p_{ik}. The functions $\xi_i^{(k)}$ represent a fundamental solution system for (1.6.35) or (1.6.36) which satisfy the initial conditions $\xi_i^{(k)}(0) = \delta_{ik}$.

In the present case we have

$$p_{11} + p_{22} = 0.$$

Therefore, the Wronskian determinant satisfies

$$W(t) = W(0) \exp\left(\int_0^\sigma p_{ii}\,dt\right) = W(0) = \text{const.}$$

The equation $W(t) = \det[\xi_i^{(k)}(t)]$ holds, and because

$$W(t) = W(0) = \text{const,}$$

it is found that

$$W(\omega) = \xi_1^{(1)}(\omega)\,\xi_2^{(2)}(\omega) - \xi_1^{(2)}(\omega)\,\xi_2^{(1)}(\omega)$$

$$= \xi_1^{(1)}(0)\,\xi_2^{(2)}(0) - \xi_1^{(2)}(0)\,\xi_2^{(1)}(0)$$

$$= \delta_{11}\,\delta_{22} - \delta_{12}\,\delta_{21} = 1.$$

The characteristic equation

$$\begin{vmatrix} a_{11} - \rho & a_{12} \\ a_{21} & a_{22} - \rho \end{vmatrix} = \begin{vmatrix} \xi_1^{(1)}(\omega) - \rho & \xi_1^{(2)}(\omega) \\ \xi_2^{(1)}(\omega) & \xi_2^{(2)}(\omega) - \rho \end{vmatrix} = 0$$

results in

$$\rho^2 - [\xi_1^{(1)}(\omega) + \xi_2^{(2)}(\omega)]\,\rho + [\xi_1^{(1)}(\omega)\,\xi_2^{(2)}(\omega) - \xi_1^{(2)}(\omega)\,\xi_2^{(1)}(\omega)] = 0,$$

which is also equal to

$$\rho^2 - a_1\rho + 1 = 0, \qquad a_1 = \xi^{(1)}(\omega) + \xi_2^{(2)}(\omega).$$

The roots $\rho_{1,2}$ of this quadratic equation are the characteristic multipliers. The absolute magnitude of these multipliers determines whether stability exists. According to Section 1.3.3, the $\rho_{1,2}$ are related to $a_1/2$ in such a way that the quantity

$$\tfrac{1}{2}[\xi_1^{(1)}(\omega) + \xi_2^{(2)}(\omega)]$$

can also be used to determine stability. In this case, one would have to know $\xi_1^{(1)}(\omega)$, $\xi_2^{(2)}(\omega)$. The latter functions are obtained according to the perturbation calculation method using (1.6.36).

First, the relationship $\xi_2^{(2)} = \dot{\xi}_1^{(2)}$ is used. We can set

$$\tfrac{1}{2}[\xi_1^{(1)}(\omega) + \xi_2^{(2)}(\omega)] = \tfrac{1}{2}[\xi_1^{(1)}(\omega) + \dot{\xi}_1^{(2)}(\omega)].$$

It is permissible to use the two particular solutions $\xi_1^{(1)}$ and $\xi_1^{(2)}$. Since $\xi_1^{(k)}(0) = \delta_{ik}$ and $\dot{\xi}_1^{(1)} = \xi_2^{(1)}$, $\dot{\xi}_1^{(2)} = \xi_2^{(2)}$, the initial conditions are given by

$$\xi_1^{(1)}(0) = 1, \qquad \dot{\xi}_1^{(1)}(0) = 0, \qquad \xi_1^{(2)}(0) = 0, \qquad \dot{\xi}_1^{(2)}(0) = 1.$$

The following trial solutions are assumed for (1.6.36):

$$\xi_1^{(1)} = \xi_{10}^{(1)} + \mu\xi_{11}^{(1)} + \mu^2\xi_{12}^{(1)} + \cdots, \qquad \xi_1^{(2)} = \xi_{10}^{(2)} + \mu\xi_{11}^{(2)} + \mu^2\xi_{12}^{(2)} + \cdots,$$

$$(1.6.37)$$

which must satisfy the differential equation $\ddot{\xi}_1^{(r)} = -p\mu_1^{(r)}$, $r = 1, 2$, as well as the initial conditions mentioned above. We require that

$$\xi_{10}^{(1)}(0) = 1, \qquad \dot{\xi}_{10}^{(1)}(0) = 0, \qquad \xi_{10}^{(2)}(0) = 0, \qquad \dot{\xi}_{10}^{(2)}(0) = 1,$$

$$\xi_{1k}^{(1)}(0) = \dot{\xi}_{1k}^{(1)}(0) = \xi_{1k}^{(2)}(0) = \dot{\xi}_{1k}^{(2)}(0) = 0, \qquad k = 1, 2, \ldots .$$

If (1.6.37) is substituted into (1.6.36), the following differential equations are obtained:

$$\ddot{\xi}_{10}^{(r)} = 0,$$

$$\ddot{\xi}_{1k}^{(r)} = -p\xi_{1,k-1}^{(r)}, \qquad r = 1, 2, \qquad k = 1, 2, 3, \dots.$$

Because of the initial conditions, the following result is obtained:

$$\xi_{10}^{(1)} = 1, \qquad \xi_{1k}^{(1)} = \int_0^t \left\{ \int_0^t -p\xi_{1,k-1}^{(1)} \, dt \right\} dt,$$

$$\xi_{10}^{(2)} = t, \qquad \dot{\xi}_{1k}^{(2)} = \int_0^t -p\xi_{1,k-1}^{(2)} \, dt, \qquad k = 1, 2, \dots. \tag{1.6.38}$$

The terms of the series expansion (1.6.37) can be obtained step by step in this way.

It is already known that the series (1.6.37) converge for any value of μ, and therefore also for $\mu = 1$. The desired fundamental solutions of the original system (1.6.35) can be obtained from (1.6.37) for $\mu = 1$.

If $\mu = 1$ and $\xi_{10}^{(1)} = 1$, $\xi_{10}^{(2)} = t$ and if (1.6.37) is substituted into $\frac{1}{2}[\xi_1^{(1)}(\omega) + \dot{\xi}_1^{(2)}(\omega)]$, the result

$$\tfrac{1}{2}[\xi_1^{(1)}(\omega) + \dot{\xi}_1^{(2)}(\omega)] = 1 + \tfrac{1}{2} \sum_{k=1}^{\infty} [\xi_{1k}^{(1)}(\omega) + \dot{\xi}_{1k}^{(2)}(\omega)] \tag{1.6.39}$$

is obtained. Based on (1.6.38) an estimation of the magnitude of the characteristic multipliers can be made, which is equivalent to a statement about stability.

Let $p(t) \leqslant 0$. It then follows from (1.6.38) that

$$\xi_{11}^{(1)} > 0, \qquad \dot{\xi}_{11}^{(2)} > 0.$$

Also, $\xi_{11}^{(2)} > 0$ because according to this $\xi_{11}^{(2)}$ increases and is zero for $t = 0$ according to the initial conditions.

The result $\xi_{1k}^{(1)} > 0$, $\dot{\xi}_{1k}^{(2)} > 0$, $\xi_{1k}^{(2)} > 0$ for $k > 1$ is obtained in a completely similar way. Owing to this fact and because (1.6.39), the final result

$$\tfrac{1}{2} \,|\, \xi_1^{(1)}(\omega) + \xi_2^{(2)}(\omega)| = \tfrac{1}{2} \,|\, \xi_1^{(1)}(\omega) + \dot{\xi}_1^{(2)}(\omega)| > 1$$

is obtained. As was shown in Section 1.3.3, this means that the two roots $\rho_{1,2}$ of the characteristic equation become real. Their magnitudes are larger or smaller than one, respectively. It can be concluded from this instability is present! In the same way it can be shown that

$$\tfrac{1}{2} \mid \xi_1^{(1)}(\omega) + \xi_2^{(2)}(\omega) \mid < 1$$

holds for $p(t) \geqslant 0$ and $\omega \int_0^\omega p \, dt \leqslant 4$ so that stability prevails.

The so-called *secular terms* occur often when the perturbation method theory is applied to nonlinear vibration problems. These are the terms of the solution series whose amplitude increases indefinitely with time. Important characteristics of the process in question can be obtained by setting up conditions for eliminating the secular terms. For example, this can be done by physical reasoning such as considering the energy balance of the mechanical process represented by the differential equation.

Example: Let us assume that the motion of a point mass suspended from a nonlinear spring is governed by the differential equation $\ddot{x} + \omega_0^2 x + \mu x^3 = 0$ and that the initial conditions are $x = A$, and $\dot{x} = 0$ for $t = 0$. If it is assumed that the magnitude of μ is sufficiently small, it is possible to use the trial solution

$$x = x_0 + \mu x_1 + \mu^2 x_2 + \cdots, \qquad \omega^2 = \omega_0^2 + \mu f_1(A) + \mu^2 f_2(A) + \cdots.$$

The initial conditions $x_0 = A$, $\dot{x}_0 = 0$, $x_k = \dot{x}_k = 0$ for $k \geqslant 1$, $t = 0$ can be specified. If this trial solution is substituted into the differential equation, the following generating solution is obtained:

$$\ddot{x}_0 + \omega^2 x_0 = 0.$$

It follows that $x_0 = A \cos \omega t$ because of the initial conditions.

If all the terms connected with μ are considered in the next step, the following is obtained:

$$\ddot{x}_1 + \omega^2 x_1 = f_1 x_0 - x_0^3 = \left(f_1 A - \frac{3A^3}{4} \right) \cos \omega t - \frac{A^3}{4} \cos 3\omega t,$$

and the general solution is

$$x_1 = c_1 \cos \omega t + c_2 \sin \omega t + \frac{\omega t}{2} \left(f_1 A - \frac{3A^3}{4} \right) \sin \omega t + \frac{A^3}{32\omega^2} \cos 3\omega t.$$

The third term is the secular term, which vanishes for

$$f_1 A - (3A^3)/4 = 0.$$

Since $A = 0$ would be trivial, the secular term vanishes if $f_1 = (3A^2)/4$. The initial conditions $x_1(0) = \dot{x}_1(0) = 0$ determine the integration constants as follows:

$$c_1 = -A^3/32\omega^2, \qquad c_2 = 0,$$

and the final result is

$$x_1 = -(A^3/32\omega^2)(\cos \omega t - \cos 3\omega t).$$

In the third step, the terms containing μ^2 result in the equation

$$\ddot{x}_2 + \omega^2 x_2 = f_1 x_1 + f_2 x_0 - 3x_0{}^2 x_1$$

$$= -\frac{3A^2}{4} \frac{A^3}{32\omega^2} (\cos \omega t - \cos 3\omega t)$$

$$+ f_2 A \cos \omega t + 3A^2 \frac{A^3}{32\omega^2} \cos^2 \omega t(\cos \omega t - \cos 3\omega t).$$

The general solution of x_2 can be calculated from this, which again contains two constants of integration and a secular term. The secular term vanishes if $f_2 = -(3A^4)/128\omega^2$. Taking the boundary conditions into account, the result obtained is

$$x_2 = -(A^5/1024\omega^4)(\cos \omega t - \cos 5\omega t).$$

If the calculation is terminated at this point, the following approximate solution for the nonlinear vibration process described by the original differential equation is obtained:

$$x = A \cos \omega t - (\mu A^3/32\omega^2)(\cos \omega t - \cos 3\omega t) - (\mu^2 A^5/1024\omega^4)(\cos \omega t - \cos 5\omega t).$$

The quantities f_1 and f_2 were determined by making the secular terms vanish. With the trial solution for ω^2, the following expression is found, which specifies the variation of frequency with amplitude:

$$\omega^2 = \omega_0{}^2 + (\mu 3A^2/4) - (\mu^2 3A^4/128\omega^2).$$

Let us consider the perturbation method theory for the case in which the functions Φ_i connected with μ appearing in the expression

$$\dot{\xi}_i = a_{ik}\xi_k + \mu\Phi_i(\xi_1, \xi_2, ..., \xi_n, \mu, t)$$

are no longer analytic. The general trial solution $\xi_i = \xi_{i0} + \mu^r \xi_{ir}$ generally used in perturbation method theory is no longer valid. The more general trial solutions must be used $\xi_i = \sum_{s=0}^{\infty} \xi_{i,s}$, whose terms can be obtained from the following iteration formula:

$$\dot{\xi}_{i,s} - a_{ik}\xi_{k,s} = \mu\Phi_i(\xi_{1,s-1}, ..., \xi_{n,s-1}, \mu, t).$$

Functional analysis must be used in every individual case to determine whether the successive approximation method for ξ_i is convergent or not.

1.6.5. *The Method of Galerkin and Ritz*

Often ordinary boundary value problems

$$D[x(t)] = F(t),$$
$$U[x(t)]_R = H(t)_R \tag{1.6.40}$$

are encountered in stability studies, where D and U are operators and F and H are functions. The subscript R refers to boundary values. The second line of (1.6.40) contains the boundary conditions.

Let us restrict ourselves to *homogeneous problems* and let us also assume that the operator D has a special form. We shall consider a case of great practical importance, which is the boundary value problem

$$L[x(t)] + N[x(t)] - cx(t) = 0,$$
$$U[x(t)]_R = 0 \tag{1.6.41}$$

in which L is a *linear and positive definite* differential operator that is *self-adjoint* with respect to the boundary conditions. N is a nonlinear differential operator in the most general case. In particular, N can also be a non-self-adjoint, linear operator or a nonlinear function.

Let us consider the following linear auxiliary problem corresponding to (1.6.41)

$$L(\varphi(t)) - \lambda \varphi(t) = 0,$$
$$U(\varphi(t))_R = 0. \tag{1.6.42}$$

It is called the *abbreviated problem*. It has the *same boundary conditions* as the "unabbreviated problem" (1.6.41), which is very important.

Because of the assumptions made for L, $\lambda = 0$ is not an eigenvalue of the abbreviated problem. This means that a Green function $G(t, \tau)$ with respect to the boundary conditions exists. Therefore $L(\varphi) = \lambda \varphi$ for φ. The expressions

$$\varphi = \lambda L^{-1}(\varphi), \tag{1.6.43}$$

or

$$\varphi = \lambda \int_a^b G(t, \tau)\, \varphi(\tau)\, d\tau \tag{1.6.44}$$

are formed. The inverse operator L^{-1} of L exists which is closely related to the Green function of L. This fact can be seen from a comparison of (1.6.43) with (1.6.44). In other words: *the integral Eq. (1.6.44) is equivalent to the problem (1.6.42).*

The abbreviated problem has an infinite number of eigenvalues λ_i , $i = 1, 2, 3,...$, which are all positive and real. The corresponding eigenfunctions are a *closed, orthogonal, and normalized* function system. The Green function is *symmetrical and positive definite* and can be expanded into the series

$$G(t, \tau) = \sum_{i=l}^{\infty} \frac{\varphi_i(t)\,\varphi_i(\tau)}{\lambda_i} , \qquad (1.6.45)$$

which is absolutely and uniformly *convergent* according to the theorem of Mercer. The eigenfunctions $\varphi_i(t)$ are orthogonal and normalized and therefore satisfy the conditions

$$(\varphi_i , \varphi_k) = \delta_{ik} ,$$
$$\{\varphi_i , L(\varphi_k)\} = [\varphi_i , \varphi_k] = \lambda_k\,\delta_{ik} . \qquad (1.6.46)$$

The following notation has been used.
The expression

$$(\varphi_i , \varphi_k) = \int_a^b \varphi_i\varphi_k\,dt$$

is the scalar product of the functions φ_i , φ_k . The expression

$$[\varphi_i , \varphi_k] = \{\varphi_i , L(\varphi_k)\} = \int_a^b \varphi_i L(\varphi_k)\,dt$$

is the "generalized scalar product" of the same functions. If the operator L is self-adjoint, it means that the relationships

$$\{\varphi_i , L(\varphi_k)\} = \{\varphi_k , L(\varphi_i)\}$$

or

$$[\varphi_i , \varphi_k] = [\varphi_k , \varphi_i] \qquad (1.6.47)$$

are satisfied.

Since the abbreviated and the unabbreviated problem satisfy the same boundary conditions, the solutions $x(t)$ of the unabbreviated problem (1.6.41) satisfy the boundary conditions of the abbreviated problem (1.6.42). Since the eigenfunctions $\varphi_i(t)$ of (1.6.42) are a closed system of functions, the expansion theorem can be applied. The series expansion

$$x(t) = \sum_{k=1}^{\infty} \rho_k\varphi_k(t), \qquad (1.6.48)$$

which is absolutely and uniformly convergent, can be formed.

The *Galerkin method* consists of substituting the series (1.6.48) into the differential equation instead of $x(t)$. All terms are multiplied by $\varphi_i(t)$ and integration is performed over the interval prescribed by the boundary values a, b of t. In this way the following algebraic and infinite system of equations is found:

$$\sum_{k=1}^{\infty} \rho_k \int_a^b \varphi_i L(\varphi_k)\, dt - c \sum_{k=1}^{\infty} \rho_k \int_a^b \varphi_i \varphi_k\, dt$$

$$= -\int_a^b \varphi_i N \left(\sum_k \rho_k \varphi_k \right) dt, \qquad i = 1, 2, 3,\ldots, \qquad (1.6.49'')$$

which determines the expansion coefficients ρ_k. It can also be written in the form

$$\sum_k \rho_k[\varphi_i, \varphi_k] - c \sum_k \rho_k(\varphi_i, \varphi_k) = -(\varphi_i, N), \qquad i = 1, 2, 3,\ldots . \qquad (1.6.49')$$

Equation (1.6.46) is then applied. Several simplifications occur and the result is

$$\rho_i = -\frac{(\varphi_i, N)}{\lambda_i - c} = -\frac{1}{\lambda_i - c} \int_a^b \varphi_i N \left(\sum_{k=1}^{\infty} \rho_k \varphi_k \right) dt, \qquad i = 1, 2, 3,\ldots .$$

$$(1.6.49)$$

Since it is not possible to solve immediately the infinite system of equations (1.6.49), the Galerkin method stipulates that the infinite series $\sum_k \rho_k \varphi_k$ on the right side of (1.6.49) is truncated at the term having n and finite order. This also makes the system of equations finite. The subscript i runs from one to n. This leads to the *Galerkin equations*

$$\rho_i^* = -\frac{1}{\lambda_i - c} \int_a^b \varphi_i N \left(\sum_{k=1}^{n} \rho_k^* \varphi_k \right) dt, \qquad i = 1, 2,\ldots, n, \qquad (1.6.50)$$

from which the approximation values ρ_i^* can be calculated. According to (1.6.48), the approximation

$$x(t) = \sum_{k=1}^{n} \rho_k^* \varphi_k(t) \qquad (1.6.51)$$

is formed from them. This is an approximation for the desired solution $x(t)$ of (1.6.41).

By letting n increase at each step of the calculation, a sequence of successive approximation is carried out. The convergence of the sequence must be investigated in each individual case.

Let us consider the simplest case where the operator N is not self-adjoint but is *linear*, so that

$$N \equiv L^*,$$

$$\int_a^b \varphi_i N \left(\sum_k \rho_k \varphi_k \right) dt \equiv \sum_k \rho_k \int_a^b \varphi_i L^*(\varphi_k) \, dt = \sum_k \rho_k \{\varphi_i , L^*(\varphi_k)\}.$$

Equation (1.6.49) then becomes

$$\rho_i(\lambda_i - c) + \sum_k \rho_k \{\varphi_i , L^*(\varphi_k)\} = 0, \qquad i = 1, 2, 3,\ldots,$$

or

$$\sum_{k=1}^{\infty} a_{ik}\rho_k = 0,$$

$$a_{ik} = (\lambda_i - c)\, \delta_{ik} + \{\varphi_i , L^*(\varphi_k)\}, \qquad i = 1, 2, 3,\ldots . \tag{1.6.52}$$

Equations (1.6.52) again represent an infinite algebraic system of equations which is now *linear* for determining the ρ_k . The ρ_k can be obtained by successive approximation according to the Galerkin method. The theory of infinite determinants of von Koch can be used to determine whether the successive approximation converges toward the unique solution of (1.6.52) and whether the Galerkin method is applicable.

Let us now consider the more general case where N is a *nonlinear* operator or is a nonlinear function.

If $L(\varphi) = \lambda\varphi$ possesses the eigenfunctions φ_i and the eigenvalues λ_i , $L(\varphi) - c\varphi = \kappa\varphi$ will have the same eigenfunctions φ_i and the eigenvalues $\kappa_i = \lambda_i - c$. If $L(\varphi)$ with the given boundary conditions has the inverse

$$L^{-1}(u) = \int_a^b G(t, \tau)u \, d\tau$$

with the Green function

$$G(t, \tau) = \sum_k \frac{\varphi_k(t)\, \varphi_k(\tau)}{\lambda_k} ,$$

then $L(\varphi) - c\varphi = L^*(\varphi)$ has the inverse

$$L^{*-1}(u) = \int_a^b G^*(t, \tau)u \, d\tau$$

for the same boundary conditions and the Green function is

$$G^*(t, \tau) = \sum_k \frac{\varphi_k(t)\, \varphi_k(\tau)}{\lambda_k - c} . \tag{1.6.53}$$

We have assumed that $c \neq \lambda_k$ holds so that G^* indeed exists. The case $c = \lambda_k$ must be examined separately which will not be done here.

It follows from $L(x) + N(x) - cx = 0$ that

$$L^*(x(t)) = -N\{x(t)\},$$

$$x(t) = -L^{*-1}\{N(x)\}, \qquad (1.6.54)$$

$$x(t) = -\int_a^b G^*(t, \tau) N\{x(\tau)\} \, d\tau.$$

Let us again consider (1.6.49) and let us rewrite this equation with another integration variable. We then obtain

$$\rho_i = -\int_a^b \frac{\varphi_i(\tau)}{\lambda_i - c} N\left(\sum_k \rho_k \varphi_k(\tau)\right) d\tau, \qquad i = 1, 2, 3, \dots .$$

Each of these ρ_i is now multiplied by $\varphi_i(t)$, and the result is summed over i.

This results in

$$\sum_i \rho_i \varphi_i(t) = -\int_a^b \sum_i \frac{\varphi_i(t) \, \varphi_i(\tau)}{\lambda_i - c} N\left(\sum_k \rho_k \varphi_k(\tau)\right) d\tau. \qquad (1.6.55)$$

When this result is compared with (1.6.48) and (1.6.53), it is found that (1.6.55) is the nonlinear integral equation (1.6.54). It follows that the successive approximation method according to the Galerkin equations is completely equivalent to the successive approximation method with the integral equation (1.6.54) if

$$x^{(n)}(t) = -\int_a^b G^{*(n)}(t, \tau) N\left(\sum_{k=1}^n \rho_k \varphi_k(\tau)\right) d\tau,$$

$$\qquad (1.6.56)$$

$$G^{*(n)}(t, \tau) = \sum_{k=1}^n \frac{\varphi_k(t) \, \varphi_k(\tau)}{\lambda_k - c} .$$

A solution of the system of Eqs. (1.6.49) can be found and the solution of (1.6.41) can be written in the form (1.6.48) if it can be shown that there is a solution for the integral equation (1.6.54). If this is the case, the Galerkin method can be used and then consists of solving the reduced system of Eqs. (1.6.50), provided that it may also be shown that the approximation (1.6.56) is close to the solution of (1.6.54). Therefore, everything has been reduced to the investigation of nonlinear integral equations. The works by Leipholz[1] and Krasnoselski[2] contain a detailed

---

[1] LEIPHOLZ, H., *Acta Mech.* 1, 339–353 (1965).

[2] KRASNOSELSKI, M. A., "Topological Methods in the Theory of Non-Linear Integral Equations." Oxford Univ. Press, London and New York, 1964.

treatment of the subject, especially a discussion of the Galerkin method convergence. These works show that the Galerkin method does indeed converge for special nonlinear operators or functions N. A very special case exists if c is equal to zero and the operator N is a *function of the independent variable t*. We then have the problem

$$L[x(t)] = f(t),$$
$$U[x(t)]_R = 0$$

(1.6.57)

and the Galerkin method can be transformed into the Ritz method. From this it can be seen *that the Ritz method is a special case of the much more general Galerkin method*.

It can be shown that obtaining a solution of (1.6.57) is the same as finding a function $x(t)$ which makes the functional

$$\Phi(x) = \{L(x), x\} - 2(x, f)$$

(1.6.58)

a minimum. If the term $\{L(x_0), x_0\}$, where x_0 is the unique solution of (1.6.57) is added and subtracted to the right side (1.6.58), the following is obtained, taking (1.6.47) and the relationship $L(x_0) = f$ into account:

$$\Phi(x) = \{L(x - x_0), x - x_0\} - \{L(x_0), x_0\}$$

or

$$\Phi(x) = [(x - x_0), (x - x_0)] - [x_0, x_0].$$

(1.6.59)

The relationship (1.6.59) shows that $\Phi(x)$ is a minimum when the first term is zero, which is the case for $x = x_0$. The solution of (1.6.57) makes the functional (1.6.58) a minimum, as we have postulated.

We may also show that the function x_0, which makes (1.6.58) a minimum is also a solution of (1.6.57). Let α be an arbitrary number, and φ an arbitrary function. Since x_0 makes $\Phi(x)$ a minimum based on our assumptions, we have

$$\Phi(x_0 + \alpha\varphi) - \Phi(x_0) \geqslant 0.$$

If we take (1.6.58) and (1.6.47) into account, the quadratic form in α may be rewritten as follows:

$$2\alpha\{L(x_0) - f, \varphi\} + \alpha^2\{L(\varphi)\varphi\} \geqslant 0.$$

It is not negative. Its discriminant must therefore satisfy

$$\{L(x_0) - f, \varphi\}^2 \leqslant 0,$$

and this may only be true for $L(x_0) = f$, i.e., x_0 is a solution of (1.6.57), which was postulated previously.

Instead of solving (1.6.57), the extremal problem $\Phi(x) = \min$ may be solved. It is assumed that L is positive definite [Eq. (1.6.57) has only *one* solution] and self-adjoint [because (1.6.47) is employed].

The *Ritz method* consists of substituting the series trial solution (1.6.48) into $\Phi(x)$, resulting in

$$\Phi(x) = \left\{\sum_k \rho_k L(\varphi_k), \sum_i \rho_i \varphi_i\right\} - 2\left(\sum_k \rho_k \varphi_k , f\right).$$

If (1.6.47) is employed, this relationship may be rewritten as follows:

$$\Phi(x) = \rho_i^2\{\varphi_i , L(\varphi_i)\} + 2 \sum_{k \neq i} \{\varphi_i , L(\varphi_k)\} \rho_i \rho_k - 2\rho_i(\varphi_i , f) - 2 \sum_{k \neq i} \rho_k(\varphi_k , f)$$

in which one arbitrary index i has been emphasized. The desired solution may be obtained by finding the minimum of $\Phi(x)$. The equation

$$\partial\Phi/\partial\rho_i = 0, \qquad i = 1, 2, 3,...,$$

is thus formed. This leads to

$$\partial\Phi/\partial\rho_i = 2 \sum_k \rho_k\{\varphi_i , L(\varphi_k)\} - 2(\varphi_i , f) = 0,$$

so that the infinite system of equations

$$\sum_{k=1}^{\infty} \rho_k[\varphi_i , \varphi_k] = (\varphi_i , f), \qquad i = 1, 2, 3,...,$$

is obtained for determining the coefficients ρ_k. In formal terms, it is completely analogous to the system (1.6.49′) if we set $c = 0$ and if the function $-f$ is substituted for the operator N. This exact assumption was employed for the transition from (1.6.41) to (1.6.57). This shows that the Ritz and Galerkin methods are closely related. The Ritz method is a special case of the Galerkin method, which is only valid for certain special assumptions.

If the Ritz method is continued, a finite system of equations is used instead of an infinite system. An iteration method using

$$\sum_{k=1}^{n} \rho_k[\varphi_i , \varphi_k] = (\varphi_i , f), \qquad i = 1, 2, 3,..., n,$$

is employed. This determines the series coefficients ρ_k of the solution (1.6.48) to any desired degree of accuracy.

The principles of mechanics may be used to derive the methods of Galerkin and Ritz. Let us first consider the most general principle, which is the principle of virtual displacements, and let us assume a deformable solid body. In a discussion of the motion or equilibrium of a body, the theory of stability assumes that the body undergoes a perturbation. In mathematical terms, this perturbation may be treated as a virtual displacement $\delta\mathbf{w}$. The *principle of virtual displacements* results in an equation specifying the time behavior of the displacement vector $\mathbf{w}$. If the magnitude of $\mathbf{w}$ is bounded for all times, the stability of the corresponding equilibrium position prevails.

The principle of virtual displacements results in the relationship

$$-\int_V \mu\ddot{\mathbf{w}}\,\delta\mathbf{w}\,dV - \delta(\Pi_i + \Pi_a) + \int_O \mathbf{k}_a\,\delta\mathbf{w}\,dO + \sum_i \mathbf{K}_{ai}\,\delta\mathbf{w}_i = 0.$$

The following notation is used: μ is the mass density, $\mathbf{w}$ is the displacement vector, Π_i is the potential of internal monogenetic forces, Π_a potential of external polygenetic forces, V the volume, O the elastic system surface, $\mathbf{k}_a$ the polygenetic external surface forces, $\mathbf{K}_{ai}$ polygenetic external single forces, $\delta\mathbf{w}_i$ the displacement vectors of the points at which the latter are applied, and the double overdots refer to double differentiation with respect to time.

Monogenetic forces may be derived from a scalar function. Conservative forces represent one example of these forces, since they may be derived from scalar potential functions, and their virtual work is the variation of the corresponding potential. Polygenetic forces, on the other hand, cannot be derived from a scalar function, and their virtual work is not the variation of a scalar function. Friction forces represent an example of polygenetic forces, as well as the so-called "follower" forces, which change their direction with the deformation of the body along which they act. The concepts of monogenetic and polygenetic forces are explained in detail in the study of Lanczos (6).

The potential energy is given by $\Pi_i + \Pi_a = A$, so that the principle of virtual displacements may also be written in the following form:

$$\delta A = -\int_V \mu\ddot{\mathbf{w}}\,\delta\mathbf{w}\,dV + \int_O \mathbf{k}_a\,\delta\mathbf{w}\,dO + \sum_{i=1}^{n} \mathbf{K}_{ai}\,\delta\mathbf{w}_i. \qquad (1.6.60)$$

If we note that $T = \int_V \frac{1}{2}\cdot\mu\mathbf{w}^2\,dV$ is the kinetic energy and that the Lagrangian L and the Hamiltonian H satisfy the following relationships,

$$L = T - A, \qquad H = T + A,$$

the following new equations may be obtained in place of (1.6.60). If

$$\delta T = \int_V \mu \dot{\mathbf{w}} \, \delta \dot{\mathbf{w}} \, dV$$

is subtracted from both sides of (1.6.60) and if multiplication by -1 is performed, it follows that

$$\delta L = \int_V \mu (\ddot{\mathbf{w}} \, \delta \mathbf{w} + \dot{\mathbf{w}} \, \delta \dot{\mathbf{w}}) \, dV - \int_O \mathbf{k}_a \, \delta \mathbf{w} \, dO - \sum_{i=1}^n \mathbf{K}_{ai} \, \delta \mathbf{w}_i \, ,$$

or

$$\delta L = \frac{d}{dt} \int_V \mu \dot{\mathbf{w}} \, \delta \mathbf{w} \, dV - \int_O \mathbf{k}_a \, \delta \mathbf{w} \, dO - \sum_{i=1}^n \mathbf{K}_{ai} \, \delta \mathbf{w}_i \, . \qquad (1.6.61)$$

On the other hand, if δT is added on both sides of (1.6.60), the result is

$$\delta H = \int_V \mu (\dot{\mathbf{w}} \, \delta \dot{\mathbf{w}} - \ddot{\mathbf{w}} \, \delta \mathbf{w}) \, dV + \int_O \mathbf{k}_a \, \delta \mathbf{w} \, dO + \sum_{i=1}^n \mathbf{K}_{ai} \, \delta \mathbf{w}_i \, . \qquad (1.6.62)$$

Let us now consider two special cases. It is first assumed that the *stability of an equilibrium position* is being investigated ($\ddot{\mathbf{w}} = 0$), and that no polygenetic forces are present ($\mathbf{k}_a = 0$, $\mathbf{K}_{ai} = 0$). Equation (1.6.60) then becomes $\delta A = 0$, $A = $ extremum, and the *principle of minimum potential energy* is obtained. Let us now consider the *stability of elastic motion* where there are no polygenetic forces (this is the case when only weight forces or "unidirectional" forces are applied). Equation (1.6.61) then becomes

$$\delta L = (d/dt) \int_V \mu \mathbf{w} \, \delta \dot{\mathbf{w}} \, dV.$$

An additional calculation results in

$$\delta \int_{t_1}^{t_2} (T - A) \, dt = 0,$$

which is the *Hamilton principle*

$$\int_{t_1}^{t_2} (T - A) \, dt = \text{extremum.}$$

Both of these special cases result in *variational problems* which can be solved by a *direct method*—that is, the Ritz method. From the standpoint of mechanics, it may be concluded that the Ritz method can be applied to stability investigations of mechanical processes if there are *no polygenetic* forces present. These polygenetic forces make it impossible to formulate variational problems. Therefore, the Ritz method has a limited range of applicability. We shall very shortly establish the fact that the Galerkin method can be obtained from (1.6.62). It has thus been determined that the Galerkin method is more general and comprehensive, because it can also be applied when polygenetic forces are present. This means that the problem is no longer a variational problem. Let us start with (1.6.62) which may be written in the following form:

$$\delta H = \delta A + \delta T = Z.$$

The quantity Z is an abbreviation for the terms on the right side of (1.6.62). In the most general case, we have

$$\delta A = -\delta T + Z. \qquad (1.6.63)$$

On the other hand, the special case of equilibrium position stability for only monogenetic forces corresponds to

$$\delta A = 0. \qquad (1.6.64)$$

Let us assume the trial solution $w = \sum_k \rho_k w_k$ of the Galerkin or Ritz method. It will be assumed from this point on that forces and displacements are directed along the same line, so that a scalar calculation may be performed. It follows from (1.6.64) that

$$(\partial A / \partial \rho_k)\, \delta \rho_k = 0.$$

Owing to the arbitary nature of the variations $\delta \rho_k$, the system of Ritz equations

$$\partial A / \partial \rho_k = 0, \qquad k = 1, 2, 3, \dots ,$$

is obtained. If the same trial solution is used, it follows from (1.6.63) that

$$\frac{\partial A}{\partial \rho_i}\, \delta \rho_i = -\frac{\partial T}{\partial \rho_i}\, \delta \rho_i + \int_V \sum_k \sum_j L_j^{(1)}(w)\, L_j^{(2)}(w_k)\, \delta \rho_k \, dV$$

$$+ \sum_k \sum_j R_j^{(1)}(w)\, R_i^{(2)}(w_k)\, \delta \rho_k ,$$

where $L_j^{(i)}$, $i = 1, 2$ are linear differential operators, and $R_j^{(i)}$, $i = 1, 2$, are linear boundary value expressions. Owing to the fact that the variations are arbitrary, the following system of equations is obtained:

$$\frac{\partial A}{\partial \rho_i} = -\frac{\partial T}{\partial \rho_i} + \int_V \sum_k \rho_k L_j^{(1)}(w_k)\, L_j^{(2)}(w_i)\, dV$$

$$+ \sum_k \rho_k R_j^{(1)}(w_k)\, R_j^{(2)}(w_i), \qquad i = 1, 2, 3, \dots ,$$

which corresponds to the *Galerkin equations* if the coordinate functions w_k of the trial solution are appropriately selected.

It is apparent that the second system of equations is more general than the Ritz system of equations, where all terms on the right side are zero. It may again be concluded that the Ritz method can only be applied when certain special assumptions have been made. This makes it possible to set terms on the right side equal to zero. On the other hand, in general it is necessary to use the Galerkin method.

The Galerkin method consists of substituting H in (1.6.62) and carrying out the variation, resulting in

$$\int_V D(w)\,\delta w\,dV + R(w,\delta w) = 0. \tag{1.6.65}$$

D is a differential operator, and R is a boundary value expression. If the trial solution $w = \sum_k \rho_k w_k$ is substituted in (1.6.65), the following is obtained:

$$\sum_j \int_V D\left[\sum_k \rho_k w_k\right] w_j\,\delta\rho_j\,dV + R\left[\sum_r \rho_r w_r, \delta\rho_j\right] = 0. \tag{1.6.66}$$

The *restricted Galerkin method* consists of selecting the coordinate functions w_k of the trial solution in such a way that *all boundary conditions* are satisfied. Thus

$$R\left[\sum_r \rho_r w_r, \delta\rho_j\right] = 0,$$

and all the boundary terms disappear. Only the following expression remains from (1.6.66):

$$\sum_j \int_V D\left[\sum_k \rho_k w_k\right] w_j\,\delta\rho_j\,dV = 0.$$

Because of the fact that the variations $\delta\rho_j$ are abitrary, the following system of equations

$$\int_V D\left[\sum_k \rho_k w_k\right] w_j\,dV = 0, \quad j = 1,2,3,..., \tag{1.6.67}$$

is obtained for determining the series coefficients ρ_k. If

$$D(w) = L(w) + N(w) - cw$$

holds, the above expression becomes the system of Eqs. (1.6.49″) if it is assumed that the integration is performed over a one-dimensional region.

The *general Galerkin method* may also be employed if the boundary conditions are very complicated, and if it is difficult to find coordinate functions w_k which satisfy all the boundary conditions. In this case, the functions w_k are used which satisfy as many, but *not all*, boundary conditions. The calculation must then start with Eq. (1.6.66), and all boundary terms are included which do not vanish, because the corresponding w_k is used. An example of this method will be presented shortly. This example will demonstrate the fact that the methods of Galerkin and Ritz are extremely important in stability investigations, especially in the mechanics of deformable, rigid bodies.

Example: Let us consider the case of a homogeneous column with constant cross section, clamped at the bottom. It is loaded by a "subtangential" compressive force (Fig. 34). This is a "follower" force

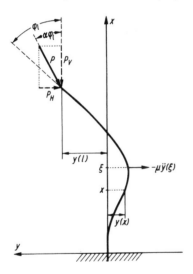

FIG. 34. Buckling column with subtangential compressive force.

which is always inclined to the vertical by the angle $\alpha\varphi_1$ as the column deforms. The quantity α is a parameter which satisfies the condition $0 < \alpha \leqslant 1$, and φ_1 is the angle between the vertical and the tangent of the column axis at its tip. Let us investigate the column *stability of the trivial equilibrium position* (straight column axis). *Transverse vibrations* are superimposed on the equilibrium position, and whether or not the deflections of the column axis remain small is determined. The load P_k is the buckling load or critical load at which the deflections increase indefinitely. If there are several such loads P_k, we shall restrict ourselves to the smallest of these loads because in practice it is usually sufficient to determine this smallest critical load.

According to the principle of d'Alembert, the inertia forces applied to the column elements are taken into account. The sectional method is applied, and the following relationship is obtained for the equilibrium of the internal and external moments

$$EJy''(x, t) = P_v[y(l, t) - y(x, t)] - P_H(l - x) - \int_x^l \mu\ddot{y}(\xi, t)(\xi - x)\, d\xi.$$

$$(1.6.68)$$

EJ is the constant bending stiffness, μ the constant linear mass density of the column, and $P_v = P \cos \alpha\varphi_1$, $P_H = \sin \alpha\varphi_1$. It can be assumed that $\alpha\varphi_1$ is a small angle, so that $\cos \alpha\varphi_1 \approx 1$ and $\sin \alpha\varphi_1 \approx \alpha\varphi_1$ holds. The calculation may be therefore continued with $P_v \approx P$, $P_H \approx P\alpha\varphi_1$. In addition, the prime indicates the derivative with respect to x, and the overdot the derivative with respect to time t.

The following partial differential equation is obtained from (1.6.68) by twofold differentiation with respect to x:

$$EJ \frac{\partial^4 y(x, t)}{\partial x^4} + P \frac{\partial^2 y(x, t)}{\partial x^2} + \mu \frac{\partial^2 y(x, t)}{\partial t^2} = 0.$$

The trial solution $y(x, t) = e^{\omega t}f(x)$ results in the following ordinary differential equation:

$$EJf^{\mathrm{IV}} + Pf'' + \mu\omega^2 f = 0. \qquad (1.6.69)$$

In addition, the boundary conditions

$$f(0) = f'(0) = f''(l) = 0, \qquad f'''(l) = (\alpha - 1)(P/EJ)f'(l) \quad (1.6.70)$$

must be taken into account. Furthermore, $\varphi_1 = f'(l)$ was employed. The boundary value problem given by (1.6.69) and (1.6.70) will now be solved by the Galerkin method. The following approximate trial solution for $f(x)$ is used:

$$f_1(x) = \rho_1 f_1(x) + \rho_2 f_2(x)$$

$$f_1(x) = 5 - \cos 2\pi(x/l) - 4 \cos \pi(x/l), \qquad (1.6.71)$$

$$f_2(x) = 28 - \cos \tfrac{3}{2}\pi(x/l) - 27 \cos(\pi/2)(x/l).$$

The functions f_1 and f_2 do *not* satisfy all the boundary conditions. The fourth condition of (1.6.70) is violated. False results would be obtained if $D(f) = EJf^{\mathrm{IV}} + Pf'' + \mu\omega^2 f$ were used in the calculation and if the Galerkin equations given by (1.6.67) were used according to

$$\int_0^l D[\rho_1 f_1(x) + \rho_2 f_2(x)] f_i(x)\, dx = 0, \qquad i = 1, 2,\dots . \qquad (1.6.72)$$

This would only be permissible if f_1 and f_2 satisfied *all* boundary conditions. Because this is not the case, the principle of virtual displacements must be used in order to find the Galerkin equation. The nonzero boundary values of the selected trial functions f_1 and f_2 must be included in the calculation.

Because $\mathbf{k}_a \equiv 0$,

$$\sum_{i=l}^{n} \mathbf{K}_{ai}\, \delta \mathbf{w}_i \equiv \mathbf{K}_{al}\, \delta \mathbf{w}_1 \equiv -P\alpha\varphi_l\, \delta w(t, l), \qquad w(t, x) \equiv y(t, x), \qquad \varphi_l \equiv y'(t, l),$$

Eq. (1.6.62) yields the following equation:

$$\delta H = \int_0^l \mu(\dot{y}\, \delta y - \ddot{y}\, \delta y)\, dx - P\alpha y'(t, l)\, \delta y(t, l). \tag{1.6.73}$$

The equation

$$H = \int_0^l \left[\frac{\mu}{2}\, \dot{y}^2 + \frac{EJ}{2}\, y''^2 - \frac{P}{2}\, y'^2 \right] dx = \int_0^l F\, dx$$

holds for an elastic column subjected to the given load. If the variation of H is carried out, the result is

$$\delta H = \int_0^l \left[\frac{\partial F}{\partial y} - \frac{\partial}{\partial x}\left(\frac{\partial F}{\partial y'} \right) + \frac{\partial^2}{\partial x^2}\left(\frac{\partial F}{\partial y''} \right) \right] \delta y\, dx$$

$$+ \left[\frac{\partial F}{\partial y'}\, \delta y + \frac{\partial F}{\partial y''}\, \delta y' - \frac{\partial}{\partial x}\frac{\partial F}{\partial y''}\, \delta y \right]_0^l + \int_0^l \frac{\partial F}{\partial \dot{y}}\, \delta y\, dx.$$

If this result and $F = (\mu\dot{y}^2 + EJy''^2 - Py'^2)/2$ are substituted into (1.6.73), the following result is obtained:

$$\int_0^l \left[(Py'' + EJy^{\mathrm{IV}})\, \delta y + \mu\dot{y}\, \delta\dot{y} \right] dx + \left[-Py'\, \delta y + EJy''\, \delta y' - EJy'''\, \delta y \right]_0^l$$

$$= \int_0^l \mu(\dot{y}\, \delta\dot{y} - \ddot{y}\, \delta y)\, dx - P\alpha y'(t, l)\, \delta y(t, l). \tag{1.6.74}$$

The trial solution $y(x, t) = e^{\omega t}f(x)$ is used and (1.6.74) becomes

$$\int_0^l (Pf'' + EJf^{\mathrm{IV}} + \mu\omega^2 f)\, \delta f\, dx + \left[-Pf'\, \delta f + EJf''\, \delta f - EJf'''\, \delta f \right]_0^l$$

$$+ P\alpha f'(l)\, \delta f(l) = 0. \tag{1.6.75}$$

In the Galerkin method, if functions f_k, $k = 1, 2$, are used, which satisfy all the conditions (1.6.70), then all terms in (1.6.75) outside of the integral sign would vanish and it would be possible to use (1.6.72).

Since one of the selected functions satisfies only the three first conditions (1.6.70), the following expression remains from (1.6.75):

$$\int_0^l (Pf'' + EJf^{IV} + \mu\omega^2 f)\,\delta f\,dx + (\alpha - 1)\,Pf'(l) - EJf'''(l)\,\delta f(l) = 0.$$

If the approximation $f^*(x) = \rho_1 f_1(x) + \rho_2 f_2(x)$ is substituted in this relationship for $f(x)$, the result is

$$\int_0^l \{[\rho_1 D(f_1) + \rho_2 D(f_2)](f_1\,\delta\rho_1 + f_2\,\delta\rho_2)\}\,dx + (\alpha - 1)\,P[\rho_1 f_1'(l)$$
$$+ \rho_2 f_2'(l)][f_1(l)\,\delta\rho_1 + f_2(l)\,\delta\rho_2] - EJ[\rho_1 f_1'''(l)$$
$$+ \rho_2 f_2'''(l)][f_1(l)\,\delta\rho_1 + f_2(l)\,\delta\rho_2] = 0.$$

By multiplying and collecting all terms containing $\delta\rho_1$ and $\delta\rho_2$, the following equations are obtained because the $\delta\rho_1$, $\delta\rho_2$ are arbitrary:

$$\rho_1 \left[\int_0^l D(f_1)f_1\,dx + (\alpha - 1)\,Pf_1'(l)f_1(l) - EJf_1'''(l)f_1(l) \right]$$
$$+ \rho_2 \left[\int_0^l D(f_2)f_1\,dx + (\alpha - 1)\,Pf_2'(l)f_1(l) - EJf_2'''(l)f_1(l) \right] = 0,$$

$$\rho_1 \left[\int_0^l D(f_1)f_2\,dx + (\alpha - 1)\,Pf_1'(l)f_2(l) - EJf_1'''(l)f_2(l) \right]$$
$$+ \rho_2 \left[\int_0^l D(f_2)f_2\,dx + (\alpha - 1)\,Pf_2'(l)f_2(l) - EJf_2'''(l)f_2(l) \right] = 0.$$

These are linear, homogeneous, algebraic equations that determine the quantities ρ_1 and ρ_2. If the calculation is continued with the functions f_1, f_2 according to (1.6.71), the following result is obtained:

$$\rho_1 \left(16\,\frac{\pi^4 EJ}{l^3} - 10P\,\frac{\pi^2}{l} + 33.5\mu\omega^2 l \right)$$
$$+ \rho_2 \left(10.5\,\frac{\pi^4 EJ}{l^3} - 21P\,\frac{\pi^2}{l} + 96\alpha P\,\frac{\pi}{l} + 78.6\mu\omega^2 l \right) = 0,$$

$$\rho_1 \left(10.5\,\frac{\pi^4 EJ}{l^3} - 21P\,\frac{\pi^2}{l} + 78.6\mu\omega^2 l \right)$$
$$+ \rho_2 \left(25.3\,\frac{\pi^4 EJ}{l^3} - 92.3P\,\frac{\pi^2}{l} + 336\alpha P\,\frac{\pi}{l} + 198.3\mu\omega^2 l \right) = 0.$$

The determinant of this system of equations must be zero if this homogeneous system is to have unique solutions ρ_1 , ρ_2. Introducing the abbreviations

$$a_1 = 33.5l, \qquad a_2 = 198.3l, \qquad b = 78.6l,$$

$$c_1 = 16\,\frac{\pi^4 EJ}{l^3} - 10P\,\frac{\pi^2}{l}, \qquad c_2 = 25.3\,\frac{\pi^4 EJ}{l^3} - 92.3P\,\frac{\pi^2}{l} + 336\alpha P\,\frac{\pi}{l},$$

$$d_1 = 10.5\,\frac{\pi^4 EJ}{l^3} - 21P\,\frac{\pi^2}{l} + 96\,\frac{\pi}{l}\,\alpha P, \qquad d_2 = 10.5\,\frac{\pi^4 EJ}{l^3} - 21P\,\frac{\pi^2}{l},$$

this condition is formulated as follows:

$$\det\left[\begin{pmatrix} c_1 & d_1 \\ d_2 & c_2 \end{pmatrix} + \omega^2 \begin{pmatrix} \mu a_1 & \mu b \\ \mu b & \mu a_2 \end{pmatrix}\right] = 0. \qquad (1.6.76)$$

Equation (1.6.76) is quadratic with respect to ω^2. Setting $\omega^2 = -\lambda$, we obtain

$$\lambda^2 - \lambda\left[\frac{a_1 c_2 + a_2 c_1}{\mu(a_1 a_2 - b^2)} - \frac{b(d_1 + d_2)}{\mu(a_1 a_2 - b^2)}\right] + \frac{c_1 c_2 - d_1 d_2}{\mu^2(a_1 a_2 - b^2)} = 0. \qquad (1.6.77)$$

The equilibrium position $y \equiv 0$ of the column is stable if the axial deflections $y(t, x)$ of the column carrying out transverse oscillations remain bounded at all times. Since $y(t, x) = e^{\omega t} f(x)$ holds, this is only the case for imaginary $\omega = \pm(-\lambda)^{1/2}$. For stability, λ must be real and satisfy $\lambda > 0$. The stability is reached when either λ becomes zero or complex.

Equation (1.6.77) describes the characteristics of λ. If

$$c_1 c_2 - d_1 d_2 = 0, \qquad (1.6.78)$$

λ can become zero and therefore (1.6.78) is a possible stability boundary. If the discriminant of (1.6.77) is positive, the roots of this equation are real. If it is negative, the roots are conjugate complex. If the discriminant is set equal to zero,

$$\text{Discr} = [a_1 c_2 + a_2 c_1 - b(d_1 + d_2)]^2 - 4(c_1 c_2 - d_1 d_2)(a_1 a_2 - b^2) = 0, \qquad (1.6.79)$$

which results in another stability boundary.

Both conditions (1.6.78) and (1.6.79) can be plotted in the P, α plane in the form of curves. This results in a stability diagram (Fig. 35). The lower limit of the instability region gives the buckling load P as a function of the parameter α.

Very often mechanical processes, the stability of which is being

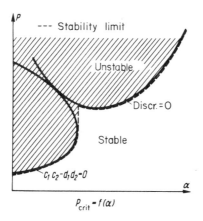

FIG. 35.   Stability map.

investigated, are described by boundary-value problems that contain partial differential equations or systems of partial differential equations. Even though the extension of stability theory from ordinary to partial differential equations would be very desirable, not enough has been done due to the mathematical difficulties. The study of Hahn (*1.2*) contains a brief discussion of this question and contains comments on the work of the Russian investigator V. J. Zubov.

In general it will again be necessary to use approximation methods which essentially are based on the Galerkin method. This again shows its importance.

1.6.6. *The Harmonic Balance Method*

The method of harmonic balance, first developed by Krylov and Bogoljubov, is an often used approximation method for the treatment of nonlinear problems of vibration and control theory. It is described in various textbooks (*2.5*, *2.12*, *2.13*) so that we can restrict ourselves to an introduction of the subject.

The harmonic balance method is often called the *method of harmonic linearization*. This harmonic linearization differs from ordinary linearization where the nonlinear function is replaced by the first term of a Taylor series (geometrically speaking, a curve is replaced by its tangent at a certain location) in that a linear approximation expression $cx(t)$ is determined for a nonlinear function $f[x(t)]$. However, its proportionality factor is no longer a true constant.

Example: Let us consider the differential equation

$$\ddot{x} + \omega_0^2 x = f(x) = x^3.$$

For its solution, the basic assumption that it is approximately harmonic is made, i.e., $x \approx b_1^* \sin \omega t$. Then $f(x)$ is a periodic function which satisfies $f(x) \approx B_1 \sin \omega t$. The relationship $\sin \omega t \approx x/b_1^*$ follows from the approximate trial solution for x. Therefore, $f(x) \approx (B_1/b_1^*)x$ holds corresponding to harmonic linearization. Apparently the factor $c = B_1/b_1^*$ has become a function of the amplitude b_1^*. The great practical importance of the harmonic balance method comes from the fact that by this method of linearization the true character of the non-linear problems is more readily apparent.

Let us continue the calculation in order to clarify the features of the method. The Fourier coefficients B_1 of $f(x)$ satisfy the relationship

$$B_1 = (2/T) \int_0^T f[x(\xi)] \sin \omega \xi \, d\xi,$$

where T is the period of the process. The result

$$cb_1^* = B_1 = (2/T) \int_0^T f[x(\xi)] \sin \omega \xi \, d\xi$$

is obtained, which will be used immediately. If $x = b_1^* \sin \omega t$ is now substituted into the linearized differential equation $\ddot{x} + \omega_0^2 x = cx$, the result is

$$b_1^*(\omega_0^2 - \omega^2) = cb_1^*.$$

The first relationship is used to obtain

$$b_1^*(\omega_0^2 - \omega^2) = (2/T) \int_0^T f[x(\xi)] \sin \omega \xi \, d\xi.$$

Because $T = 2\pi/\omega, f[x(\xi)] = x^3(\xi) = b_1^{*3} \sin^3 \omega \xi$,

$$b_1^*(\omega_0^2 - \omega^2) = (\omega/\pi) \int_0^{2\pi/\omega} b_1^{*3} \sin^4 \omega \xi \, d\xi$$

is obtained. After integration and transformation, the following equation is obtained:

$$\omega_0^2 - \omega^2 = \tfrac{3}{4} b_1^{*2},$$

which is well known in the theory of vibrations and gives the dependence of amplitude b_1^* on the frequency ω. The approximate solution of the problem has thus been found.

Up to the present we have discussed the simple vibration problem

$$\ddot{x} + \omega_0^2 x = f(x). \tag{1.6.80}$$

ω_0 was a given constant and $f(x)$ is a nonlinear restoring function that satisfies

$$\int_0^T f[x(\xi)] \, d\xi = 0, \tag{1.6.81}$$

thus greatly simplifying the problem. In addition, let us assume that (1.6.80) has a periodic solution that is twice differentiable and has a period T. The boundary conditions corresponding to (1.6.80) are therefore

$$x(0) = x(T), \qquad \dot{x}(0) = \dot{x}(T). \tag{1.6.82}$$

The harmonic balance method can be extended to more general problems, for example, the problem

$$D[x(t)] = f(x),$$

where D is a linear differential operator of higher order with constant coefficients. It can also be applied to

$$\ddot{x} + \omega_0^2 x = f(x, \dot{x}).$$

Let us now extend the method in another direction. The following *linear substitution problem* corresponding to (1.6.80) is introduced:

$$\ddot{x} + \omega_0^2 x = c_1 x + c_2 \dot{x}. \tag{1.6.83}$$

The general trial solution is

$$x = a_1^* \cos \omega t + b_1^* \sin \omega t. \tag{1.6.84}$$

If the latter is substituted into the right side of (1.6.80), $f(x)$ becomes a periodic function that can be expanded in Fourier series. Only the first term of the series is retained. Owing to the assumption (1.6.81) there is not a constant term in this series. The following result is thus obtained:

$$f(x) = f(a_1^* \cos \omega t + b_1^* \sin \omega t) \approx A_1 \cos \omega t + B_1 \sin \omega t. \tag{1.6.85}$$

This expression should coincide with the right side of the substitution problem so that

$$A_1 \cos \omega t + B_1 \sin \omega t = c_1 x + c_2 \dot{x} \tag{1.6.86}$$

must hold. If the trial solution (1.6.84) is again substituted on the right side of (1.6.86), the following system of equations is obtained:

$$A_1 = c_1 a_1{}^* + c_2 b_1{}^* \omega,$$
$$B_1 = c_1 b_1{}^* - c_2 a_1{}^* \omega.$$

(1.6.87)

The Fourier coefficients A_1 and B_1 satisfy the relationships

$$A_1 = (2/T) \int_0^T f(x) \cos \omega \xi \, d\xi, \qquad B_1 = (2/T) \int_0^T f(x) \sin \omega \xi \, d\xi,$$

and (1.6.87) becomes

$$(2/T) \int_0^T f(x) \cos \omega \xi \, d\xi = c_1 a_1{}^* + c_2 b_1{}^* \omega,$$
$$(2/T) \int_0^T f(x) \sin \omega \xi \, d\xi = c_1 b_1{}^* - c_2 a_1{}^* \omega.$$

(1.6.88)

Substitution of the trial solution (1.6.84) into the linear substitution problem (1.6.83) results in

$$a_1{}^*(\omega_0{}^2 - \omega^2) \cos \omega t + b_1{}^*(\omega_0{}^2 - \omega^2) \sin \omega t$$
$$= c_1 a_1{}^* \cos \omega t + c_1 b_1{}^* \sin \omega t - c_2 a_1{}^* \omega \sin \omega t + c_2 b_1{}^* \omega \cos \omega t.$$

This is only true if the equations

$$a_1{}^*(\omega_0{}^2 - \omega^2) = c_1 a_1{}^* + c_2 b_1{}^* \omega,$$
$$b_1{}^*(\omega_0{}^2 - \omega^2) = c_1 b_1{}^* - c_2 a_1{}^* \omega$$

(1.6.89)

are satisfied. Comparison of (1.6.88) and (1.6.89) shows that

$$a_1{}^*(\omega_0{}^2 - \omega^2) = (2/T) \int_0^T f(x) \cos \omega \xi \, d\xi,$$
$$b_1{}^*(\omega_0{}^2 - \omega^2) = (2/T) \int_0^T f(x) \sin \omega \xi \, d\xi$$

(1.6.90)

must hold. The following fundamental equations of the harmonic balance method are found using the new variables $a_1 = (T/2)^{1/2} a_1{}^*$:

$$a_1 = -\frac{1}{\omega^2 - \omega_0{}^2} \int_0^T f[x(\xi)](2/T)^{1/2} \cos \omega \xi \, d\xi,$$
$$b_1 = -\frac{1}{\omega^2 - \omega_0{}^2} \int_0^T f[x(\xi)](2/T)^{1/2} \sin \omega \xi \, d\xi.$$

(1.6.91)

The most important application of the method is the determination of periodic solutions of mechanical processes and the investigation of their stability.

Example[1]: Let us assume that a mechanical process is described by the nonlinear differential equation

$$\ddot{x} - (\alpha - \beta \mid x \mid) \mid \dot{x} \mid \dot{x} + \omega_0^2(x + \gamma x^3) = 0.$$

Let us investigate whether a periodic solution exists and whether it is stable.

The following trial solution $x = a_1^* \cos \omega t$ is used for this purpose. For purposes of simplicity, $b_1^* = 0$ has been assumed (in contrast to the first example where $a_1^* \equiv 0$, $b_1^* \neq 0$), so that the following relationships are obtained for the coefficients c_1 and c_2 of the linear substitution problem (1.6.83):

$$c_1 = (2/a_1^*T) \int_0^T f[x, \dot{x}] \cos \omega \xi \, d\xi,$$

$$c_2 = -(2/a_1^*\omega T) \int_0^T f[x, \dot{x}] \sin \omega \xi \, d\xi.$$

The following substitutions must be made:

$$x \approx a_1^* \cos \omega \xi, \qquad \dot{x} \approx -\omega a_1^* \sin \omega \xi, \qquad T = 2\pi/\omega,$$

as well as

$$f[x, \dot{x}] = (\alpha - \beta \mid x \mid) \mid \dot{x} \mid \dot{x} - \omega_0^2 \gamma x^3.$$

The result after integration is

$$c_1 = -\frac{3}{4} \omega_0^2 \gamma a_1^{*2}, \qquad c_2 = \frac{8}{3} \frac{\alpha \omega}{\pi} a_1^* - \frac{\beta \omega}{\pi} a_1^{*2},$$

and therefore (1.6.83) becomes

$$\ddot{x} + \omega_0^2 x = -\frac{3}{4} \omega_0^2 \gamma a_1^{*2} x + a_1^* \frac{\omega}{\pi} \left(\frac{8}{3} \alpha - \beta a_1^* \right) \dot{x}.$$

The linear substitution problem is given by the vibration equation

$$\ddot{x} + a_1^* \frac{\omega}{\pi} \left(\beta a_1^* - \frac{8}{3} \alpha \right) \dot{x} + \omega_0^2 \left(1 + \frac{3}{4} \gamma a_1^{*2} \right) x = 0.$$

---

[1] This example is from MAGNUS, K., "Vibrations," p. 125. Teubner, Stuttgart, 1961.

When the damping coefficient $d = a_1^*(\omega/\pi)(\beta a_1^* - (8/3)\alpha)$ vanishes, the periodic solution is an undamped oscillation. This is the case for $a_1^* = a_{1st}^* \equiv (8/3)(\alpha/\beta)$. Simultaneously, the stationary amplitude a_{1st}^* of this undamped oscillation has been found. Its frequency can be obtained from the coefficients of the third term of the linear vibration equation

$$\omega^2 = \omega_0^2(1 + \tfrac{3}{4}\gamma a_{1st}^{*2}) \quad \text{or} \quad \omega = \omega_0(1 + \tfrac{3}{4}\gamma a_{1st}^{*2})^{1/2}.$$

If the amplitude of the approximate solution satisfies

$$a_1^* > 8\alpha/3\beta = a_{1st}^*,$$

the damping coefficient d becomes positive and the oscillation is damped. Conversely if $a_1^* < a_{1st}^*$, d becomes negative and we have an induced oscillation case. This is because stability in the large and instability in the small prevail, and because the periodic solution with the amplitude a_{1st}^* is described by a stable *limit cycle* in the phase diagram.

The harmonic balance method can also be used to determine the existence and stability of periodic solutions of systems of differential equations such as

$$\dot{\xi}_i = F_i(\xi_1,...,\xi_n), \qquad i = 1, 2,..., n, \tag{1.6.92}$$

which correspond to nonlinear variational problems.

Let us replace (1.6.92) by the linear substitution system

$$\dot{\xi}_i = c_{1ik}\xi_k + c_{2ik}\dot{\xi}_k. \tag{1.6.93}$$

The method is used to determine the coefficients. If, in particular, $F_i(\xi_1,...,\xi_n) = \sum_k F_{ik}(\xi_k)$ holds, the following result is obtained:

$$c_{1ik} = (1/\pi a_k^*) \int_0^{2\pi} F_{ik}(a_k^* \sin\varphi) \sin\varphi \, d\varphi,$$

$$c_{2ik} = (1/\pi\omega a_k^*) \int_0^{2\pi} F_{ik}(a_k^* \sin\varphi) \cos\varphi \, d\varphi. \tag{1.6.94}$$

The c_{jik} are related to "description functions," which are used in the theory of nonlinear control systems.

The trial solution $\xi_k = a_k^* e^{\lambda t}$ and the linear substitution system (1.6.93) results in the following homogeneous, algebraic system of equations:

$$a_k^*(c_{1ik} + c_{2ik}\lambda - \delta_{ik}\lambda) = 0, \qquad i = 1, 2,..., n.$$

It only has nontrivial solutions for the $a_k{}^*$ if its determinant is equal to zero. The requirement

$$\det(c_{1ik} + c_{2ik}\lambda - \delta_{ik}\lambda) = 0$$

must be satisfied, which leads to the characteristic equation

$$a_k\lambda^k = 0, \qquad k = 1, 2, ..., n,$$

of the problem.

It is therefore possible to apply the stability investigation method of Section 1.3 to the characteristic equation. This leads to results that make it possible to estimate the stability of the problem at hand.

If, for example, the Hurwitz criterion is applied to the coefficients a_k of the characteristic equation given above, the so-called *A-curve method* of Magnus is obtained.

Application of Stability Theory to Selected Problems of Mechanics

2.1. CELESTIAL MECHANICS PROBLEMS

2.1.1. *Stability of Central Force Motion*

The equations of motion of central force motion of a point mass are given by

$$m(\ddot{r} - r\dot{\varphi}^2) = f(r),$$
$$r\ddot{\varphi} + 2\dot{r}\dot{\varphi} = (1/r)(d/dt)(r^2\dot{\varphi}) = 0. \tag{2.1.1}$$

If the attracting force directed to the center satisfies Newton's laws, we have

$$f(r) = -(mM\gamma/r^2) = -(mK/r^2). \tag{2.1.2}$$

r and φ are the polar coordinates of motion, m and M are the masses of the two bodies, and γ and K are certain constants.

The Lagrangian of central force motion is given by

$$L = (m/2)(\dot{r}^2 + r^2\dot{\varphi}^2) - \int_r^\infty f(r) \, dr. \tag{2.1.3}$$

If the terms and formulas of Hamiltonian mechanics are not known to the reader, he is referred to the works of Frank and of Mises (7).

Because $dL/d\varphi = 0$, φ is a "cyclic" coordinate, one of the Lagrangian equations

$$\frac{dL}{d\varphi} - \frac{d}{dt}\frac{\partial L}{\partial \dot{\varphi}} = 0$$

results in $d/dt(\partial L/\partial \dot{\varphi}) = 0$, or finally $dL/\partial \dot{\varphi} = $ const. The generalized momenta are given by

$$p_r = \partial L/\partial \dot{r} = m\dot{r}, \qquad p_\varphi = \partial L/\partial \dot{\varphi} = mr^2\dot{\varphi}.$$

Owing to the relationship $\partial L/\partial \dot{\varphi} = $ const, we have

$$r^2\dot{\varphi} = h = \text{const.} \qquad (2.1.4)$$

This result can easily be derived from the second equation of motion (2.1.1).

In central force motion the radius vector r represents the "noncyclic" coordinate and the angle φ is the "cyclic" coordinate. Let us investigate the *stability of quasistatic motions*. Quasistatic motions, which are similar to equilibrium, are motions in which the noncyclic coordinate r remains constant and the cyclic coordinate φ increases linearly with time. These motions are circular motions for which $r = r_0 = $ const holds. Therefore, according to (2.1.4), $\dot{\varphi} = h/r_0^2 = $ const and therefore $\varphi = ht/r_0^2 + C$ holds. Therefore, r and φ satisfy the conditions mentioned above.

Let us now carry out a perturbation of the quasistatic circular motion in such a way that the noncyclic coordinate r varies while the momentum p_φ of the cyclic coordinate φ remains constant. This is similar to the perturbation of equilibrium positions. Consequently, let us investigate the *"conditional" stability of circular motions*. We have

$$r = r_0 + \xi, \qquad p_\varphi = mr^2\dot{\varphi} = mh = \text{const,}$$

that is, $h = $ const.

Let us first apply the *method of small oscillations*. Substitution of the quantities $r = r_0 + \xi, r_0 = $ const, $h = r_0^2\dot{\varphi} = $ const into the first equation of motion (2.1.1) yields the following result where ξ is the variation of r. If the terms of

$$\ddot{\xi} - h^2/(r_0 + \xi)^3 - (1/m)f(r_0 + \xi) = 0 \qquad (2.1.5)$$

are expanded into a series, the following result is obtained:

$$\ddot{\xi} - \frac{h^2}{r_0^3}\left(1 - \frac{3\xi}{r_0} + \frac{6\xi^2}{r_0^2}\cdots\right) - \frac{1}{m}\left[f(r_0) + \left(\frac{df(r)}{dr}\right)'_{r=r_0}\xi\right.$$
$$\left. + \frac{1}{2}\left(\frac{d^2f(r)}{dr^2}\right)_{r=r_0}\xi^2\cdots\right] = 0. \qquad (2.1.6)$$

The first equation of (2.1.1) yields the result

$$-r_0\dot{\varphi}^2 - (1/m)f(r_0) = 0$$

for $r = r_0$. Since the relationship $r_0\dot{\varphi}^2 = h^2/r_0^3$ is obtained from $r_0^2\dot{\varphi} = h$, this result is equal to

$$-h^2/r_0^3 - (1/m)f(r_0) = 0. \tag{2.1.7}$$

After substitution of (2.1.7) into (2.1.6) and neglecting the higher-order terms, we obtain

$$\ddot{\xi} + \left[\frac{3h^2}{r_0^4} - \frac{1}{m}\left(\frac{df(r)}{dr}\right)_{r=r_0} \right]\xi = 0. \tag{2.1.8}$$

Therefore, a *linear* variational equation has been obtained which can be rewritten as follows because of (2.1.7):

$$\ddot{\xi} - \left[\frac{3f(r_0)}{mr_0} + \frac{1}{m}\left(\frac{df(r)}{dr}\right)_{r=r_0} \right]\xi = 0 \tag{2.1.9}$$

or with the abbreviation

$$\alpha = \frac{3f(r_0)}{mr_0} + \frac{1}{m}\left(\frac{df(r)}{dr}\right)_{r=r_0} \tag{2.1.10}$$

it can be rewritten as follows:

$$\ddot{\xi} - \alpha\xi = 0. \tag{2.1.11}$$

With $\xi = \xi_1$, $\dot{\xi} = \xi_2$, Eq. (2.1.11) finally results in a system of differential equations

$$\dot{\xi}_1 = \xi_2, \qquad \dot{\xi}_2 = \alpha\xi_1 \tag{2.1.12}$$

or

$$\dot{\xi}_i = a_{ik}\xi_k, \qquad i, k = 1, 2.$$

The system coefficients are $a_{11} = 0$, $a_{12} = 1$, $a_{21} = \alpha$, $a_{22} = 0$ and the characteristic equation follows from

$$\det(a_{ik} - \lambda\,\delta_{ik}) = \begin{vmatrix} -\lambda & 1 \\ \alpha & -\lambda \end{vmatrix} = \lambda^2 - \alpha = 0.$$

The *characteristic exponents* $\lambda_{1,2}$ satisfy

$$\lambda_{1,2} = \pm\alpha^{1/2}.$$

Theorem 2 of Section 1.3.1 states that instability occurs if $\alpha > 0$ because one of the characteristic exponents has a positive real part. In this case, according to Theorem 5 of Section 1.5.3, instability can be predicted with certainty from the linear variational equations. If it is assumed that

$$f(r) = -mK/r^n,$$ (2.1.13)

the instability condition is

$$\alpha = \frac{3f(r_0)}{mr_0} + \frac{1}{m}\left(\frac{df(r)}{dr}\right)_{r=r_0} = -\frac{3K}{r_0^{n+1}} + \frac{nK}{r_0^{n+1}} > 0.$$ (2.1.14)

Equation (2.1.14) shows that central force motions with a central force of the form (2.1.13) are unstable for $n > 3$.

If the linear variational equations were used to determine whether stability is possible for $\alpha \leqslant 0$, a vanishing double root would be found for $\alpha = 0$. Two imaginary roots with zero real part would be found for $\alpha < 0$. Therefore, $\alpha \leqslant 0$ corresponds to the *critical case*, and the question of stability can no longer be decided with certainty from the linear variational equations. It is then necessary to use the nonlinear equation (2.1.5).

The substitution $\xi = \xi_1$, $\dot{\xi} = \xi_2$, and Eq (2.1.5) result in the system

$$\dot{\xi}_1 = \xi_2, \qquad \dot{\xi}_2 = h^2/(r_0 + \xi_1)^3 + (1/m)f(r_0 + \xi_1).$$ (2.1.15)

The *Lyapunov function* is assumed to have the following form:

$$V(\xi_1, \xi_2) = \frac{1}{2}\xi_2^2 + \frac{1}{2}\frac{h^2}{(r_0 + \xi_1)^2} - \frac{1}{m}F(r_0 + \xi_1) - \frac{1}{2}\frac{h^2}{r_0^2} + \frac{1}{m}F(r_0),$$ (2.1.16)

where $F(x) = \int f(x)\, dx$. Apparently $V(0, 0) = 0$ due to the second equation in (2.1.15) and also $\dot{V} \equiv 0$. The condition for which the function V is definite in the vicinity of the origin must be investigated in order to apply the Lyapunov stability Theorem 1 of Section 1.5.3. $V(\xi_1, 0)$ is expanded with respect to ξ_1 around the origin, which results in

$$V(\xi_1, 0) = V(0, 0) + \left(\frac{\partial V}{\partial \xi_1}(\xi_1, 0)\right)_{\xi_1=0} \xi_1 + \frac{1}{2}\left(\frac{\partial^2 V}{\partial \xi_1^2}(\xi_1, 0)\right)_{\xi_1=0} \xi_1^2$$

$$+ \frac{1}{3!}\left(\frac{\partial^3 V}{\partial \xi_1^3}(\xi_1, 0)\right)_{\xi_1=0} \xi_1^3 + \cdots.$$

$V(\xi_1, \xi_2)$ is positive definite in the vicinity of the origin if $V(\xi_1, 0)$ is. Because $V(0, 0) = 0$ the sign of $V(\xi_1, \xi_2)$ depends only on the first- and second-order ξ_1 terms sufficiently close to the origin. Because of (2.1.7) we have

$$\left(\frac{\partial V}{\partial \xi_1}(\xi_1, 0)\right)_{\xi_1=0} = -\frac{h^2}{r_0^3} - \frac{1}{m}f(r_0) = 0,$$

so that the first-order ξ_1 term is zero. Therefore, the sign of $V(\xi_1, \xi_2)$ is determined only by the properties of the term $\frac{1}{2}\{\partial^2 V(\xi_1, 0)/\partial\xi_1^2\}_{\xi_1=0}\, \xi_1^2$. In order for $V(\xi_1, \xi_2)$ to be positive definite, the coefficient $\{\partial^2 V(\xi_1 0)/\partial\xi_1^2\}_{\xi_1=0}$ must always be positive.

Therefore

$$\left(\frac{\partial^2 V}{\partial\xi_1^2}(\xi_1, 0)\right)_{\xi_1=0} = \frac{3h^2}{r_0^4} - \frac{1}{m}\left(\frac{df(x)}{dx}\right)_{\xi_1=0} > 0$$

is assumed. For $f(x) = -mK/x^n$ this means that

$$3h^2/r_0^4 - nK/r_0^{n+1} > 0.$$

Because of (2.1.7), we have $h^2/r_0^3 = -(1/m)f(r_0) = K/r_0^n$, and the last inequality becomes

$$3K/r_0^{n+1} - nK/r_0^{n+1} > 0.$$

Therefore, it can be concluded that $n < 3$ must hold for stability of circular motion considered as a quasistatic central force motion. Therefore, circular motion is stable for the Newtonian law $n = 2$ with respect to sufficiently small "conditional" perturbations (perturbations of the radius without perturbations of p_φ).

It will now be shown that point masses moving along circles can only have *orbital stability*.

The equation $r = r_0 = \text{const}$ holds. $\dot\varphi = h/r_0^2 = \text{const}$ then follows from (2.1.4) and therefore $v = r_0\dot\varphi = h/r_0 = \text{const}$. If the initial conditions of the unperturbed motion are r_0, v_0 at the time $t = 0$, $r \equiv r_0$, $v \equiv v_0$ holds at all times.

Let us consider the motion of a point mass along a circular trajectory with the initial conditions $r_0{}^* = r_0 + \epsilon_{r0}$, $v_0 = v_0{}^* + \epsilon_{v0}$ as an example of perturbed motion. Since $r^* = \text{const}$, $v^* = \text{const}$ also holds for this motion, $r^* = r + \epsilon_r = r_0{}^*$, $v^* = v + \epsilon_v = v_0{}^*$ holds always so that $r^* - r = r_0{}^* - r_0 = \epsilon_r = \epsilon_{r0}$, $v^* - v = v_0{}^* - v_0 = \epsilon_v = \epsilon_{v0}$. If the initial perturbations ϵ_{r0}, ϵ_{v0} have been selected small enough, the perturbations ϵ_r, ϵ_v will also remain as small. In particular, the difference ϵ_r between the radii of the circles is always small enough so that the circles are as close as desired. This implies orbital stability.

On the other hand, there is no stability in the sense of Lyapunov. This is an example that shows that the concept of stability is a function of the selected coordinates. In order to show this, let us consider the problem in Cartesian coordinates.

The unperturbed motion satisfies

$$x = r_0\cos(v_0t/r_0), \qquad y = r_0\sin(v_0t/r_0),$$

and the perturbed motion satisfies

$$x^* = r_0^* \cos(v_0^* t/r_0^*), \qquad y^* = r_0^* \sin(v_0^* t/r_0^*).$$

The equation $x^* - x = r_0^* \cos(v_0^* t/r_0^*) - r_0 \cos(v_0 t/r_0)$ holds, and let us assume that $v_0^*/r_0^* - v_0/r_0 = \epsilon$. Thus

$$x^* - x = r_0^* \cos(v_0 t/r_0 + \epsilon t) - r_0 \cos(v_0 t/r_0)$$

is obtained.

Let us consider the motion at a certain time $t = \pi/\epsilon$. At this instant $x^*(\pi/\epsilon) - x(\pi/\epsilon) = -(r_0^* + r_0) \cos(v_0 \pi/r_0 \epsilon)$ holds and also

$$y^*(\pi/\epsilon) - y(\pi/\epsilon) = -(r_0^* + r_0) \sin(v_0 \pi/r_0 \epsilon).$$

The Euclidian distance of the image points $P[x^*(t), y^*(t)]$, $Q[x(t), y(t)]$, which is given by

$$d[P(t), Q(t)] = \{[x^*(t) - x(t)]^2 + [(y^*(t) - y(t)]^2\}^{1/2},$$

has the value $d[P(\pi/\epsilon), Q(\pi/\epsilon)] = r_0^* + r_0$ at the time $t = \pi/\epsilon$. This is a finite, large, and constant value which cannot be made as small as desired regardless of how small the initial perturbations $x_0^* - x_0$, $y_0^* - y_0$ have been selected. Therefore, it is not possible to make the distance $d(P, Q)$ arbitrarily small at *all* times by making the initial perturbations sufficiently small. The Lyapunov stability condition is not satisfied and there is no Lyapunov stability.

2.1.2. *Stability Investigations of the Three-Body Problem*

Let us consider the restricted three-body problem for which the mass of one body (for example, the satellite) is made so *small* that it can be neglected with respect to the mass of the other two bodies (for example, earth and moon). In addition, let us consider a *plane problem* for which the small body always moves in the plane common to all three bodies. We shall justify these assumptions in the following.

The acceleration **b** satisfies the following equation in the inertial system, according to Newton's law

$$\mathbf{b} = -k \left(\frac{m_E \mathbf{r}_{ES}}{|\mathbf{r}_{ES}|^3} + \frac{m_M \mathbf{r}_{MS}}{|\mathbf{r}_{MS}|^3} \right). \tag{2.1.17}$$

m_E and m_M are the masses of the earth and the moon, respectively. The definitions of $\mathbf{r}_{ES}$ and $\mathbf{r}_{MS}$ follow from Fig. 36. It is useful to use the synodic reference system (x, y, z) to describe the three-body problem from this point on. It rotates around the z axis (Fig. 36) with the angular rate $\omega = 2.66 \times 10^{-6}$ sec^{-1}.

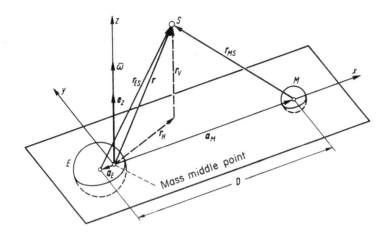

FIG. 36.   Three-body problem: E, earth; M, moon; S, satellite.

As a result of the laws of relative motion, the following equation is obtained in the synodic system:

$$\mathbf{b} = \partial \mathbf{v}/\partial t + 2\boldsymbol{\omega} \times \mathbf{v} + \boldsymbol{\omega} \times (\boldsymbol{\omega} \times \mathbf{r}) = \partial \mathbf{v}/\partial t + 2\boldsymbol{\omega} \times \mathbf{v} - \omega^2 \mathbf{r}_H .$$

$$(2.1.18)$$

The following units are selected: the length unit $|\mathbf{a}_M|$, the time unit $T/2\pi$ (where T is a sidereal month), and the mass unit $m_E + m_M$. All other notation can be obtained from Fig. 36.

In these units we have $\omega = 1$, $k = 1$ so that the following equation for the motion of the satellite is obtained from (2.1.17) and (2.1.18):

$$\partial \mathbf{v}/\partial t + 2\mathbf{e}_z \times \mathbf{v} = \operatorname{grad} U. \qquad (2.1.19)$$

Here $\mathbf{e}_z$ is the unit vector in the z direction, and we have

$$U = \frac{1}{2}\mathbf{r}_H{}^2 + \frac{1-\mu}{|\mathbf{r}_{ES}|} + \frac{\mu}{|\mathbf{r}_{MS}|}, \qquad (2.1.20)$$

where $\mu \approx m_M/m_E = 1/82$.

Let us first show that the motion of the satellite remains in the earth–moon plane (x, y) if the initial values $\mathbf{r}_0$, $\mathbf{v}_0$ are vectors in this plane. In this way we will show that the restriction to *plane motion* is a reasonable assumption for the three-body problem.

Scalar multiplication of (2.1.19) with $\mathbf{v}$ results in

$$\mathbf{v}(\partial \mathbf{v}/\partial t - \operatorname{grad} U) = 0. \qquad (2.1.21)$$

This means that the vector $\mathbf{c} = \partial \mathbf{v}/\partial t - \text{grad } U = 2\mathbf{v} \times \mathbf{e}_z$ is perpendicular to the vector $\mathbf{v}$ and lies in the x, y plane. If the initial values $\mathbf{r}_0$, $\mathbf{v}_0$ of a motion are selected so that they are vectors in the x, y plane, grad U is also a vector in the x, y plane at this time, as well as $\partial \mathbf{v}/\partial t = \mathbf{c} + \text{grad } U$. Therefore, $\mathbf{v}$ and $\mathbf{r}$ remain in the x, y plane for all times. The motion is plane as predicted.

Equation (2.1.21) can be written in the form

$$\mathbf{v} \, (\partial \mathbf{v}/\partial t) = \mathbf{v} \text{ grad } U = dU/dt.$$

Integration of this expression yields the Jacobi integral

$$\mathbf{v}^2 = 2U - J, \tag{2.1.22}$$

where J is the Jacobi constant.

In the following discussion of the plane three-body problem, the equipotential curves $2U = C_i$ corresponding to different values of the constants C_i, $i = 1, 2, 3,...$, are plotted. These curves are called *Hill limit curves* of zero relative velocity or Darwin curves. Figure 37 shows these curves, which correspond to limits of motion. This can be seen as follows: Let us assume that the satellite begins its motion from the position $\mathbf{r}_a$ with the velocity $\mathbf{v}_a$ (Fig. 38). From Eq. (2.1.22) it follows that the Jacobi constant of this motion is $J = C_a - \mathbf{v}_a^2$ because the satellite began its motion along the equipotential curve $2U = C_a$. When the satellite reaches the curve $2U = C_i$, the square of the magnitude of the velocity is $\mathbf{v}_i^2 = C_i - J = \mathbf{v}_a^2 + C_i - C_a$. The equipotential curve $2U = J$ represents a limit to the motion because $\mathbf{v}_0^2 = J - J = 0$ holds along it. In general, however, the satellite will not remain in relative equilibrium at the limit curve but instead will move away from it again. For $\mathbf{v}_0 = 0$ (which holds along the limit curve), it follows from (2.1.19) that $\partial \mathbf{v}_0/\partial t = \text{grad } U$. If grad $U = 0$ does not also hold by coincidence, $\partial \mathbf{v}_0/\partial t \neq 0$ must hold along the limiting curve and therefore $\mathbf{b} \neq 0$ must hold. Consequently there is a force acting on the satellite so that this one is repulsed from the limit curves.

The condition grad $U = 0$ leads to singular points of the surface $2U = \text{const.}$ The level curves of the surface are the Hill limit curves $2U = C_i$, $i = 1, 2, 3,...$, in the x, y plane. The singular points in the x, y plane correspond to the *libration points* L_k, $k = 1, 2, 3, 4, 5$. The coordinates of these points are the Lagrange particular integrals of the three-body problem. Let us consider the case where the satellite remains in relative equilibrium at one of the libration points where $\mathbf{v} = 0$, $\partial \mathbf{v}/\partial t = 0$. In addition, the coordinates of the libration point satisfy the condition grad $U = 0$. All the terms in (2.1.19) are simultaneously zero and the system of differential equations is satisfied.

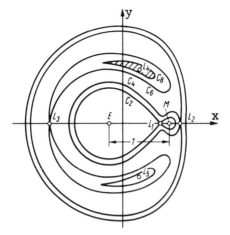

FIG. 37. Darwin curves.

Libration points:	L_1	L_2	L_3	L_4	L_5
Distance from					
$E = r_i$:	0.85	1.17	0.99	1.0	1.0
Distance from					
$M = \rho_i$:	0.15	0.17	1.99	1.0	1.0

Let us now investigate the stability of the points L_1 to L_5 in the three-body problem for the earth, moon, and satellite.

For this three-body problem we have

$$2U = x^2 + y^2 + \frac{2(1-\mu)}{|\,r_{ES}\,|} + \frac{2\mu}{|\,r_{MS}\,|} = C, \qquad |\,r_{ES}\,| = [(x-x_E)^2 + y^2]^{1/2},$$
$$|\,r_{MS}\,| = [(x-x_M)^2 + y^2]^{1/2},$$

and grad $U = 0$ is given by

$$x - (1-\mu)\frac{x-x_E}{|\,r_{ES}\,|^3} - \mu\frac{x-x_M}{|\,r_{MS}\,|^3} = 0,$$

$$y - (1-\mu)\frac{y}{|\,r_{ES}\,|^3} - \mu\frac{y}{|\,r_{MS}\,|^3} = 0. \tag{2.1.23}$$

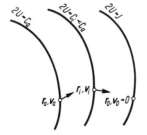

FIG. 38. Hill limit curves.

The libration points L_1 to L_3 lie along the x axis where $y = 0$. The second of Eqs. (2.1.23) is identically satisfied. The x coordinates of the points of the L_i, $i = 1, 2, 3$, can therefore be calculated from the first of Eqs. (2.1.23) with $y = 0$, which results in

$$x - (1 - \mu)(x - x_E)^{-2} - \mu(x - x_M)^{-2} = 0. \qquad (2.1.24)$$

Since the origin of the x, y axis system is at the center of gravity of the earth and moon, $x_E = -\mu$, $x_M = 1 - \mu$. If the distance between M and L_2 is called ρ, then $x = 1 - \mu + \rho$. If this is substituted into (2.1.24), the following equation is obtained for determining ρ:

$$\rho^5 + (3 - \mu)\rho^4 + (3 - 2\mu)\rho^3 - \mu\rho^2 - 2\mu\rho - \mu = 0.$$

For $\mu = 1/82$, the one real positive solution is $\rho = 0.17$. The locations of L_1, L_3 along the x axis can also be calculated in this way. The results are given in the table of Fig. 37.

Figure 39 shows the location of the points L_4, L_5. It may be seen, for example, that L_4 has coordinates

$$x = \tfrac{1}{2} - \mu, \qquad y = \tfrac{1}{2}\sqrt{3}. \qquad (2.1.25)$$

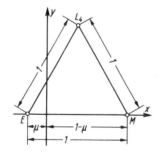

FIG. 39.   Position of the libration point L_4.

If the libration point locations are known, the Gaussian curvature

$$K = (U_{xx}U_{yy} - U_{xy}^2)/(1 + U_x^2 + U_y^2)^2 \qquad (2.1.26)$$

of the surface U (and therefore also of $2U$) at these points can be found. From

$$U = \frac{1}{2}(x^2 + y^2) + \frac{1 - \mu}{|\mathbf{r}_{ES}|} + \frac{\mu}{|\mathbf{r}_{MS}|},$$

$$|\mathbf{r}_{ES}| = [(x - x_E)^2 + y^2]^{1/2}, \qquad |\mathbf{r}_{MS}| = [(x - x_M)^2 + y^2]^{1/2},$$

it follows that

$$U_{xx} = \left[1 - (1-\mu)\frac{1}{|\mathbf{r}_{ES}|^3} - \mu\frac{1}{|\mathbf{r}_{MS}|^3}\right]$$
$$+ 3\left[(1-\mu)\frac{(x-x_E)^2}{|\mathbf{r}_{ES}|^5} + \mu\frac{(x-x_M)^2}{|\mathbf{r}_{MS}|^5}\right],$$

$$U_{yy} = \left[1 - (1-\mu)\frac{1}{|\mathbf{r}_{ES}|^3} - \mu\frac{1}{|\mathbf{r}_{MS}|^3}\right] \qquad (2.1.27)$$
$$+ 3y^2\left[(1-\mu)\frac{1}{|\mathbf{r}_{ES}|^5} + \mu\frac{1}{|\mathbf{r}_{MS}|^5}\right],$$

$$U_{xy} = 3y\left[(1-\mu)\frac{x-x_E}{|\mathbf{r}_{ES}|^5} + \mu\frac{x-x_M}{|\mathbf{r}_{MS}|^5}\right].$$

$y = 0$ holds for L_1 to L_3. The following relationships are obtained for these points from (2.1.27):

$$U_{xy} = 0, \quad U_{yy} = \left[1 - (1-\mu)\frac{1}{|x-x_E|^3} - \mu\frac{1}{|x-x_M|^3}\right], \quad U_{xx} = 3 - 2U_{yy}.$$

The sign of K is determined by the sign of

$$U_{xx}U_{yy} = 3U_{yy} - 2U_{yy}^2$$

because $U_{xy} = 0$. For example, for L_1 where $x - x_E = 0.85$, $x - x_M = -0.15$ and because $\mu = 1/82$, we have

$$U_{yy} = 1 - \frac{0.988}{0.85^3} - \frac{0.012}{0.15^3} = -4.15.$$

Since U_{yy} is negative, our question is already answered, because K must also be negative. Therefore, L_1 is a *hyperbolic* point of U. The same method shows that L_2, L_3 are also hyperbolic points of U.

Equation (2.1.25) is satisfied for L_4. Furthermore, the coordinates of this point satisfy $|\mathbf{r}_{ES}| = |\mathbf{r}_{MS}| = 1$, $x - x_E = -(x - x_M) = \frac{1}{2}$ (see Fig. 39) and

$$U_{xx} = \frac{3}{4}, \quad U_{yy} = \frac{9}{4}, \quad U_{xy} = \frac{3}{2}\sqrt{3}\left(\frac{1}{2}-\mu\right). \qquad (2.1.28)$$

We then have

$$U_{xx}U_{yy} - U_{xy}^2 = \frac{3}{4}\frac{9}{4} - \frac{3\cdot9}{4}\left(\frac{1}{2}-\mu\right)^2 = \frac{27}{16}\left[1 - 4\left(\frac{1}{2}-\mu\right)^2\right]. \qquad (2.1.29)$$

Since $\mu = 1/82$ and $4(\frac{1}{2} - \mu)^2 < 1$, we have $U_{xx}U_{yy} - U_{xy}^2 > 0$ and therefore also $K > 0$ so that L_4 is an *elliptic* point of U. The same is true for point L_5 that is a mirror image of L_4 with respect to the x axis.

Let us now show the manner in which the curvature of U at the singular points (libration points) leads to a first approximation of the stability of these points. It will be shown that the points L_1 to L_3 are unstable according to the definition of Lyapunov, because they are hyperbolic points of U with $K < 0$. It will be seen that the points L_4, L_5 are hyperbolic points of U with $K > 0$ which are stable in the first approximation.

The following calculation is made to prove this postulate. The scalar form of Eq. (2.1.19) is

$$\ddot{x} - 2\dot{y} = U_x,$$
$$\ddot{y} + 2\dot{x} = U_y. \tag{2.1.30}$$

The libration points L_i, $i = 1, 2, 3, 4, 5$, which have the coordinates $x_i = a_i$, $y_i = b_i$, satisfy grad $U = 0$ and therefore

$$U_x(a_i, b_i) = U_y(a_i, b_i) = 0 \qquad \text{for all } i. \tag{2.1.31}$$

According to (2.1.30), (2.1.31) they also satisfy

$$\ddot{a}_i - 2\dot{b}_i = 0,$$
$$\ddot{b}_i + 2\dot{a}_i = 0 \qquad \text{for all } i. \tag{2.1.32}$$

The variations ξ, η are introduced in order to investigate the stability of the L_i. We have

$$x = a_i + \xi,$$
$$y = b_i + \eta. \tag{2.1.33}$$

The variational equations for the stability investigation are obtained from (2.1.30) by substitution of (2.1.33). This results in

$$\ddot{a}_i - 2\dot{b}_i + \ddot{\xi} - 2\dot{\eta} = U_x(a_i + \xi, b_i + \eta),$$
$$\ddot{b}_i + 2\dot{a}_i + \ddot{\eta} + 2\dot{\xi} = U_y(a_i + \xi, b_i + \eta). \tag{2.1.34}$$

Because of (2.1.32) and (2.1.31), and making use of the Taylor expansions of U_x and U_y, we obtain

$$\ddot{\xi} - 2\dot{\eta} = U_{xx}(a_i, b_i)\,\xi + U_{xy}(a_i, b_i)\,\eta + X(\xi, \eta),$$
$$\ddot{\eta} + 2\dot{\xi} = U_{xy}(a_i, b_i)\,\xi + U_{yy}(a_i, b_i)\,\eta + Y(\xi, \eta). \tag{2.1.35}$$

This is a nonlinear system of equations, the nonlinear terms of which are given by X and Y.

Linearized equations corresponding to (2.1.35) given by

$$\ddot{\xi} - 2\dot{\eta} = U_{xx}^{(i)}\xi + U_{xy}^{(i)}\eta,$$
$$\ddot{\eta} + 2\dot{\xi} = U_{xy}^{(i)}\xi + U_{yy}^{(i)}\eta, \tag{2.1.36}$$

are used in the stability investigation, where

$$U_{xx}^{(i)} = U_{xx}(a_i, b_i), \qquad \text{etc.}$$

The trial solutions

$$\xi = ae^{\lambda t}, \qquad \eta = be^{\lambda t} \tag{2.1.37}$$

substituted into (2.1.36) result in the system

$$a\lambda^2 - 2b\lambda = aU_{xx}^{(i)} + bU_{xy}^{(i)},$$
$$b\lambda^2 + 2a\lambda = aU_{xy}^{(i)} + bU_{yy}^{(i)}. \tag{2.1.38}$$

In order that this system of equations may have nontrivial solutions for a and b, we must set

$$\begin{vmatrix} \lambda^2 - U_{xx}^{(i)} & -2\lambda - U_{xy}^{(i)} \\ 2\lambda - U_{xy}^{(i)} & \lambda^2 - U_{yy}^{(i)} \end{vmatrix} = 0.$$

This leads to the characteristic equation

$$\lambda^4 + (4 - U_{xx}^{(i)} - U_{yy}^{(i)})\lambda^2 + [U_{xx}^{(i)}U_{yy}^{(i)} - (U_{xy}^{(i)})^2] = 0. \tag{2.1.39}$$

Because of (2.1.26) and (2.1.31), the Gaussian curvature K of the surface U takes on the following values at the libration points:

$$K^{(i)} = U_{xx}^{(i)}U_{yy}^{(i)} - (U_{xy}^{(i)})^2, \qquad i = 1, 2, 3, 4, 5. \tag{2.1.40}$$

Therefore, Eq. (2.1.39) can be transformed into

$$\lambda^4 + (4 - U_{xx}^{(i)} - U_{yy}^{(i)})\lambda^2 + K^{(i)} = 0. \tag{2.1.41}$$

Let us introduce the further abbreviation

$$4 - U_{xx}^{(i)} - U_{yy}^{(i)} = \beta^{(i)}. \tag{2.1.42}$$

The characteristic equation finally becomes

$$\lambda^4 + \beta^{(i)}\lambda^2 + K^{(i)} = 0. \tag{2.1.43}$$

The roots of this equation are given by

$$\lambda_1 = \left[-\frac{\beta^{(i)}}{2} + \left(\frac{\beta^{(i)2}}{4} - K^{(i)}\right)^{1/2}\right]^{1/2}, \qquad \lambda_2 = -\lambda_1,$$

$$\lambda_3 = \left[-\frac{\beta^{(i)}}{2} - \left(\frac{\beta^{(i)2}}{4} - K^{(i)}\right)^{1/2}\right]^{1/2}, \qquad \lambda_4 = -\lambda_3. \tag{2.1.44}$$

At the libration points for L_1, L_2, L_3—that is, for $i = 1, 2, 3$, $K^{(i)} < 0$ holds. According to (2.1.44) $\lambda_1 > 0$ and is real, so that instability follows from (2.1.37). According to the theorems derived previously, in this case the result of the stability calculation using the linear equations (2.1.36) is also valid for the nonlinear case (2.1.35). This means that for the condition $K^{(i)} < 0$ the libration points L_1, L_2, L_3 are unstable not only in terms of the definition of small oscillations but also in terms of the Lyapunov definition. The negative Gaussian curvature K of the surface U, i.e., the hyperbolic character of U of the points L_1, L_2, L_3 derived from it, has indeed revealed the instability of these points.

For the libration points L_4, L_5, i.e., for $i = 4, 5$, we find $\beta^{(i)} = 1$ from (2.1.42). This can easily be calculated for L_4, where $U_{xx} = 3/4$, $U_{yy} = 9/4$ according to (2.1.28). This also holds for L_5. The characteristic equation (2.1.43) can therefore be simplified as follows:

$$\lambda^4 + \lambda^2 + K^{(i)} = 0, \qquad i = 4, 5. \tag{2.1.45}$$

Instead of (2.1.44), the following simplification can be made:

$$\lambda_1 = \{\tfrac{1}{2}[-1 + (1 - 4K^{(i)})^{1/2}]\}^{1/2}, \qquad \lambda_2 = -\lambda_1$$

$$\qquad\qquad\qquad\qquad\qquad\qquad\qquad\qquad \text{for} \quad i = 4, 5. \tag{2.1.46}$$

$$\lambda_3 = \{\tfrac{1}{2}[-1 - (1 - 4K^{(i)})^{1/2}]\}^{1/2}, \qquad \lambda_4 = -\lambda_3$$

From (2.1.46) it may be seen that all λ_k, $k = 1, 2, 3, 4$, are purely imaginary. Because of (2.1.37), small oscillation stability in the first approximation prevails if

$$\tfrac{1}{4} > K^{(i)} > 0, \qquad i = 4, 5, \tag{2.1.47}$$

is satisfied.

Since we are considering the earth–moon–satellite problem, we have $\mu = 1/82$ so that the following result is obtained from (2.1.40), (2.1.28) for $i = 4$, i.e., from (2.1.29)

$$K^{(4)} = \frac{27}{16}\left[1 - 4\left(\frac{1}{2} - \frac{1}{82}\right)^2\right] \approx 0.08.$$

Condition (2.1.47) is satisfied at the elliptical point L_4 because of the positive and sufficiently small Gaussian curvature $K^{(4)}$ of the surface

U at this point. Therefore, stability in the first approximation does indeed prevail at this point. This also holds for L_5 .

Since the λ_k , $k = 1, 2, 3, 4$, are purely imaginary for (2.1.47), the critical case according to the definition of Lyapunov prevails. This means that the stability of L_4 , L_5 obtained from the linearized equations (2.1.36) does not necessarily apply to the general case based on the nonlinear equations (2.1.35). In order to obtain results in this case, the stability would have to be investigated again, and the nonlinear terms would have to be taken into account. A first step in this direction was made by Leontovic.[1]

2.2. PROBLEMS OF THE MECHANICS OF RIGID BODIES AND OF SYSTEMS

2.2.1. *Stability of Vibrations*

Vibration problems and the associated stability problems have been treated extensively in the literature. For example, we should mention the books by Klotter (*2.4*), Minorsky (*2.1*), Kauderer (*2.10*), and Magnus (*2.14*). The following limited discussion is given in order to show the way in which the approximate methods derived in Section 1.6 can be applied. Let us restrict ourselves to *nonlinear vibrations of one-mass systems*. This makes it possible to deal with the most important questions of modern stability theory without any great difficulty.

Let us assume a vibration process given by the nonlinear differential equation (Fig. 40)

$$\ddot{x} + \omega_0{}^2 x + \mu\Phi(x, \dot{x}) = G(t). \tag{2.2.1}$$

ω_0 is a given constant, μ a small parameter, Φ a nonlinear function, and G is a periodic function.

The first possibility is to discuss problem (2.2.1) in the *phase plane*. The phase curves can be obtained with the isocline method, for example. The method can only be recommended when $G \equiv 0$, that is, when the mechanical process is *autonomous*. If, in addition, Φ is only a function of x, the phase curve can be determined exactly by integrating (2.2.1)

$$\dot{x} = \sqrt{2}[E - U(x)]^{1/2}, \tag{2.2.2}$$

where

$$U(x) = \tfrac{1}{2}\omega_0{}^2 x^2 + \mu \int \Phi(x)\, dx.$$

The phase curves are levels of the energy surface $E = \tfrac{1}{2}\dot{x}^2 + U(x)$ in the $\dot{x}, x$ plane. The course of these curves makes it possible to determine

[1] LEONTOVIC, A. M., *Dokl. Akad. Nauk SSSR* **143**, 525–528 (1962).

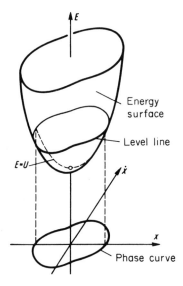

FIG. 40. Connection between energy
surface and phase curves.

the stability of stationary points (equilibrium positions) or limit cycles (periodic solutions). If the phase curve equation depends on the parameter, the stability investigation consists of determining the change in the properties of the stationary points or limit cycles as the parameter is varied. The "critical" parameter values are of special interest, because the transition from stable to unstable occurs there.

Example: Let us consider $\ddot{x} + \omega_0^2 x + \mu x^3 = 0$. The phase curves are determined from $\dot{x} = (1/\sqrt{2})(4E - 2\omega_0^2 x^2 - \mu x^4)^{1/2}$ (Fig. 41). The

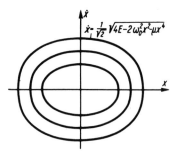

FIG. 41. Phase diagram.

phase diagram shows that the stability of the stationary solution $\dot{x} = x = 0$, as well as that of the periodic solutions, do not change if the parameters ω_0 or μ are changed. The periodic solutions have orbital stability with respect to perturbations of the initial conditions.

The approximation methods of Section 1.6 can also be used to investigate the stability of the solutions of the problem (2.2.1). Let us now consider this method.

Let us reconsider the *method of small oscillations*. It is assumed that (2.2.1) has a periodic solution, which is essentially given by $x_s = A \cos \omega t$. In order to investigate its stability, $x = x_s + \xi$ is substituted into (2.2.1), and the result is

$$\ddot{\xi} + \omega_0^2 \xi + \mu \Phi(x_s + \xi, \dot{x}_s + \dot{\xi}) = G(t) - \ddot{x}_s \, \omega_0^2 x_s^2.$$

Linearization of this equation results in

$$\ddot{\xi} + \omega_0^2 \xi + \mu \left[\frac{\partial \Phi(x, \dot{x})}{\partial x} \right]_{\substack{x=x_s \\ \dot{x}=\dot{x}_s}} \xi + \mu \left[\frac{\partial \Phi(x, \dot{x})}{\partial \dot{x}} \right]_{\substack{x=x_s \\ \dot{x}=\dot{x}_s}} \dot{\xi}$$

$$= G(t) - \ddot{x}_s - \omega_0^2 x_s^2 - \mu \Phi(x_s, \dot{x}_s).$$

The right side is zero according to our assumption and the following variational equation remains:

$$\ddot{\xi} + \omega_0^2 \xi + \mu \left\{ \left[\frac{\partial \Phi}{\partial x} \right]_{\substack{x=x_s \\ \dot{x}=\dot{x}_s}} \xi + \left[\frac{\partial \Phi}{\partial \dot{x}} \right]_{\substack{x=x_s \\ \dot{x}=\dot{x}_s}} \dot{\xi} \right\} = 0. \qquad (2.2.3)$$

This is a linear differential equation with variable, *periodic* coefficients. The remainder of the stability investigation follows Section 1.3.3.

Example: Let us again consider $\ddot{x} + \omega_0^2 x + \mu x^3 = 0$. The variational equation is $\ddot{\xi} + \omega_0^2 \xi + 3\mu A^2 (\cos^2 \omega t) \xi = 0$. It can be changed into the following standard form of the Mathieu differential equation by the substitutions of

$$\lambda = (1/\omega^2)(\omega_0^2 + 3\mu A^2/2), \qquad h = (-1/\omega^2)(3\mu A^2/4), \qquad \tau = \omega t$$

$$d^2\xi/d\tau^2 + (\lambda - 2h \cos 2\tau) \xi = 0.$$

In Section 1.6.4 the approximate solution

$$x = A \cos \omega t \quad \text{of} \quad \ddot{x} + \omega_0^2 x + \mu x^3 = 0$$

was found. The relationship $\omega_0^2 = \omega^2 + \frac{3}{4}\mu A^2$ was a first approximation. Therefore, $\lambda = 1 - h$ would show that this approximation lies along a limiting curve of the Strutt diagram (Fig. 16). It thus follows that x_s was a solution, as can be determined from the phase diagram (Fig. 41), and that it must have orbital stability. It can easily be shown that it is not stable according to the definition of Lyapunov. This is a first indication that the concept of *orbital stability* is very important in vibration problems, because often the Lyapunov stability definition is too restrictive in practical applications.

It is better to use the *harmonic linearization* in such a case, because the linearization described above is often not permissible and may lead to the critical case where the real parts of the characteristic exponents vanish, which means that it certainly is questionable. Harmonic linearization is carried out by first obtaining the linear substitution equations using the harmonic balance method as described in Section 1.6.6. The coefficients of the substitution equation are then used to determine stability. This was done for the nonlinear differential equation

$$\ddot{x} - (\alpha - \beta \mid x \mid) \mid \dot{x} \mid \dot{x} + \omega_0^2 (x + \gamma x^3) = 0.$$

It was immediately possible to conclude the stability of the periodic solution from the behavior of the damping coefficients of the substitution equation.

Example: Let us again consider the differential equation $\ddot{x} + \omega_0^2 x + \mu x^3 = 0$, which will be transformed into the substitution equation $\ddot{x} + \omega_0^2 x = cx$. From Section 1.6.6 it follows that

$$c = -\frac{\mu \omega}{A\pi} \int_0^{2\pi/\omega} A^3 \sin^4 \omega \xi \, d\xi = -\frac{3\mu A^2}{4},$$

so that $\ddot{x} + (\omega_0^2 + 3\mu A^2/4) x = 0$ is obtained. This substitution equation corresponds to an oscillation with amplitude A and the frequency $\omega = [\omega_0^2 + 3\mu A^2/4]^{1/2}$ which depends on the amplitude. Apparently the phase curves always have orbital stability because the stability limit $\omega = 0$ can never be reached. On the other hand, oscillations corresponding to adjacent values of A have different frequencies.

This means that image points moving along adjacent phase curves having orbital stability can be displaced by a finite amount at certain times, even though the phase curves may be as close to each other as desired. Therefore, Lyapunov stability does not exist. We have again shown the properties of this differential equation in another way.

It is possible that the harmonic balance method will be suitable for *special stability criteria*, which can be different in each case and are adjusted to the conditions at hand. We will continue this method by means of two examples.

Example 1:   Let us first consider a *forced vibration* with the differential equation

$$\ddot{x} + \omega_0^2 x + \mu x^3 = Z \cos \omega t.$$

It can be written in the following form with the notation $L = d^2/dt^2 + \omega_0{}^2$ and $f(x, t) = Z \cos \omega t - \mu x^3$: $L(x) = f(x, t)$. The solution is $x = L^{-1}[f(x, t)]$. The inverse of the operator L is given by

$$L^{-1}(\varphi) = -\frac{2}{T} \int_0^t \sum_k \frac{\cos k\omega(t - \xi)}{k^2\omega^2 - \omega_0{}^2} \varphi(\xi) \, d\xi$$

and the nonlinear integral equation

$$x(t) = -\frac{2}{T} \int_0^t \sum_{k=1}^\infty \frac{\cos k\omega(t - \xi)}{k^2\omega^2 - \omega_0{}^2} [Z \cos \omega\xi - \mu x^3(\xi)] \, d\xi$$

is obtained where $T = 2\pi/\omega$. According to the harmonic balance method, only the first term in the sum under the integral sign is retained and therefore $k = 1$. The approximation $x = A \cos \omega t$ is assumed for x. This results in

$$A \cos \omega t = -\frac{\omega}{\pi(\omega^2 - \omega_0{}^2)} \left[\int_0^{2\pi/\omega} Z \cos^2 \omega\xi \, d\xi - \mu \int_0^{2\pi/\omega} A^3 \cos^4 \omega\xi \, d\xi \right] \cos \omega t.$$

The integration is carried out and the coefficients of $\cos \omega t$ are set equal, which results in the relationship

$$Z = (3\mu/4) A^3 - A(\omega^2 - \omega_0{}^2). \qquad (2.2.4)$$

For constant Z this relationship represents the resonance curve that gives the dependence of the amplitude A on the square of the exciting frequency ω (Fig. 42). As the figure shows, it is possible to have a jump in the amplitude from A_u to A_0 for $\omega = \omega_{gr}$, which indicates that the

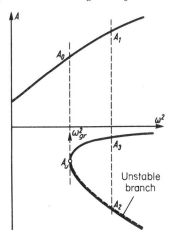

FIG. 42.   Resonance curve.

stability boundary has been reached. Three amplitudes are possible
for $\omega > \omega_{gr}$. A decision must be made as to which of the amplitudes
correspond to stable or unstable solutions. This is done in the following
way.

The resonance curve is a level line of the surface $Z = Z(A, \omega^2)$
generated by (2.2.4). Since $Z = $ const holds for the resonance curve, the
following equation must also be satisfied:

$$dZ = (\partial Z/\partial A)\, dA + (\partial Z/\partial \omega^2)\, d\omega^2 = 0.$$

This equation, together with the relationship $\partial Z/\partial \omega^2 = -A$ according
to (2.2.4), results in the following equation:

$$dA/d\omega^2 = A(\partial Z/\partial A)^{-1}. \tag{2.2.5}$$

Obviously $\partial Z/\partial A > 0$ must hold for stable solutions, which means
that the magnitude of the forcing function must increase when A is
increased. Conversely, $\partial Z/\partial A < 0$ for unstable solutions. If this is taken
into account in (2.2.5), it follows that the stability boundary is given by
$\partial Z/\partial A = 0$, that is, by $dA/d\omega^2 = \infty$. The unstable branch of the resonance curve must start at the point A_u where the resonance curve has a
vertical tangent. The stable branches follow from (2.2.5) and $\partial Z/\partial A > 0$.
In Fig. 42 they are the branches with positive slope. The unstable
branch is the one with negative slope, because $\partial Z/\partial A < 0$.

Example 2: Let us now consider the differential equation
$\ddot{x} + \omega_0^2 x = av^2(\varphi - \dot{x}/v)$. It corresponds to the *bending oscillations of
a wing*. The flight velocity is v and the wing simultaneously twists by an
amount φ. The critical flight velocity v_k is to be determined at which
flutter oscillations, and therefore instability phenomena, of the wing
begin.[1]

This example is used to show that stability can be determined using the
approximate trial solution $x = A \cos \omega_0 t$ of the harmonic balance
method and by a consideration of the *energy balance*.

The amount of damping energy dissipated over a period is given by
$E_D = \int_0^T av\dot{x}^2\, dt$.

The energy supplied by the excitation over a period is given by
$E_Z = -\int_0^T av^2\varphi\dot{x}\, dt$.

The amplitude of the torsional oscillation φ is different from that of the
bending oscillation x and there is a phase displacement between them.

---

[1] This example was given by MAGNUS (*2.14*, p. 126).

Therefore, $\varphi = bA \cos(\omega_0 t - \psi)$, and the following expressions are calculated:

$$E_D = \int_0^{2\pi} av\omega_0^2 A^2 \sin^2 \omega_0 t \, d(\omega_0 t) = av\omega_0^2 A^2 \pi,$$

$$E_Z = \int_0^{2\pi} av^2 b\omega_0 A^2 \sin \psi \sin^2 \omega_0 t \, d(\omega_0 t) = av^2 b\omega_0 A^2 \sin \psi \pi.$$

The oscillation can only build up to flutter of the wing for $E_Z > E_D$, that is, for $v > \omega_0/b \sin \psi$. The stability boundary is given by $E_Z = E_D$, which means that the critical flight velocity is $v = \omega_0/b \sin \psi$.

The question of *orbital stability* can also be decided by the *variation of constants* method described in Section 1.6.3. Let us consider the phase plane $\dot{A}$, A and the phase curves in this plane (Fig. 43). The time param-

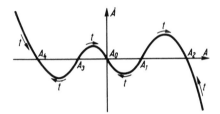

FIG. 43. Phase curve of the vibration amplitude.

eter t increases along the phase curve in the right direction in the upper half-plane. In the lower half-plane, it increases toward the left. The points $\dot{A} = 0$ are *equilibrium points* for the amplitude A that correspond to *periodic solutions*. Points such as A_4, A_0, and A_2 are *stable*, and the corresponding phase curves of the periodic solutions in the $\dot{x}$, x plane have orbital stability if the time in the $\dot{A}$, A plane increases in the direction toward these points. On the other hand, the points A_3, A_1 are *unstable* because time increases in the direction away from them. It can be seen that $d\dot{A}/dA < 0$ holds in the region of stable points and $d\dot{A}/dA > 0$ holds in the region of unstable points. If $[\dot{A}]_M = 0$ is identically satisfied, which is the case for the example $\ddot{x} + \omega_0^2 x + hx^3 = 0$, then the mechanical system has orbital stability of the neutral type.

In the variation of constants method, the average value $[\dot{A}]_M$ is substituted for $\dot{A}$ and the stability is determined from the expression $d[\dot{A}]_M/dA$. This was already carried out for the example mentioned above.

Example: Let us consider the *Van der Pol equation*

$$\ddot{x} - (\alpha - \beta x^2) \dot{x} + x = 0$$

to show the exact calculation. Let us proceed according to Section 1.6.3. For $x = \xi_1$, $\dot{x} = \xi_2$, the differential equation is transformed into the system

$$\dot{\xi}_1 = \xi_2, \qquad \dot{\xi}_2 = (\alpha - \beta \xi_1^2) \xi_2 - \xi_1.$$

The trial solution $\xi_1 = A \cos(t + \vartheta) = A \cos \psi$, $\xi_2 = -A \sin \psi$ is substituted into the system of differential equations, where A and ϑ are allowed to be variables. The following result is then obtained:

$$\dot{\xi}_1 = \dot{A} \cos \psi - A \sin \psi(1 + \dot{\vartheta}) = \xi_2 = -A \sin \psi,$$

$$\dot{\xi}_2 = -\dot{A} \sin \psi - A \cos \psi(1 + \dot{\vartheta}) = (\alpha - \beta \xi_1^2) \xi_2 - \xi_1,$$

$$= (\alpha - \beta A^2 \cos^2 \psi)(-A \sin \psi) - A \cos \psi,$$

from which it follows that

$$\dot{A} \cos \psi - A\dot{\vartheta} \sin \psi = 0,$$

$$-\dot{A} \sin \psi - A\dot{\vartheta} \cos \psi = -A(\alpha - \beta A^2 \cos^2 \psi) \sin \psi.$$

These are the determining equations for $\dot{A}$, $\dot{\vartheta}$, from which the following expression is calculated:

$$\dot{A} = A(\alpha - \beta A^2 \cos^2 \psi) \sin^2 \psi.$$

The final result is

$$[\dot{A}]_M = \frac{1}{2\pi} \int_0^{2\pi} A(\alpha - \beta A^2 \cos^2 \psi) \sin^2 \psi \, d\psi = \frac{\alpha A}{2} \left(1 - \frac{\beta}{\alpha} \frac{A^2}{4}\right).$$

The stationary values of A are $A = 0$ and $A_{st} = 2(\alpha/\beta)^{1/2}$. The first value is a trivial one. The second value is the amplitude of a possible periodic solution. The properties of the expression

$$\frac{d[\dot{A}]_M}{dA} = \frac{\alpha}{2} - \frac{3}{8} \beta A^2$$

are determined for A_{st}. We have

$$\{d[\dot{A}]_M/dA\}_{A=A_{st}} = -\alpha,$$

which we shall assume to be negative for positive α. It follows that the corresponding periodic solution has *orbital stability*.

Let us now consider the application of the *perturbation method*. The calculation of *parameter-dependent stability regions* will be shown by means of an example.

Example: Let us consider the differential equation

$$\ddot{x} + \lambda^2[1 + \mu f(t)] x = 0,$$

where λ^2 and μ are parameters and the function $f(t)$ is *periodic* with period ω. It is assumed that this function can be expanded into the series $f_i(t)\mu^i$, $i = 0, 1, 2, 3,...$, according to powers of μ.

The substitutions $\xi_1 = x$, or $\xi_2 = \dot{x}$ transform the differential equation into the system $\dot{\xi}_1 = \xi_2$, $\dot{\xi}_2 = -\lambda^2(1 + \mu f)\xi_1$ or $\dot{\xi}_i = p_{ik}\xi_k$ $(i, k = 1, 2)$, respectively, where $p_{11} = 0$, $p_{12} = 1$, $p_{21} = -\lambda^2(1 + \mu f)$, $p_{22} = 0$. These are the same conditions as in the example in Section 1.6.4 and these results can be used here. The magnitude of $\frac{1}{2}[\xi_1^{(1)}(\omega) + \dot{\xi}_1^{(2)}(\omega)]$ determines stability, where $\xi_1^{(1)}$, $\xi_1^{(2)}$ are fundamental solutions of the system that satisfy the initial conditions:

$$\xi_1^{(1)}(0) = 1, \qquad \dot{\xi}_1^{(1)}(0) = 0, \qquad \xi_1^{(2)}(0) = 0, \qquad \dot{\xi}_1^{(2)}(0) = 1.$$

Since $\xi_1 = x$, we can also use the old notation and can use the magnitude of $K = \frac{1}{2}[x_1(\omega) + \dot{x}_2(\omega)]$ to determine stability, where x_1 and x_2 are two solutions of $\ddot{x} + \lambda^2[1 + \mu f(t)] x = 0$ which satisfy the corresponding initial conditions. There is stability for $|K| < 1$ and instability for $|K| > 1$. The regions of stability and instability are separated by the condition $|K| = 1$. Let us assume that the convergent series expansion

$$x_i = x_{ik}\mu^k, \qquad i = 1, 2, \qquad k = 0, 1, 2, 3,...,$$

is possible for $|\mu| < \epsilon$. Substituting these series into the differential equation, taking $f = f_k\mu^k$, $k = 0, 1, 2, 3,...$, into account and setting the coefficients of equal powers of μ equal to zero, we obtain the following system of differential equations:

$$\ddot{x}_{i0} = -\lambda^2 x_{i0}, \qquad \ddot{x}_{ik} = -\lambda^2\left(x_{ik} + \sum_{s=0}^{k-1} f_{k-s-1}x_{is}\right), \qquad i = 1, 2, \; k = 1, 2, 3,...,$$

with the initial conditions

$$x_{10}(0) = 1, \qquad \dot{x}_{10}(0) = 0,$$

$$x_{20}(0) = 0, \qquad \dot{x}_{20}(0) = 1, \qquad i = 1, 2, \qquad k = 1, 2, 3,....$$

$$x_{ik}(0) = \dot{x}_{ik}(0) = 0,$$

In particular, $x_{10} = \cos \lambda t$, $x_{20} = (1/\lambda) \sin \lambda t$ so that

$$K = \cos \lambda\omega + \frac{1}{2}[x_{1k}(\omega) + \dot{x}_{2k}(\omega)] \mu^k, \qquad k = 1, 2, 3,.... \qquad (2.2.6)$$

For $K = (-1)^n$, $\lambda = (n + \alpha)\,\pi/\omega$, Eq. (2.2.6) results in a sequence of α values that prescribe the stability region boundaries in the λ, μ plane. This sequence is given by the roots $\alpha_k(\mu)$ of the equation

$$(-1)^n = (-1)^n \left(1 - \frac{\alpha^2 \pi^2}{2!} + \frac{\alpha^4 \pi^4}{4!} - \cdots\right)$$

$$+ \tfrac{1}{2}[x_{11}(\omega) + \dot{x}_{21}(\omega)]_{\lambda=(n+\alpha)\pi/\omega} \cdot \mu + \cdots .$$

The boundaries between the stability and instability regions can also be interpreted as level lines of the surface $K(\lambda, \mu)$ in the λ, μ plane (Fig. 44), which helps to clarify the calculation.

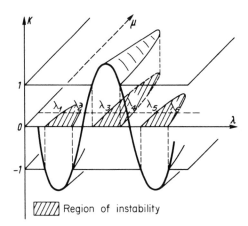

FIG. 44. Stability map: $\lambda_k(\mu) = [n + \alpha_k(\mu)]\pi/\omega$.

////// Region of instability

2.2.2. Stability of Control Systems

Stability investigations are important in control theory. In the design and optimization of control loops one of the important questions is to design the control system in such a way that it will be stable for all cases, at least in the large. The reader is referred to the books on control theory by Lefschetz (1.4), Letov (1.10), Solodovnikov (2.5), Popov (2.12), and Oppelt (2.11). In this book it will only be possible to give a short discussion of the topic and to show a few applications of stability theory to control theory. First, a linear problem is discussed using the methods of Section 1.3.3. A nonlinear problem will be discussed, which will be linearized using the harmonic balance method, and the method outlined in Section 1.3 will be applied. Finally, a nonlinear problem will be investigated using the Lyapunov stability theorems.

Example 1:   The *linear control loop* shown in the block diagram of Fig. 45 is given. The equation of motion for the dynamic behavior of the control loop is

$$T_{s2}\ddot{x} + \dot{x} = s_1\dot{y}_s + y_s \qquad (2.2.7)$$

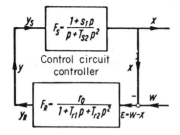

FIG. 45.   Block diagram of a linear control loop.

and the equation of motion for the controller is

$$T_{r2}\ddot{y}_R + T_{r1}\dot{y}_R + y_R = r_0(w - x). \qquad (2.2.8)$$

In addition, $y_R = -y_s$, so that the following differential equation of the control loop is obtained from (2.2.7) and (2.2.8):

$$T_{s2}T_{r2}x^{\text{IV}} + (T_{r1}T_{s2} + T_{r2})\,\dddot{x} + (T_{r1} + T_{s2})\,\ddot{x} + (1 + r_0s_1)\,\dot{x} + r_0x$$
$$= r_0w + r_0s_1\dot{w}. \qquad (2.2.9)$$

It is assumed that $w = w_0 = \text{const}$ holds for the command variable. The stationary solution $x_0 = w_0$ is obtained from (2.2.9) and the stability of the solution will now be discussed. For this purpose, the substitution $x = x_0 + \xi$ is made into (2.2.9). The following variational equation is obtained:

$$T_{s2}T_{r2}\xi^{\text{IV}} + (T_{r1}T_{s2} + T_{r2})\,\dddot{\xi} + (T_{r1} + T_{s2})\,\ddot{\xi} + (1 + r_0s_1)\,\dot{\xi} + r_0\xi = 0. \qquad (2.2.10)$$

It is assumed that the magnitudes of the system parameters T_{s2}, T_{r1}, T_{r2} are given by

$$T_{s2} = 0.10, \qquad T_{r1} = 0.15, \qquad T_{r2} = 0.01$$

in the appropriate dimensions. The parameters s_1 and r_0 must be selected in such a way that the stationary solution x_0 is stable. Without first deriving a system of first-order differential equations, we find the following characteristic equation with the trial solution $x = e^{\lambda t}$ and the numerical values of T_{s2}, T_{r1}, T_{r2} given above:

$$\lambda^4 + 25\lambda^3 + 250\lambda^2 + a_3\lambda + a_4 = 0, \qquad (2.2.11)$$

where $a_3 = (1 + r_0 s_1) \times 10^3$ and $a_4 = r_0 \times 10^3$. The *Hurwitz matrix* is

$$\mathbf{H} = \begin{bmatrix} 25 & 1 & 0 & 0 \\ a_3 & 250 & 25 & 1 \\ 0 & a_4 & a_3 & 250 \\ 0 & 0 & 0 & a_4 \end{bmatrix}.$$

In order that stability may prevail, the principal minors of this matrix must all be positive. This leads to the conditions

$$a_3 > 0, \qquad a_4 > 0, \qquad 6250 a_3 - a_3{}^2 - 625 a_4 > 0.$$

The stability diagram is shown in Fig. 46. The missing system values

$$r_0 = a_4 \times 10^{-3} = 5,$$
$$s_1 = (a_3 \times 10^{-3} - 1)(1/r_0) = 2/5 = 0.4$$

are obtained by selecting a point from the stability region which is not cross hatched (for example, P with $a_3 = 3 \times 10^3$, $a_4 = 5 \times 10^3$). If these values are used, the control loop is stable with respect to perturbations of the stationary solution.

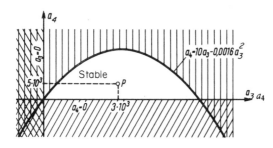

FIG. 46.   Stability map in the parameter plane.

Example 2:   The *stability of periodic solutions* is often of interest. Let us consider the control loop represented by the block diagram shown in Fig. 47. The following equations are found from the block diagram:

$$x_1 = z - y, \qquad x_2 = F_L x_1, \qquad y = F_{NL} x_2.$$

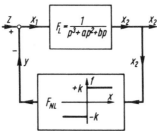

FIG. 47.   Block diagram of a nonlinear control loop.

The frequency response of the control loop is calculated as follows:

$$F = F_L/(1 + F_L F_{NL}),$$

where $x_2 = Fz$. In order to investigate the stability of oscillations that occur in the nonlinear system without any external influences, it is sufficient to consider the relationship

$$(1 + F_L F_{NL}) x_2 = 0.$$

For

$$F_L = 1/(p^3 + ap^2 + bp) \text{ and } F_{NL} x_2 \equiv f(x_2) \text{ in } (p^3 + ap^2 + bp) x_2 + f(x_2) = 0,$$

it is transformed into the differential equation

$$\dddot{x}_2 + a\ddot{x}_2 + b\dot{x}_2 + f(x_2) = 0 \tag{2.2.12}$$

because $p^n = d^n/dt^n$. In this example, $f(x)$ is given by the *nonlinear function* $f(x) = k \operatorname{sgn} x$.

The *harmonic linearization method* is applied to (2.2.12) and $f(x_2) = cx_2$ with

$$c = (1/\pi A) \int_0^{2\pi} f(A \sin \varphi) \sin \varphi \, d\varphi = 4k/\pi A$$

is obtained. The harmonic linearized substitution equation

$$\dddot{x}_2 + a\ddot{x}_2 + b\dot{x}_2 + cx_2 = 0 \tag{2.2.13}$$

can be used for the rest of the investigation. The book by Popov and Paltov (*2.13*) contains a very detailed treatment of this equation.

A periodic solution with amplitude A and frequency ω exists if $\lambda = i\omega$ satisfies the characteristic equation (2.2.13). Therefore

$$(i\omega)^3 + a(i\omega)^2 + bi\omega + c = 0,$$

which leads to

$$X(A, \omega) + iY(A, \omega) = 0. \tag{2.2.14}$$

The functions

$$X(A, \omega) = -a\omega^2 + c, \qquad Y(A, \omega) = -\omega^3 + b\omega$$

depend on A as well as ω, because at least some of the coefficients in (2.2.13) are functions of A. In our example, this holds true for the coefficient c. Equation (2.2.14) decomposes into the two determining equations

$$X(A, \omega) = 0, \qquad Y(A, \omega) = 0,$$

which must result in positive and real values for A and ω if the system is capable of any oscillations at all.

We obtain

$$-a\omega^2 + (4k/\pi A) = 0, \qquad -\omega^3 + b\omega = 0$$

with the solutions

$$\omega^2 = b, \qquad A = 4k/\pi ab. \tag{2.2.15}$$

It is possible for oscillations to exist for $a > 0$, $b > 0$, provided they are stable. The stability is determined as follows. If the $(n-1)$th Hurwitz determinant H_{n-1} satisfies the condition

$$H_{n-1} = 0,$$

the system is on the oscillation boundary. Let us suppose that the amplitude of the oscillation is changed by ΔA. If this oscillation is stable, then a damped process must result for $\Delta A > 0$ and a process with negative damping must result for $\Delta A < 0$. Therefore, the following is a necessary condition for stability of the periodic solution with the amplitude A_0

$$(\partial H_{n-1}/\partial A)_{A=A_0} > 0.$$

Another necessary condition is that all Hurwitz determinants are positive with the exception of H_{n-1}.

In our case this amounts to the condition that the coefficients of (2.2.13) are positive (which is implied from the above condition $a > 0$, $b > 0$, because c is greater than zero in all cases) and that

$$\frac{\partial H_2}{\partial A} = \frac{\partial}{\partial A} (ab - c) = \frac{4k}{\pi A^2} > 0,$$

which is also satisfied in all cases.

It is therefore possible to conclude that for $a > 0$, $b > 0$ it is possible to have stable oscillations with the parameters (2.2.15). The *A-curve method* (Section 1.6.6) gives the same result. The R curve is given by $H_{n-1} \equiv H_2 = 0$ in the parameter plane b, c (Fig. 48). It follows from

$$H_2 = ab - c = 0$$

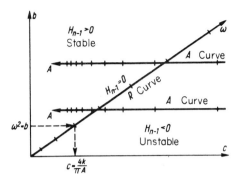

FIG. 48. The A-curve method.

that the R curve is the line

$$b = (1/a)\,c$$

passing through the origin. Along the oscillation boundary the square of the frequency satisfies

$$\omega^2 = a_n(H_{n-3}/H_{n-2}).$$

In our case $a_n = a_3 = c$, $H_{n-2} = H_1 = a$, $H_{n-3} = H_0 = 1$, so that

$$\omega^2 = c/a$$

is obtained. Because $H_{n-1} = H_2 = ab - c = 0$, we have $c/a = b$ so that the well-known relationship $\omega^2 = b$ is obtained in this way. It is therefore possible to plot the ω scale along the R curve shown in Fig. 48.

Since the parameter b is independent of A, the A curves are horizontal lines, and the A scales are given by $A = 4k/\pi c$. Since the A curves intersect the R curve first in the unstable region and then the stable region as A is increased, $\partial R/\partial A \equiv \partial H_2/\partial A > 0$ holds. The oscillations corresponding to the intersection points of the A curves with the R curves are all stable. These are the same stability conditions found earlier by calculation.

The amplitude $A = 4k/\pi c$ can be read from Fig. 48 for any $\omega^2 = b$. Because $c = ab$, $A = 4k/\pi ab$. There is a stable oscillation with a certain amplitude A for any arbitrary frequency ω. Thus, all the results of the previous calculation have been illustrated graphically.

Example 3: Finally let us consider the application of the *Lyapunov theory* to a *nonlinear* problem of control theory (Fig. 49). Let us assume that the differential equations of the control loop are given by

$$\ddot{x} + a\dot{x} + by = c,$$
$$\dot{y} = F(\sigma), \tag{2.2.16}$$
$$\sigma = ex + g\dot{x} + h\ddot{x} - ky.$$

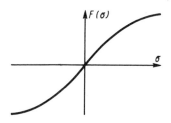

FIG. 49. Course of the nonlinear
function of a control loop.

The coefficients are constants $(a, b, c > 0)$ and $F(\sigma)$ is a nonlinear function that satisfies

$$F(0) = 0, \qquad \sigma F(\sigma) > 0 \qquad \text{for} \quad \sigma \neq 0.$$

By means of the transformation $q_1 = x$, $q_2 = \dot{x}$, $q_3 = y$, (2.2.16) is transformed into the system

$$\dot{q}_1 = q_2,$$
$$\dot{q}_2 = -aq_2 - bq_3 + c, \qquad\qquad (2.2.17)$$
$$\dot{q}_3 = F[eq_1 + (g - ah) q_2 - (k + bh) q_3 + ch].$$

The following stationary solution is obtained for $\dot{q}_i = 0$ the stability of which will now be investigated:

$$q_1{}^0 = kc/be, \qquad q_2{}^0 = 0, \qquad q_3{}^0 = c/b. \qquad\qquad (2.2.18)$$

The trial solution $\xi_i = q_i - q_i{}^0$, $\dot{\xi}_i = \dot{q}_i$ leads to the following variational equations from (2.2.17):

$$\dot{\xi}_1 = \xi_2,$$
$$\dot{\xi}_2 = -(a\xi_2 + b\xi_3), \qquad\qquad (2.2.19)$$
$$\dot{\xi}_3 = F[e\xi_1 + (g - ah) \xi_2 - (k + bh) \xi_3].$$

The following *positive definite* function V is selected as the *Lyapunov function*:

$$V = A \left(\frac{a}{b} \xi_2 + \dot{\xi}_3\right)^2 + B\xi_3{}^2 + C \int_0^{\sigma^*} F(\sigma^*)\, d\sigma^*, \qquad\qquad (2.2.20)$$
$$\sigma^* = e\xi_1 + (g - ah) \xi_2 - (k + bh) \xi_3,$$

with the constants A, B, C which are assumed to be greater than zero. Let us now form

$$\frac{dV}{dt} = \frac{\partial V}{\partial \xi_1} \dot{\xi}_1 + \frac{\partial V}{\partial \xi_2} \dot{\xi}_2 + \frac{\partial V}{\partial \xi_3} \dot{\xi}_3,$$

which becomes, making use of (2.2.19) and (2.2.20)

$$\frac{dV}{dt} = -\frac{2aA}{b^2}(a\xi_2 + b\xi_3)^2 - C(k + bh)F^2(\sigma^*)$$

$$+ \left[\frac{2aA}{b} + eC - aC(g - ah)\right]F(\sigma^*)\,\xi_2$$

$$+ [2A + 2B - bC(g - ah)]\,F(\sigma^*)\,\xi_3\,.$$

If we set

$$(2aA/b) + C(e - ag + a^2h) = 0$$

$$2(A + B) = bC(g - ah), \qquad (2.2.21)$$

$$k + bh > 0,$$

the remaining terms are

$$dV/dt = -(2aA/b^2)(a\xi_2 + b\xi_3)^2 - C(k + bh)F^2(\sigma^*),$$

which is a *negative semidefinite* function. The stationary solution (2.2.18) is therefore stable according to Theorem 1 of Section 1.5.3 for the assumptions (2.2.21).

If the first two conditions of (2.2.21) are solved for A and B which are assumed to be positive, it follows that

$$A = \frac{bC}{2a}(ag - e - a^2h) > 0, \qquad B = \frac{bC(g - ah) - 2A}{2} = \frac{beC}{2a} > 0.$$

These conditions together with the third condition of (2.2.21) and the fact that $a, b > 0$ leads to the conditions

$$ag > e + a^2h,$$

$$e > 0,$$

$$k + bh > 0.$$

The solution (2.2.18) is stable if the parameters of the control loop satisfy these conditions.

2.2.3. *Stability of the Gyroscope*

Numerous important gyroscope problems of various types occur in technology. The stability of the gyroscope itself or of the gyroscope as a stabilizing element are problems which are often encountered.

Let us describe the gyroscope motion in the rotating body-fixed reference system x, y, z along the principal axes of inertia. The following vector equation of motion is obtained from the angular momentum theorem:

$$\partial \mathbf{D}/\partial t + \boldsymbol{\omega} \times \mathbf{D} = \mathbf{M}. \tag{2.2.22}$$

$\mathbf{D} = (D_x, D_y, D_z)$ is the angular momentum vector, $\boldsymbol{\omega} = (\omega_x, \omega_y, \omega_z)$ the rotation vector, and $\mathbf{M} = (M_x, M_y, M_z)$ the torque vector. The angular momentum components satisfy

$$D_x = A\omega_x, \qquad D_y = B\omega_y, \qquad D_z = C\omega_z,$$

where A, B, C are the principal moments of inertia. Therefore, the components of (2.2.22) are

$$A\dot{\omega}_x + (C - B)\,\omega_y\omega_z = M_x,$$
$$B\dot{\omega}_y + (A - C)\,\omega_z\omega_x = M_y, \tag{2.2.23}$$
$$C\dot{\omega}_z + (B - A)\,\omega_x\omega_y = M_z.$$

Example 1: Let us first consider a special case of the gyroscope motion without forces, so that $\mathbf{M} \equiv 0$. Therefore, Eq. (2.2.23) results in

$$A\dot{\omega}_x + (C - B)\,\omega_y\omega_z = 0,$$
$$B\dot{\omega}_y + (A - C)\,\omega_z\omega_x = 0, \tag{2.2.24}$$
$$C\dot{\omega}_z + (B - A)\,\omega_x\omega_y = 0.$$

One *stationary motion* is given by $\omega_{x0} = \text{const}$, $\omega_{y0} = \omega_{z0} = 0$ because these values satisfy the system (2.2.24). Let us investigate the stability of the motion by setting $\omega_x = \omega_{x0} + \xi_1$, $\omega_y = \xi_2$, $\omega_z = \xi_3$. The following nonlinear variational equations are obtained from (2.2.24):

$$A\dot{\xi}_1 + (C - B)\,\xi_2\xi_3 = 0,$$
$$B\dot{\xi}_2 + (A - C)(\omega_{x0} + \xi_1)\,\xi_3 = 0, \tag{2.2.25}$$
$$C\dot{\xi}_3 + (B - A)(\omega_{x0} + \xi_1)\,\xi_2 = 0.$$

If the investigation is carried out by the *small oscillation method*, it is necessary to linearize the problem by setting products of the variations equal to zero. Equation (2.2.25) is therefore simplified as follows:

$$\dot{\xi}_1 = 0,$$
$$\dot{\xi}_2 = [(C - A)/B]\,\omega_{x0}\xi_3, \tag{2.2.26}$$
$$\dot{\xi}_3 = [(A - B)/C]\,\omega_{x0}\xi_2.$$

The linearized system (2.2.26) has the characteristic equation

$$\begin{vmatrix} -\lambda & 0 & 0 \\ 0 & -\lambda & [(C-A)/B]\,\omega_{x0} \\ 0 & [(A-B)/C]\,\omega_{x0} & -\lambda \end{vmatrix} = 0,$$

$$\lambda \left[\lambda^2 + \frac{(A-C)(A-B)}{BC}\,\omega_{x0}^2 \right] = 0.$$

The roots of this equation are

$$\lambda_1 = 0, \qquad \lambda_{2,3} = \pm\omega_{x0}[(C-A)(A-B)/BC]^{1/2}.$$

If $C < A < B$ or $C > A > B$, then one of the roots λ_i has a positive real part. According to Theorem 5 of Section 1.5.3, the instability of the stationary solution can be predicted with certainty from the linearized equations (2.2.26). This stationary motion consists of rotation about the intermediate principal axis of inertia.

Two imaginary roots are obtained for $A < B < C$ or $A > B > C$, which are rotations around the axis corresponding to the smallest or largest principal moment of inertia, respectively. Since $\lambda_1 = 0$ always holds, this is the *critical case*. It is now impossible to determine stability or instability from the linearized equations (2.2.26), and we must return to the nonlinear variational equations (2.2.25).

The following positive definite *Lyapunov function* is used in the case $A < B < C$ (rotation about the axis corresponding to the *smallest* principal moment of inertia):

$$V = B(B-A)\,\xi_2^2 + C(C-A)\,\xi_3^2 + [B\xi_2^2 + C\xi_3^2 + A(\xi^2 + 2\xi_1\omega_{x0})]^2.$$

It is easy to see that $\dot{V} \equiv 0$, so that the stationary solution is stable according to Theorem 1 of Section 1.5.3.

The following positive definite Lyapunov function is used for the rotation around the axis corresponding to the largest principal moment of inertia for $A > B > C$:

$$V = B(A-B)\,\xi_2^2 + C(A-C)\,\xi_3^3 + [B\xi_2^2 + C\xi_3^2 + A(\xi_1^2 + 2\xi_1\omega_{x0})]^2.$$

$\dot{V} \equiv 0$ also holds, so that the stationary motion is stable.

Example 2: Let us now consider a rocket controlled by an internal rotor. This system can be considered as a *gyroscope with rotation rate-dependent self-excitation*.[1]

---

[1] LEIPHOLZ, H., A contribution to the problem of a gyroscope with rotation rate dependent self-excitation. *Ing. Arch.* 32, 255–285 (1963).

The equations of motion are

$$A\dot{\omega}_x + (C - B)\,\omega_y\omega_z = 0,$$
$$B\dot{\omega}_y + (A - C)\,\omega_z\omega_x = -D\omega_z,$$
$$C\dot{\omega}_z + (B - A)\,\omega_x\omega_y = D\omega_y,$$

(2.2.27)

where A, B, C, and D are constant parameters of the mechanical system. One *stationary solution* of (2.2.27) is given by

$$\omega_{x0} = [D/(C - A)], \qquad \omega_{y0} = 0, \qquad \omega_{z0} = 0.$$

Let us investigate the stability of this solution. The equations $\omega_x = \omega_{x0} + \xi_1$, $\omega_y = \xi_2$, $\omega_z = \xi_3$ are substituted into (2.1.54) and the following nonlinear variational equations are obtained:

$$A\dot{\xi}_1 + (C - B)\,\xi_2\xi_3 = 0,$$
$$B\dot{\xi}_2 + (A - C)(\omega_{x0} + \xi_1)\,\xi_3 = -D\xi_3,$$
$$C\dot{\xi}_3 + (B - A)(\omega_{x0} + \xi_1)\,\xi_2 = D\xi_2.$$

(2.2.28)

There is no point in linearizing these equations because it is not possible to determine stability with certainty from the linearized equations, as is easily derived. Therefore, the system (2.2.28) is used, which has the following *integrals:*

$$B\xi_2{}^2 + C\xi_3{}^2 + A(\omega_{x0} + \xi_1)^2 = \text{const},$$
$$B^2\xi_2{}^2 + C^2\xi_3{}^2 + A^2[(\omega_{x0} + \xi_1) + D/A]^2 = \text{const}.$$

By using combinations of these integrals, it is possible to *obtain a Lyapunov function.* Let us select

$$V = B(K - B)\,\xi_2{}^2 + C(K - C)\,\xi_3{}^2 + A(K - A)\,\xi_1{}^2 + 2A[\omega_{x0}(K - A) - D]\,\xi_1.$$

$\dot{V} = 0$ holds, because V is made up of integrals of (2.2.28). V becomes positive definite if

$$K > B, \qquad K > C, \qquad K > A, \qquad K = (A\omega_{x0} + D)/\omega_{x0}$$

holds simultaneously. These conditions are equivalent to

$$D > 0, \qquad \omega_{x0} < D/(C - A), \qquad \omega_{x0} < D/(B - A).$$

(2.2.29)

These are the stability conditions for the stationary motion given above. If (2.2.29) are satisfied, V is positive definite and the motion is stable according to Theorem 1 of Section 1.5.3.

Example 3:   Let us now investigate the stability of a *monorail*. The *Scherl* system contains a vertical gyroscopic pendulum. Figure 50 shows a diagram of the system. Θ_W, M_W are the moments of inertia and mass of the carriage Θ_K, M_K for the moments of inertia and mass of the pendulum including the frame, and D is the constant eigen angular momen-

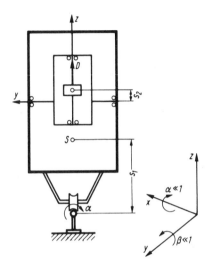

FIG. 50.   Monorail.

tum of the gyroscope. In addition, it is assumed that the rotation angles and their derivatives α, β, $\dot\alpha$, $\dot\beta$ are small quantities. Under these assumptions, the angular momentum vector

$$\mathbf{D} = (\Theta_W\dot\alpha + D\beta, \Theta_K\dot\beta - D\alpha, D)$$

and the torque vector

$$\mathbf{M} = (M_W g s_1 \alpha, M_K g s_2 \beta, 0)$$

and $d\mathbf{D}/dt = \mathbf{M}$ lead to the following scalar equations:

$$\Theta_W\ddot\alpha + D\dot\beta = M_W g s_1 \alpha,$$
$$\Theta_K\ddot\beta - D\dot\alpha = M_K g s_2 \beta. \qquad (2.2.30)$$

The characteristic equation of the system is

$$\begin{vmatrix} \Theta_W\lambda^2 - M_W g s_1 & D\lambda \\ -D\lambda & \Theta_K\lambda^2 - M_K g s_2 \end{vmatrix} = 0,$$

or

$$\Theta_W\Theta_K\lambda^4 + (D^2 - \Theta_W M_K g s_2 - \Theta_K M_W g s_1)\,\lambda^2 + M_W M_K g^2 s_1 s_2 = 0.$$

All the Hurwitz determinants of this fourth-order equation are identically zero, as can easily be calculated. Therefore, the necessary condition for asymptotic stability and for the applicability of the linearized equations (2.2.30), which would imply stability (that is, roots with negative reals parts only) can never be satisfied. Therefore, a moment generator must be added to the system, which introduces a moment around the y axis proportional to α. It is also necessary to have damping about the x axis proportional to the angular rate $\dot{\alpha}$. Differential equations (2.2.30) then become

$$\Theta_W \ddot{\alpha} + D\dot{\beta} + K\dot{\alpha} = M_W g s_1 a,$$

$$\Theta_K \ddot{\beta} - D\dot{\alpha} = M_K g s_2 \beta + R\alpha, \tag{2.2.31}$$

with the characteristic equation

$$\begin{vmatrix} \Theta_W \lambda^2 + K\lambda - M_W g s_1 & D\lambda \\ -R - D\lambda & \Theta_K \lambda^2 - M_K g s_2 \end{vmatrix} = 0.$$

They can also be written

$$\Theta_M \Theta_K \lambda^4 + K\Theta_K \lambda^3 + (D^2 - \Theta_W M_K g s_2 - \Theta_K M_W g s_1) \lambda^2$$
$$+ (DR - M_K g s_2 K) \lambda + M_W M_K g^2 s_1 s_2 = 0.$$

The following conditions must be satisfied for asymptotic stability according to the *Hurwitz criterion*:

$$D^2 > M_W g s_1 \Theta_K + M_K g s_2 \Theta_W,$$

$$DR > M_K g s_2 K,$$

$$K\Theta_K (D^2 - M_W g s_1 \Theta_K - M_K g s_2 \Theta_W)(DR - M_K g s_2 K)$$
$$- \Theta_W \Theta_K (DR - M_K g s_2 K)^2 - K^2 \Theta_K^2 M_W M_K g^2 s_1 s_2 > 0.$$

As can be seen, all three conditions can be simultaneously satisfied by a sufficiently large eigen angular momentum D of the gyroscope after all other parameters of the system have been specified. The characteristic equation then has only roots with negative real parts. Stability prevails and the stability can be determined with the linearized equations (2.2.31).

Example 4:   Finally, let us consider a *gyroscopic pendulum with an erection motor* (Fig. 51). There are friction moments about the x and y axes proportional to the angular rates $\dot{\alpha}, \dot{\beta}$. The supporting moment is about the y axis and is a nonlinear function of the angle α. Weight forces are neglected. The quantities $\alpha, \beta, \dot{\alpha}, \dot{\beta}$ are again assumed to be small, and

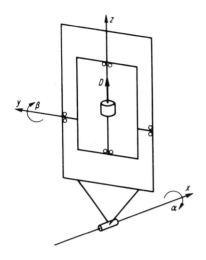

Fig. 51. Gyroscopic pendulum
with erection motor.

the rotor has the eigen angular momentum D. The equations of motion
are obtained in exactly the same way as in the previous example from

$$\mathbf{D} = (\Theta_A \dot{\alpha} + D\beta, \Theta_J \dot{\beta} - D\alpha, D),$$

$$\mathbf{M} = (-K_x \dot{\alpha}, F(\alpha) - K_y \dot{\beta}, 0),$$

and $d\mathbf{D}/dt = \mathbf{M}$. They are given by

$$\Theta_A \ddot{\alpha} + D\dot{\beta} + K_x \dot{\alpha} = 0,$$
$$\Theta_J \ddot{\beta} - D\dot{\alpha} + K_y \dot{\beta} = F(\alpha). \qquad (2.2.32)$$

Θ_A is the moment of inertia of the entire system with respect to the
x axis, Θ_J the moment of inertia of the system consisting of the rotor and
the inner frame with respect to the y axis, and K_x, K_y are the damping
coefficients.

Equation (2.2.32) can be transformed into a third-order equation with
respect to α

$$\Theta_A \Theta_J \dddot{\alpha} + (\Theta_J K_x + \Theta_A K_y) \ddot{\alpha} + (D^2 + K_x K_y) \dot{\alpha} + DF(\alpha) = 0.$$

The abbreviations

$$(1/\Theta_A \Theta_J)(\Theta_J K_x + \Theta_A K_y) = a, \quad (1/\Theta_A \Theta_J)(D^2 + K_x K_y) = b, \quad (1/\Theta_A \Theta_J) D = e$$

and substitutions $x = \alpha$ and $eF(\alpha) = f(\alpha)$ lead to

$$\dddot{x} + a\ddot{x} + b\dot{x} + f(x) = 0,$$

which is exactly the differential equation (2.2.12) of Section 2.2.2. It is not necessary to discuss this equation again. The existence of stable oscillations of a supported gyroscopic pendulum can be determined according to the discussion in Section 2.2.2.

2.2.4. *Stability of Aircraft, Rockets, Satellites*

Example 1:   Let us first consider the center-of-gravity motion of an aircraft moving in the x, z plane as shown in Fig. 52. The aircraft is

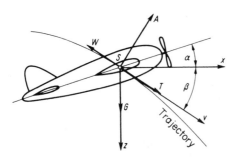

FIG. 52.   Forces on the aircraft.

symmetrical with respect to this plane. The velocity of the aircraft center of gravity is v, the weight of the aircraft is $G = mg$, and the angle between the trajectory and the x axis is β.

The equations of motion of the center of gravity with respect to the axis parallel and perpendicular to the trajectory are given by

$$m\dot{v} = -W + T + mg \sin \beta,$$
$$mv\dot{\beta} = -A + mg \cos \beta. \tag{2.2.33}$$

W is the drag, T the component of the propulsion force in the flight direction, and A is the lift, which includes the component of the propulsion force perpendicular to the trajectory. It is assumed that $T = W$. The first equation in (2.2.33) can be integrated by first multipling by v and noting that $v \sin \beta = \dot{z}$. After integration the result is

$$v^2 = 2gz + c, \tag{2.2.34}$$

where c is a constant of integration.

In addition, let us assume that the lift A is proportional to the square velocity, $A = kv^2$. For $T = W$, the system $\beta = 0$, $v = v_0 = \text{const}$, $A = mg$ is a solution. Therefore, $mg = kv_0^2$ must hold and the final result is

$$A = mg(v^2/v_0^2). \tag{2.2.35}$$

Furthermore

$$\frac{d\beta}{dt} = \frac{d\beta}{dz}\frac{dz}{ds}\frac{ds}{dt} = v\frac{d\beta}{dz}\sin\beta. \qquad (2.2.36)$$

Substitution of (2.2.34), (2.2.35), and (2.2.36) into the second equation of (2.2.23) results in

$$(2gz + c)(d\beta/dz)\sin\beta = -(g/v_0^2)(2gz + c) + g\cos\beta.$$

It follows from $q_1 = \beta$, $q_2 = 2gz + c$, $dz = dq_2/2g$ that

$$\frac{dq_1}{dq_2} = \frac{\cos q_1 - 1/v_0^2 \cdot q_2}{2q_2\sin q_1} \qquad (2.2.37)$$

or

$$\dot{q}_1 = (1/\alpha)[\cos q_1 - (1/v_0^2)\,q_2], \qquad \dot{q}_2 = (1/\alpha)\,2q_2\sin q_1 ,$$

where α is a proportionality factor.

There are stationary solutions to Eq. (2.2.37) for $\dot{q}_1 = \dot{q}_2 = 0$, which are obtained from the equations

$$\cos q_1 - (1/v_0^2)\,q_2 = 0, \qquad 2q_2\sin q_1 = 0.$$

As was mentioned previously, one of the solutions is

$$q_{1s} = \beta = 0, \qquad q_{2s} = 2gz_s + c = v_0^2. \qquad (2.2.38)$$

This corresponds to constant velocity at a constant altitude $z_s = (v_0^2 - c)/2g$. Let us discuss the stability of this motion.

Let us carry out the discussion in the *phase plane*. According to Section 1.4, the following abbreviated linear variational equations may be used in the immediate vicinity of the stationary point:

$$\dot{\xi}_i = (\partial F_i/\partial q_k)_{q_s} \cdot \xi_k .$$

This a simple stationary point.

In our case,

$$F_1(q_1 , q_2) = (1/\alpha)(\cos q_1 - 1/v_0^2 \cdot q_2),\, F_2(q_1 , q_2) = (1/\alpha)\,2q_2\sin q_1$$

and therefore

$$\left(\frac{\partial F_1}{\partial q_1}\right)_{q_s} = \frac{1}{\alpha}(-\sin q_1)_{\substack{q_1=0\\q_2=v_0^2}} = 0, \qquad \left(\frac{\partial F_1}{\partial q_2}\right)_{q_s} = -\left(\frac{1}{\alpha v^2}\right)_{\substack{q_1=0\\q_2=v_0^2}} = -\frac{1}{\alpha v_0^2},$$

$$\left(\frac{\partial F_2}{\partial q_1}\right)_{q_s} = \frac{1}{\alpha}(2q_2\cos q_1)_{\substack{q_1=0\\q_2=v_0^2}} = \frac{2v_0^2}{\alpha}, \qquad \left(\frac{\partial F_2}{\partial q_2}\right)_{q_s} = \frac{1}{\alpha}(2\sin q_1)_{\substack{q_1=0\\q_2=v_0^2}} = 0,$$

which means that it is indeed such a point, because not all the derivatives vanish simultaneously. Therefore

$$\dot{\xi}_1 = -(1/\alpha v_0{}^2)\, \xi_2, \qquad \dot{\xi}_2 = (2v_0{}^2/\alpha)\, \xi_1$$

can be used. The integration of this equation results in

$$\xi_1{}^2 + (1/2v_0{}^4)\, \xi_2{}^2 = \text{const},$$

which is the equation for the phase curves surrounding the stationary point. Since they are ellipses, the stationary point is a stable center. It follows that stationary flight at constant altitude is stable for the assumptions made above.

Example 2: Let us now consider the motion of a *rocket* without propulsion. The rocket is symmetrical with respect to the ζ axis and rotates at the angular rate ω (Fig. 53). As soon as the axis of rotation ζ deviates from the direction z of the velocity vector v by the angle ϑ, the drag moment W affects the rocket. Let us determine whether the special motion of the rocket given by $\vartheta = 0$ is stable and determine whether it is insensitive to small disturbances which result from the moment caused by W.

An orthogonal axis system is used to describe the rocket motion and the deflection of the rocket axis by the angle ϑ. This system is comprised of the ζ axis, the node axis ξ and the orthogonal axis η (Fig. 54). The rotation of this system is given by the angular rates

$$\omega_\xi = \dot{\vartheta}, \qquad \omega_\eta = \dot{\psi} \sin \vartheta, \qquad \omega_\zeta = \dot{\psi} \cos \vartheta. \qquad (2.2.39)$$

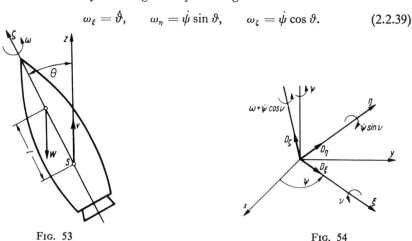

FIG. 53   FIG. 54

FIG. 53.   Rocket stabilized by angular momentum.

FIG. 54.   Axis angles and angular rates.

The angular momentum vector has the components

$$D_\xi = A\dot\vartheta, \qquad D_\eta = A\dot\psi \sin\vartheta, \qquad D_\zeta = C(\omega + \dot\psi \cos\vartheta), \qquad (2.2.40)$$

where A, A, C are the corresponding moments of inertia.

The equations of motion are obtained from the scalar components of the vector equation

$$\mathbf{M} = \partial\mathbf{D}/\partial t + \boldsymbol{\omega} \times \mathbf{D}$$

and from $\mathbf{M} = (Wl \sin\vartheta, 0, 0)$. The result is

$$Wl \sin\ddot\vartheta = A\ddot\vartheta + C(\omega + \dot\psi \cos\vartheta)\dot\psi \sin\vartheta - A\dot\psi^2 \sin\vartheta \cos\vartheta,$$

$$0 = A(d/dt)(\dot\psi \sin\vartheta) + A\dot\vartheta\dot\psi \cos\vartheta - C\dot\vartheta(\omega + \dot\psi \cos\vartheta), \quad (2.2.41)$$

$$0 = C(d/dt)(\omega + \dot\psi \cos\vartheta).$$

Two integrals are used to reduce the system. The first integral is obtained from the third line of (2.2.41):

$$\omega + \dot\psi \cos\vartheta = n = \text{const.} \qquad (2.2.42)$$

The second one is the energy integral

$$E = T + U = \tfrac{1}{2}Cn^2 + \tfrac{1}{2}A(\dot\vartheta^2 + \dot\psi^2 \sin^2\vartheta) + Wl \cos\vartheta, \qquad (2.2.43)$$

which is transformed in the following way. We have $M_z = 0$, and the corresponding angular momentum component satisfies

$$D_z = D_\zeta \cos\vartheta + D_\eta \sin\vartheta = Cn \cos\vartheta + A\dot\psi \sin^2\vartheta = \text{const.} \qquad (2.2.44)$$

It follows that

$$\dot\psi = (D_z - Cn \cos\vartheta)/A \sin^2\vartheta,$$

which is substituted into Eq. (2.2.43). The result is

$$E - \frac{Cn^2}{2} = \frac{A\dot\vartheta^2}{2} + \frac{(D_z - Cn \cos\vartheta)^2}{2A \sin^2\vartheta} + Wl \cos\vartheta. \qquad (2.2.45)$$

By means of the substitutions

$$(2/A)(E - (Cn^2/2)) = \alpha,$$

$$2Wl/A = \beta, \qquad D_z/A = \gamma, \qquad Cn/A = D, \qquad \cos\vartheta = u, \qquad (2.2.46)$$

Eq. (2.2.45) becomes

$$\dot{u}^2 = (1 - u^2)(\alpha - \beta u) - (\gamma - Du)^2 = f(u). \qquad (2.2.47)$$

This last relationship makes it possible to describe the rocket motion in the $\dot{u}$, u plane by using the phase curves $\dot{u} = [f(u)]^{1/2}$. The characteristics of the "potential function" $f(u)$ determine the course of any phase curve. A motion which is physically possible only results if $f(u)$ is positive, because $\dot{u}$ must be real (Fig. 55). The stationary solution for

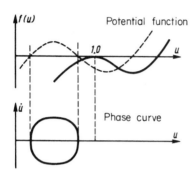

FIG. 55.   Potential function and phase diagram.

$\dot{u} \equiv 0$ is obtained from (2.2.47). It is given by $u = 1$ because this makes the first term on the right side of (2.2.47) equal to zero. However, for $u = 1$ we also have $\vartheta = 0$ and therefore $D_z = C_n$ according to (2.2.44), which means that $\gamma = D$ according to (2.2.46). This means that the second term on the right side of (2.2.47) is also equal to zero and therefore $f(u) = 0$, which must be the case for the stationary solution. The motion of the rocket corresponding to $u = 1$, $\vartheta = 0$, the stability of which is being investigated, is exactly the stationary solution of (2.2.47). The *phase diagram* (Fig. 55) shows that $f(u)$ must have a *maximum* at $u = 1.0$ for stability to prevail. This guarantees that a disturbed motion will occur with u values (ϑ values) that remain sufficiently close to the stationary values $u = 1$ ($\vartheta = 0$).

$f(u)$ has a maximum at $u = 1$ if the conditions $f'(1) = 0$ and $f''(1) < 0$ are satisfied.

We have $df/du = -2u(\alpha - \beta u) - \beta(1 - u^2) + 2D(\gamma - Du)$. This expression is zero for $u = 1$ provided that $\alpha = \beta$. This is because we can make use of the fact that $\gamma = D$ holds for $u = 1$.

The other condition at $u = 1$ is

$$d^2f/du^2 = -2(\alpha - \beta u) + 2\beta u + 2u\beta - 2D^2 < 0.$$

We make use of the relationship $\alpha = \beta$ so that

$$f''(1) = 4\beta - 2D^2 < 0.$$

It follows that

$$\beta < \tfrac{1}{2}D^2.$$

Because of (2.2.46), this inequality becomes

$$n > (2/C)(WlA)^{1/2}. \qquad\qquad (2.2.48)$$

The condition (2.2.48) causes the function $f(u)$ to be stable—that is, to have a maximum at $u = 1(\vartheta = 0)$. If the rocket rotates around the z axis, which coincides with the ζ axis in this case, with an angular rate n satisfying (2.2.48), the rocket motion is stable.

Example 3: The *flutter of airfoils* is an important problem in flight dynamics. Flutter is an instability phenomenon which can occur at certain critical flight velocities. This problem really belongs in the field of aeroelasticity. However, it is possible to discuss a simplified version of the problem in this section without losing sight of its characteristic features.

Let us consider the problem shown in Fig. 56. The wing carries out

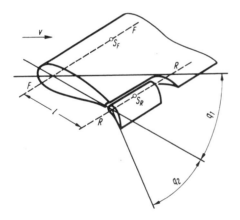

FIG. 56.   Flutter of an airfoil.

torsional oscillations around the axis F whose amplitude is the angle q_1. The rudder is deflected by the angle q_2 about the axis R. The centers of gravity S_F and S_R are located along the axis. J_F is the moment of inertia of the wing with respect to the axis F and J_R the moment of inertia of the rudder with respect to the axis R, m_R is the mass of the rudder, and l is the distance between the axes F and R. The expression for kinetic energy of the system is

$$T = \tfrac{1}{2}(J_F\dot{q}_1{}^2 + J_R\dot{q}_2{}^2 + m_R l^2 \dot{q}_1{}^2). \qquad\qquad (2.2.49)$$

The generalized forces applied to the system are

$$Q_1 = -(k - c_{11}v^2)\, q_1 - c_{12}v^2 q_2\,, \qquad Q_2 = c_{21}v^2 q_1 - c_{22}v^2 q_2\,, \qquad (2.2.50)$$

These forces are restoring moments, aerodynamic moments, and rotational spring moments about the axis R. All coefficients in (2.2.50) are constant and positive. v is the velocity of the wing.

Substitution of (2.2.49) and (2.2.50) into the Lagrange equations

$$\frac{d}{dt}\left(\frac{\partial T}{\partial \dot{q}_k}\right) - \frac{\partial T}{\partial q_k} = Q_k$$

results in the system of differential equations

$$\begin{aligned}
(J_F + m_R l^2)\, \ddot{q}_1 &= (c_{11}v^2 - k)\, q_1 - c_{12}v^2 q_2\,, \\
J_R \ddot{q}_2 &= c_{21}v^2 q_1 - c_{22}v^2 q_2\,.
\end{aligned} \qquad (2.2.51)$$

One stationary solution is given by $q_{10} = q_{20} = 0$, and the stability of this solution will now be investigated. The following system of variational equations is obtained from (2.2.51) by means of the trial solution $q_i = q_{i0} + \xi_i$:

$$\begin{aligned}
(J_F + m_R l^2)\, \ddot{\xi}_1 &= (c_{11}v^2 - k)\, \xi_1 - c_{12}v^2 \xi_2\,, \\
J_R \ddot{\xi}_2 &= c_{21}v^2 \xi_1 - c_{22}v^2 \xi_2\,,
\end{aligned}$$

The characteristic equation is

$$\begin{vmatrix} (J_F + m_R l^2)\, \lambda^2 + (k - c_{11}v^2) & c_{12}v^2 \\ -c_{21}v^2 & J_R \lambda^2 + c_{22}v^2 \end{vmatrix} = 0.$$

It is given by

$$(J_F + m_R l^2)\, J_R \lambda^4 + [(J_F + m_R l^2)\, c_{22}v^2 + J_R(k - c_{11}v^2)]\, \lambda^2$$
$$+ (k - c_{11}v^2)\, c_{22}v^2 + c_{12}c_{21}v^4 = 0.$$

The roots of this equation are

$$\begin{aligned}
\lambda_{1,2,3,4} = \pm\Bigg\{ &-\frac{(J_F + m_R l^2)\, c_{22}v^2 + J_R(k - c_{11}v^2)}{2(J_F + m_R l^2)\, J_R} \\
&\pm\Bigg[\frac{[(J_F + m_R l^2)\, c_{22}v^2 + J_R(k - c_{11}v^2)]^2}{4(J_F + m_R l^2)^2\, J_R^2} \\
&-\frac{(k - c_{11}v^2)\, c_{22}v^2 + c_{12}c_{21}v^4}{(J_F + m_R l^2)\, J_R}\Bigg]^{1/2}\Bigg\}^{1/2}.
\end{aligned} \qquad (2.2.52)$$

By means of the abbreviations

$$\frac{(J_F + m_R l^2) c_{22} v^2 + J_R(k - c_{11} v^2)}{(J_F + m_R l^2) J_R} = a,$$

$$\frac{(k - c_{11} v^2) c_{22} v^2 + c_{12} c_{21} v^4}{(J_F + m_R l^2) J_R} = b,$$

(2.2.53)

Eq. (2.2.52) becomes

$$\lambda_{1,2,3,4} = \pm \left[\tfrac{1}{2} a \pm (\tfrac{1}{4} a^2 - b)^{1/2}\right]^{1/2}.$$

(2.2.54)

The characteristic exponents are all purely imaginary if the radicand of the outer square root is negative and real. In this case, there are four different and simple roots which have zero real parts, which means that stability prevails according to Theorem 3 of Section 1.3.1. On the other hand, if the radicand of the outer square root is positive or complex, roots λ having a positive real part exist. The stationary solution is then unstable according to Theorem 2 of Section 1.3.1.

Therefore, the following two conditions

$$\tfrac{1}{4} a^2 - b > 0, \qquad \tfrac{1}{2} a > (\tfrac{1}{4} a^2 - b)^{1/2}$$

(2.2.55)

must be satisfied for stability.

Let us consider the first condition. According to (2.2.53)

$$[(J_F + m_R l^2) c_{22} v^2 + J_R(k - c_{11} v^2)]^2$$
$$-4(J_F + m_R l^2) J_R[(k - c_{11} v^2) c_{22} v^2 + c_{12} c_{21} v^4] > 0.$$

This can also be written in the form

$$[(J_F + m_R l^2) c_{22} v^2 - J_R(k - c_{11} v^2)]^2 > 4(J_F + m_R l^2) J_R c_{12} c_{21} v^4.$$

This inequality is satisfied by

$$(J_F + m_R l^2) c_{22} v^2 - J_R(k - c_{11} v^2) > 2v^2[(J_F + m_R l^2) J_R c_{21} c_{21}]^{1/2} \quad (2.2.56)$$

as well as by

$$J_R(k - c_{11} v^2) - (J_F + m_R l^2) c_{22} v^2 > 2v^2[(J_F + m_R l^2) J_R c_{12} c_{21}]^{1/2}. \quad (2.2.57)$$

If the first condition in (2.2.55) is satisfied, the second one is also satisfied provided that b is positive. This is true for $k > c_{11} v^2$ according to (2.2.53), which is always satisfied for the velocity region under consideration. Therefore, the question of stability is determined solely from (2.2.56) and (2.2.57). Therefore, according to (2.2.56), flutter cannot occur for

$$v^2 > \frac{J_R k}{(J_F + m_R l^2)\, c_{22} + J_R c_{11} - 2[(J_F + m_R l^2)\, J_R c_{12} c_{21}]^{1/2}} = v^2_{\text{crit},0}$$

and according to (2.2.57), cannot occur for

$$v^2 < \frac{J_R k}{(J_F + m_R l^2)\, c_{22} + J_R c_{11} + 2[(J_F + m_R l^2)\, J_R c_{12} c_{21}]^{1/2}} = v^2_{\text{crit},u}.$$

On the other hand, flutter can be expected for velocities in the range

$$v_{\text{crit},u} < v < v_{\text{crit},0}$$

This velocity range must be avoided.

Example 4:   As a final example, let us consider the stability of motion of a *satellite*. We will follow a method developed by Magnus.[1]

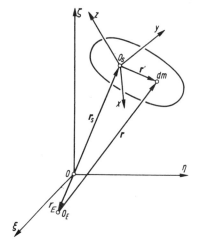

Fig. 57.   Satellite motion.

Let O be the common center of mass of the earth and the satellite, O_E the center of mass of the earth, O_S the center of mass of the satellite, dm a mass element of the satellite, m_E the mass of the earth, m_S the mass of the satellite, $m = m_E + m_S$ the total mass. We will employ the space-fixed orthogonal coordinate system ξ, η, ζ and the principal axis coordinate system x, y, z fixed on the satellite. The vectors $\mathbf{r}$, $\mathbf{r}_E$, $\mathbf{r}_S$, $\mathbf{r}'$ and $\mathbf{r}_a = \mathbf{r}_S - \mathbf{r}_E = \mathbf{r} - \mathbf{r}'$ are defined in Fig. 57 and are used to

---

[1] MAGNUS, K., The rod satellite in radially symmetric field of gravitation. *Z. Flugwiss.* pp. 233–241 (1963); contributions to the investigation of the rotational motion of rigid-body satellites along circular orbits. *Jahrb. Wiss. Ges. Luft Raumfahrt* pp. 174–180 (1963).

formulate Newton's law

$$\ddot{\mathbf{r}}_S + \gamma \frac{m_E}{m_S} \int \frac{\mathbf{r}}{r^3}\, dm = 0. \qquad (2.2.58)$$

γ is the gravitational constant. The integration must be carried out over the satellite. The angular momentum theorem with respect to O_S and the mowing orthogonal coordinate system x, y, z is

$$\frac{\partial \mathbf{D}_S}{\partial t} + \boldsymbol{\omega} \times \mathbf{D}_S = -\gamma m_E \int \frac{\mathbf{r}' \times \mathbf{r}}{r^3}\, dm. \qquad (2.2.59)$$

$\mathbf{D}_S = \int [\mathbf{r}' \times (\boldsymbol{\omega} + \mathbf{r}')]\, dm$ is the angular momentum of the satellite and $\boldsymbol{\omega}$ is the rotation vector of the satellite with respect to the space-fixed system.

We have

$$\mathbf{r}' = \mathbf{e}_x x + \mathbf{e}_y y + \mathbf{e}_z z,$$
$$\mathbf{r}_a = r_a(\mathbf{e}_x \gamma_x + \mathbf{e}_y \gamma_y + \mathbf{e}_z \gamma_z). \qquad (2.2.60)$$

$\mathbf{r}_a$ is the vertical direction and $\gamma_x, \gamma_y, \gamma_z$ are the direction cosines with respect to the body-fixed system. It is important to note that $|\mathbf{r}'| \ll |\mathbf{r}_a|$. The following series expansion holds:

$$\frac{1}{r^3} = \frac{1}{r_a^3} - 3(\mathbf{r}_a \mathbf{r}') \frac{1}{r_a^5} - \frac{3}{2} \frac{r'^2}{r_a^5} \cdots . \qquad (2.2.61)$$

Only the first three terms are retained, and the following ones are suppressed because they are small. It follows that

$$\int \frac{\mathbf{r}}{r^3}\, dm = \int \frac{\mathbf{r}_a + \mathbf{r}'}{r^3}\, dm = \int \frac{\mathbf{r}_a}{r_a^3}\, dm - 3 \int \frac{\mathbf{r}_a(\mathbf{r}_a \mathbf{r}')}{r_a^5}\, dm$$
$$-\frac{3}{2} \int \frac{\mathbf{r}_a r'^2}{r_a^5}\, dm \cdots$$
$$+ \int \frac{\mathbf{r}'}{r_a^3}\, dm - 3 \int \frac{\mathbf{r}'(\mathbf{r}_a \mathbf{r}')}{r_a^5}\, dm$$
$$-\frac{3}{2} \int \frac{\mathbf{r}' r'^2}{r_a^5}\, dm \cdots .$$

We have $\int \mathbf{r}_a\, dm = m_S \mathbf{r}_a$, $\int \mathbf{r}'\, dm = 0$, and we will neglect all terms which contain r_a^4 and higher powers of r_a in the denominator. The expression

$$\int \frac{\mathbf{r}}{r^3}\, dm \approx \frac{m_S \mathbf{r}_a}{r_a^3} \qquad (2.2.62)$$

remains. Using the relationships

$$\mathbf{r}_S = (m_E/m)\,\mathbf{r}_a, \qquad r_S = (m_E/m)\,r_a, \tag{2.2.63}$$

the following approximation for Newton's law (2.2.58) is obtained:

$$\ddot{\mathbf{r}}_S + \gamma(m_E{}^3/m^2 r_S{}^3)\,\mathbf{r}_S = 0. \tag{2.2.64}$$

The angular momentum theorem can also be simplified. Because of
(2.2.61) we have

$$\int \frac{\mathbf{r}' \times \mathbf{r}}{r^3}\, dm = \int \frac{\mathbf{r}' \times (\mathbf{r}_a + \mathbf{r}')}{r^3}\, dm$$

$$= -\mathbf{r}_a \times \int \frac{\mathbf{r}'}{r_a{}^3}\, dm - 3 \int \frac{(\mathbf{r}' \times \mathbf{r}_a)(\mathbf{r}_a\mathbf{r}')}{r_a{}^5}\, dm\cdots,$$

if small terms are again suppressed. The first is zero so that

$$\int \frac{\mathbf{r}' \times \mathbf{r}}{r^3}\, dm \approx -\frac{3}{r_a{}^5} \int (\mathbf{r}' \times \mathbf{r}_a)(\mathbf{r}_a\mathbf{r}')\, dm$$

$$= -\frac{3}{r_a{}^5} \left[\int \mathbf{r}'(\mathbf{r}_a\mathbf{r}')\, dm\right] \times \mathbf{r}_a \tag{2.2.65}$$

remains. The following expression is calculated from (2.2.60):

$$\int \mathbf{r}'(\mathbf{r}_a\mathbf{r}')\, dm = P\mathbf{r}_a - r_a(\mathbf{e}_x\gamma_x A + \mathbf{e}_y\gamma_y B + \mathbf{e}_z\gamma_z C). \tag{2.2.66}$$

A, B, C are the principal moments of inertia:

$$A = \int (y^2 + z^2)\, dm, \qquad B = \int (z^2 + x^2)\, dm, \qquad C = \int (x^2 + y^2)\, dm,$$

and $P = \int(x^2 + y^2 + z^2)\, dm$ is polar moment of inertia. Making use of
(2.2.60) and (2.2.66), we can rewrite Eq. (2.2.65) as follows:

$$\int \frac{\mathbf{r}' \times \mathbf{r}}{r^3}\, dm \approx -\frac{3}{r_a{}^3} [\mathbf{e}_x\gamma_y\gamma_z(C - B) + \mathbf{e}_y\gamma_z\gamma_x(A - C) + \mathbf{e}_z\gamma_x\gamma_y(B - A)]. \tag{2.2.67}$$

Substitution of (2.2.67) into the theorem of angular momentum (2.2.59),
together with (2.2.63), results in

$$\partial \mathbf{D}_S/\partial t + \boldsymbol{\omega} \times \mathbf{D}_S$$
$$= (3\gamma m_E{}^4/m^3 r_S{}^3)[\mathbf{e}_x\gamma_y\gamma_z(C - B) + \mathbf{e}_y\gamma_z\gamma_x(A - C) + \mathbf{e}_z\gamma_x\gamma_y(B - A)].$$

Using the abbreviation

$$K = 3\gamma m_E^4 / m^3 r_S^3 \approx 3\gamma m_E^3 / m^2 r_S^3 \qquad (2.2.68)$$

and with $\boldsymbol{\omega} = (\omega_x, \omega_y, \omega_z)$, the scalar form of this equation is

$$A\dot{\omega}_x + (C - B)\,\omega_y\omega_z = K\gamma_y\gamma_z(C - B),$$
$$B\dot{\omega}_y + (A - C)\,\omega_z\omega_x = K\gamma_z\gamma_x(A - C),$$
$$C\dot{\omega}_z + (B - A)\,\omega_x\omega_y = K\gamma_x\gamma_y(B - A).$$

If the satellite is assumed to have symmetry so that $A = B$, and introducing the notation

$$a = (C/A) - 1,$$

the following equations are obtained from the theorem of angular momentum:

$$\dot{\omega}_x + a\omega_y\omega_z = aK\gamma_y\gamma_z\,,$$
$$\dot{\omega}_y - a\omega_z\omega_x = -aK\gamma_z\gamma_x\,, \qquad (2.2.69)$$
$$\dot{\omega}_z = 0.$$

Let us now determine the particular solutions of (2.2.64) and (2.2.69) for the motion of the satellite. Such a solution can be found for (2.2.64), for example, i.e., for $\ddot{\mathbf{r}}_S + \frac{1}{3}K\mathbf{r}_S = 0$ using the trial solution $|\,\mathbf{r}_S\,| = \text{const}$. In this case, $K = \text{const}$ so that the vector $\mathbf{r}_S$ rotates in a plane with the angular velocity $\Omega = (K/3)^{1/2}$. This means that the center of gravity of the satellite moves along a plane circular trajectory. A particular solution of (2.2.69) for the rotation of the satellite about its center of gravity is given by

$$\omega_{z0} = \text{const}, \qquad \omega_{x0} = \omega_{y0} = 0, \qquad \gamma_{z0} = 0.$$

It corresponds to the rotation of the satellite around the z axis which is always perpendicular to the circular trajectory plane for this solution.

The three angles φ, ψ, δ are used to uniquely determine the position of the satellite in space, assuming that $|\,\delta\,| < \pi/2$ (Fig. 58). Let us select the space-fixed orthogonal coordinate system in such a way that the ξ, ζ plane is the plane of motion of the satellite. Because $\mathbf{r}_S \approx \mathbf{r}_a$, the vertical vector $\mathbf{r}_a$, which is a radius of the orbit, rotates uniformly around the origin of the fixed axes at the angular rate Ω. Therefore, $\gamma_{x0} = \cos(\varphi_0 - \Omega t)$, $\gamma_{y0} = -\sin(\varphi_0 - \Omega t)$.

If the satellite rotates around the z axis at the rate ω_{z0}, the body-fixed

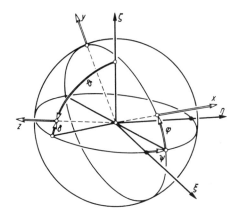

FIG. 58.   Reference axes and angles.

axis system x, y, z will also rotate. Figure 58 shows that $\varphi_0 = \int \omega_{z0} \, dt$ must hold.

The particular solution in question, which satisfies the equations of motion and kinematic equations, can be summarized as follows. The satellite's center of gravity describes a circular motion around the earth's center at the angular rate $\Omega = (K/3)^{1/2}$. The radius r_a of this orbit is constant. In addition, the satellite rotates about an axis through the center of gravity z, which remains perpendicular to the plane of the circle. This motion is described by

$$\omega_{x0} = \omega_{y0} = 0 \qquad \omega_{z0} = \text{const},$$

$$\delta_0 = \psi_0 = 0, \qquad \varphi_0 = \int \omega_{z0} \, dt, \qquad (2.2.70)$$

$$\gamma_{x0} = \cos(\varphi_0 - \Omega t), \qquad \gamma_{y0} = -\sin(\varphi_0 - \Omega t), \qquad \gamma_{z0} = 0.$$

Let us determine whether solution (2.2.70) is stable. We shall consider the perturbed motion

$$\delta = \delta_0 + \delta^*, \qquad \psi = \psi_0 + \psi^*, \qquad \varphi = \varphi_0 + \varphi^*,$$

i.e., the z axis has been raised by δ^* and the ξ, ζ plane is no longer the plane of motion. δ^*, ψ^*, ϕ^* are the variations. Because $\delta_0 = \psi_0 = 0$, we may also write $\delta = \delta^*$, $\psi = \psi^*$. For small variations we have

$$\omega_x = \dot{\psi} \sin \varphi - \dot{\delta} \cos \varphi, \qquad \gamma_x = \cos(\varphi - \Omega t),$$

$$\omega_y = \dot{\psi} \cos \varphi + \dot{\delta} \sin \varphi, \qquad \gamma_y = -\sin(\varphi - \Omega t), \qquad (2.2.71)$$

$$\omega_z = \dot{\varphi}, \qquad \gamma_z = \psi \cos \Omega t + \delta \sin \Omega t.$$

If (2.2.71) is substituted into the first equation of (2.2.69) (because the second equation gives the same result and the third equation gives $\dot{\varphi} = \text{const}$), we obtain the variational equation

$$\ddot{\psi} \sin \varphi - \ddot{\delta} \cos \varphi + \dot{\varphi}(1 + a)(\dot{\psi} \cos \varphi + \dot{\delta} \sin \varphi)$$
$$= -aK \sin(\varphi - \Omega t)(\dot{\psi} \cos \Omega t + \dot{\delta} \sin \Omega t).$$

Introducing the new variables

$$\alpha = \psi \cos \Omega t + \delta \sin \Omega t, \qquad \beta = \psi \sin \Omega t - \delta \cos \Omega t,$$

we obtain

$$\cos(\varphi - \Omega t)\{\dot{\alpha}[\dot{\varphi}(1 + a) - 2\Omega] + \ddot{\beta} + \beta[\Omega\dot{\varphi}(1 + a) - \Omega^2]\}$$
$$+ \sin(\varphi - \Omega t)\{\ddot{\alpha} + \alpha[\Omega\dot{\varphi}(1 + a) - \Omega^2 + aK] - \dot{\beta}[\dot{\varphi}(1 + a) - 2\Omega]\} = 0.$$

Since this equation must always hold, the following two differential equations are obtained for α and β:

$$\ddot{\alpha} + c_{11}\alpha - c_{12}\dot{\beta} = 0,$$
$$\ddot{\beta} + c_{22}\beta + c_{12}\dot{\alpha} = 0, \tag{2.2.72}$$

with the constant coefficients

$$c_{11} = \Omega\dot{\varphi}(1 + a) - \Omega^2 + aK,$$
$$c_{12} = \dot{\varphi}(1 + a) - 2\Omega,$$
$$c_{22} = \Omega\dot{\varphi}(1 + a) - \Omega^2.$$

The quantity $\dot{\varphi}$ is constant because of the third equations in (2.2.71) and (2.2.69). The characteristic equation of (2.2.72) follows from

$$\begin{vmatrix} \lambda^2 + c_{11} & -\lambda c_{12} \\ \lambda c_{12} & \lambda^2 + c_{22} \end{vmatrix} = 0$$

and is given by

$$\lambda^4 + \lambda^2(c_{22} + c_{11} + c_{12}^2) + c_{11}c_{22} = 0. \tag{2.2.73}$$

The existence of four imaginary roots (2.2.73) is a necessary condition for stability. This condition is satisfied if λ^2, calculated from (2.2.73), is negative and real for all cases. This corresponds to conditions

$$c_{22} + c_{11} + c_{12}^2 > 0, \qquad c_{11}c_{22} > 0, \qquad (c_{22} + c_{11} + c_{12}^2)^2 - 4c_{11}c_{22} > 0. \tag{2.2.74}$$

The stability conditions (2.2.74) can be written in the following form, making use of $K = 3\Omega^2$ and the new quantities $p = C/A = (a + 1)$ and $\nu = \dot{\varphi}/\Omega$:

$$\nu^2 p^2 - 2\nu p + 3a + 2 > 0,$$

$$\nu^2 p^2 + \nu p(3a - 2) + (1 - 3a) > 0,$$

$$\nu^4 p^4 - 4\nu^3 p^3 + (4 + 6a)\,\nu^2 p^2 - 24a\nu p + 9a^2 + 24a > 0.$$

They have been evaluated in the form of a *stability diagram* in Fig. 59.

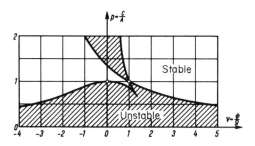

FIG. 59. Stability diagram for satellites.

This diagram shows that for $C = 0$ or $p = 0$, the rod satellite is always unstable with respect to the particular solution (2.2.76). On the other hand, a satellite having a flattened moment of inertia ellipsoid ($p > 1$) is always stable for $\nu > 1$. Satellites with a long moment of inertia ellipsoid ($p < 1$) may be stabilized with respect to the motion (2.2.70) for sufficiently large ν, i.e., for sufficiently large angular momentum with respect to the symmetry axis (z axis) perpendicular to the orbit.

2.3. PROBLEMS OF ELASTOMECHANICS

As outlined in Section 1.1, we will now discuss stability problems in elastomechanics in connection with the fundamental concepts contained in the stability theory discussed previously. A well-developed and independent theory exists for elastomechanics, which can be found in the literature. In addition to Refs. (*1*) and (*2*), the reader is also referred to several important works (*8*), which by no means exhausts the list. It is not necessary to repeat the methods and examples of these works. Instead we will carefully analyze their contents. Modern studies on the stability of elastic bodies frequently question the validity and applicability of the old concepts and methods. The theory of elastic-body stability is often reformulated and classified within the overall mechanical theory of stability.

The modern theory of elastic-body stability is as follows. The behavior of a special state of the elastic body (for example, an equilibrium shape considered as the stationary solution of the fundamental equations of elastostatics) is investigated as a function of a *system parameter* (for example, load, temperature, etc.). This behavior is expressed by means of a *characteristic quantity* ω, which determines whether stability or instability prevails. Let us assume that the functional dependence of the parameter is given by $\omega = \omega(p)$. By plotting this law, we obtain a diagram from which it is possible to determine whether the corresponding state is stable or unstable. Let us assume that the curve $\omega = \omega(p)$ passes into the instability region of ω for certain critical values of p. It follows that the state of the elastic body is no longer stable for such parameter values. The term *eigenvalue curve* $\omega = \omega(p)$ is often used, because the parameter is very often an eigenvalue in the mathematical treatment of the problem. In addition, *structure stability* must be investigated according to the concepts introduced in Section 1.1.2, because the stability of a special state is influenced by a perturbation of parameters.

In practice, the *lower stability boundary* is usually of interest, which is the smallest value $p_{\text{crit,min}}$ of the parameter p, assumed to be positive, for which instability first occurs. The behavior of the elastic body after this lower bound has been exceeded is called its "postcritical" behavior. Euler investigated the postcritical behavior of elastic rods. The works by Tricomi (9), Wang (10), or Langhaar (8.5) contain modern investigations of this problem. Nonlinear equations and the corresponding special methods must be used. At present, a great deal of work is being done on the postcritical behavior of plates and shells. Koiter (11) has reviewed this subject.

The stability behavior of bodies in the nonelastic range, as well as disturbances caused by undesirable inaccuracies of elastic constructions, are also of great interest. The latter problem leads to the theory of statistics.

We shall not deal further with these problems. Instead, we will discuss a linearized theory of bucking of *ideal and elastic* bodies. We will attempt to determine the lowest critical value of the system parameter (the so-called Euler value). Let us first consider elastostatic problems. The stability problems of elastokinetics will be considered in a concluding chapter.

2.3.1. *Stability Criteria of Elastostatics*

Two criteria have been used up to the present to determine stability regions and stability boundaries: These are the *energy criterion* and the *static criterion.* The application of these criteria is called the *energy method* and *equilibrium method.* The *energy method* uses the fact that the principle of virtual displacements can be extended to elastic bodies with infinite degrees of freedom. The relationship $\delta\Pi = 0$ holds in the equilibrium state (and in statics this is always the special state for which the stability is investigated) provided that the system is conservative or, more generally, monogenetic according to Lanczos (6). The potential is $\Pi = \Pi_i + \Pi_a$ (where Π_i is the deformation energy and Π_a the potential of the external loads). This relationship follows immediately from the principle of virtual work. The total potential Π takes on a stationary value in the equilibrium position. Let us assume that $\Pi = \Pi(h)$, i.e., the potential depends on a system parameter that specifies the position. Therefore, $\delta\Pi(h_0) = 0$ holds for the parameter value h_0 corresponding to equilibrium. In order to investigate the effect of h on Π, h_0 is varied and

$$\Pi(h_0 + \delta h) = \Pi(h_0) + (\partial\Pi/\partial h)_{h_0}\,\delta h + \tfrac{1}{2}(\partial^2\Pi/\partial h^2)_{h_0}\,\delta h^2 + \cdots$$

$$= \Pi(h_0) + \delta\Pi(h_0) + \tfrac{1}{2}\delta^2\Pi(h_0) + \cdots$$

is assumed. This can immediately be transformed to the relationship

$$\Pi(h_0 + \delta h) = \Pi(h_0) + \tfrac{1}{2}\delta^2\Pi(h_0) + \cdots.$$

In the vicinity of the equilibrium position, i.e., for small variations δh, the behavior of $\Pi(h_0 + \delta h)$ is given by

$$(\partial^2\Pi/\partial h^2)_{h_0}\,\delta h^2 = \delta^2\Pi(h_0).$$

The following results are obtained analogous to *Dirichlet's* theorem, which is valid for a point system, and from a consideration of the energy relationships.

If $\delta^2\Pi < 0$, energy is released when h_0 is varied. The body can move away from the equilibrium position that is unstable. On the other hand, if $\delta^2\Pi > 0$, energy is consumed when h_0 is varied. The body cannot leave the equilibrium position by itself. The equilibrium is stable. The stability boundary is given by $\delta^2\Pi = 0$ and the equilibrium is neutral. The stability criterion therefore amounts to the requirement $\delta^2\Pi > 0$. For *stability*, $\delta^2\Pi$ must be *positive definite* and Π must have a *minimum* in the equilibrium position.

Often the stability boundary is determined from a condition $\Delta\Pi_i = \Delta A_a$, where $\Delta\Pi_i$ and ΔA_a are the increases of the potential of the internal forces and the work of the external forces for variation of the equilibrium position. For example, Timoshenko (*1.6*) used this criterion successfully. Because $\Delta A_a = -\Delta\Pi a$, this is identical to

$$\Delta(\Pi_i + \Pi_a) = \Delta\Pi = 0,$$

and because $\Delta\Pi = \delta\Pi + \frac{1}{2}\delta^2\Pi \equiv \frac{1}{2}\delta^2\Pi$, this again amounts to $\delta^2\Pi = 0$.

There is another variation of the energy criterion advanced by E. Trefftz. He postulated that $\delta^2\Pi_a \equiv 0$ for loads that only depend linearly on h. The discussion can be restricted to $\delta^2\Pi_i$. What is true for $\delta^2\Pi$ can now be stated for $\delta^2\Pi_i$.

In the *equilibrium method* it is assumed that because of the neutral equilibrium state on the stability boundary, *neighboring* equilibrium positions can exist for the same load conditions. If we assume the trivial equilibrium position in our discussion, for which displacements are zero for the chosen coordinate system, the application of the equilibrium method postulates that *nontrivial equilibrium positions* exist. If they occur, they can be determined by the methods of statics. In this case, this may be called the *static stability criterion*, because the critical parameter value $p_{\text{crit,min}}$, follows from the proof that nontrivial equilibrium positions exist. Instability would occur after this parameter value is exceeded.

In the case of *conservative or monogenetic* problems, which are usually treated without being explicitly mentioned as such, the energy and static methods lead to the *same statement*. This may be shown as follows:

The energy method specifies $\delta^2\Pi = 0$ at the stability boundary.

The static method assumes that in addition to the trivial equilibrium position h_0 there is a nontrivial equilibrium position $h_1 = h_0 + \Delta h$. It must satisfy $\delta\Pi(h_1) = 0$ because it is an equilibrium position, and $\delta\Pi(h_0) = 0$ holds for the trivial equilibrium position.

We have

$$\Pi(h_1) = \Pi(h_0 + \Delta h) = \Pi(h_0) + \delta\Pi(h_0) + \frac{1}{2}\delta^2\Pi(h_0) + \cdots$$

and therefore because of $\delta\Pi(h_0) = 0$,

$$\Pi(h_1) = \Pi(h_0) + \frac{1}{2}\delta^2\Pi(h_0) + \cdots \equiv \Pi(h_0) + \frac{1}{2}(\partial^2\Pi/\partial h^2)_{h_0}(\Delta h)^2 + \cdots.$$

Furthermore

$$\delta\Pi(h_1) = \delta\Pi(h_0 + \Delta h) = [\partial\Pi(h_1)/\partial(\Delta h)]\,\delta(\Delta h)$$

is calculated and by substitution of $\Pi(h_1)$ we have

$$\delta\Pi(h_1) = (\partial^2\Pi/\partial h^2)_{h_0}\,\Delta h\,\delta(\Delta h).$$

The condition $\delta \Pi(h_1) = 0$ which follows from the existence of the nontrivial solution therefore implies

$$(\partial^2 \Pi / \partial h^2)_{h_0} \, \Delta h = 0$$

and therefore the condition

$$\delta^2 \Pi(h_0) \equiv (\partial^2 \Pi / \partial h^2)_{h_0} \, (\Delta h)^2 = 0,$$

which correspond to the energy criterion. It has thus been shown that the same conclusions may be reached for both criteria.

Finally, it should be mentioned that it is not necessary to be restricted to only *one* load parameter p or one position parameter h. In more complicated cases it can become necessary to calculate with several parameters, which is immediately possible. The practical application of the energy and static criteria will be demonstrated in the following sections by means of several examples. The first serious objections to the universal validity of the two classical stability criteria mentioned above were brought forth by H. Ziegler. He showed that the static criterion can only be automatically used for conservative problems. This is because there are nonconservative problems (for example, the Beck problem[1]), for which there are no nontrivial equilibrium positions. The assumptions for the criterion are therefore not satisfied. This result makes it necessary to distinguish two classes of problems in the stability theory of elastomechanics. One class encompasses the conservative problems that are monogenetic in the broader sense of the word. These are the problems for which the internal as well as the external forces can be derived from a potential. Only for these problems can the principle of virtual displacements be transformed into the simple form $\delta \Pi = 0$, which is the foundation for the static as well as for the energy criteria, as we have seen. Since the vast majority of problems in practice are of this type, it was always assumed that these conditions were satisfied until incorrect results made it necessary to reconsider the problems with greater attention. The other class encompasses the nonconservative problems, which are polygenetic in the broad sense. These are the problems for which the external forces cannot be derived from a potential (the internal forces of elastic bodies always have a potential). The "follower" forces are examples for this, which occur in problems in aeroelasticity and which occur more often in modern technology.

There are just as many objections to the energy criterion. This is

---

[1] BECK, M., The buckling load of a tangentially loaded beam clamped on one side. *Z. Angew. Math. Phys.* **3**, 255–228, 476–477 (1952).

because the positive definite property of the expression $\delta^2\Pi$ according to the *Dirichlet* theorem does not necessarily imply stability of equilibrium according to the more rigorous definition of Lyapunov for two- or three-dimensional continua. Lyapunov's definition of stability stipulates that the displacements are sufficiently and arbitrarily small at all times and at every point of the continuum. These doubts were already formulated by E. Hellinger. Shield and Green[1] showed that the objections to the energy criterion are justified.

Ziegler recommends a third stability criterion in cases where the energy criterion can definitely not be applied and where the static criterion cannot be applied with certainty because of the absence of a potential for the external forces. This is the *kinetic criterion*, which consists of the following: small displacements from the equilibrium position of the elastic body under consideration are carried out. Then the disturbed motion which occurs is discussed. If it remains bounded for all times, stability of the equilibrium position is implied.

The introduction of this third criterion and the class of nonconservative problems immediately leads to questions that will now be discussed. (1) What mathematical methods must be applied in the nonconservative case (the Ritz method does not qualify because $\delta\Pi = 0$ no longer holds)? (2) Can the static criterion continue to hold for nonconservative problems (the energy criterion certainly does not hold)? (3) Does the mass distribution of the elastic body and its internal damping influence the stability, because there is now a motion to be considered? (4) Does linearization make sense?

2.3.2. *Application of General Methods to Elastostatic Problems*

Stability problems are by nature kinetic problems. This is because, according to the general theory developed in Part 1 of the book, they consist of observing the motion about the equilibrium position of interest, which is caused by a perturbation. Therefore, we will use this kinetic point of view in the following. In spite of this, we will discuss elastostatic stability problems when we are dealing with stability of equilibrium positions. On the other hand, later on we will discuss elastokinetic stability problems that involve the stability of stationary motions. Making use of the ideas of Ziegler, we will use kinetic methods and kinetic stability criteria in the investigation of elastostatic problems, which follow the fundamental principles of a general theory.

Let us investigate the *stability of an equilibrium position*, which will be

[1] SHIELD, R. T., and GREEN, A. E., On certain methods in the stability theory of continuous systems. *Arch. Ratl. Mech. Anal.* 12, 354–360 (1963).

called the *original state*. For simplicity, a reference system is assumed
for which all displacements $\mathbf{w}(x, y, z, t)$ are zero in the equilibrium
position. The original state is disturbed so that a small motion of the
body occurs. According to the general theory, the corresponding equations of motion are the variational equations. Their solution leads to a
decision regarding the stability of the original state. Even though the
displacements themselves are substituted in the variational equations in
the following, it should be remembered that these are really variations
of the displacements of the original state. This effect cannot be seen
because of the selection of the reference system.

In order to obtain variational equations, the principle of virtual
displacements is used. According to Hellinger, the latter is at least a
permissible axiom in the field of continuum mechanics. The perturbation consists of an infinitesimal displacement which is compatible with
the constraints of the body. The work equations which follow from the
principle are then formulated, and the equations of motion and the
boundary conditions can be obtained from them.

Example: Let us consider the *buckling column* clamped at one end
with a tangential (follower) point load P_t and a vertical (constant
direction) point load P_v (Fig. 60). The disturbance consists of a lateral

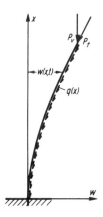

FIG. 60. Buckling rod with
unidirectional force P_v and
follower force P_t.

deflection of the column from its trivial equilibrium position. It will
then carry out small oscillations around the original state (vertical column
axis). The points of the column axis undergo the transverse displacements $w(x, t)$. For these displacements, the principle of virtual displacements yields

$$\int_0^l \mu w_{tt}\, \delta w\, dx + \delta \Pi_i + (P_v + P_t)\, \delta u(l) + P_t w_x(l)\, \delta w(l) = 0. \quad (2.3.1)$$

μ is the linear mass density, l the length, Π_i the potential energy of the internal forces of the column, w and u are the horizontal and vertical displacements of the column axis. Subscripts correspond to the partial derivative with respect to the independent variable t (time) and x (abscissa of the column axis). If it is assumed that the axis of the column is incompressible, the extension is given by

$$\epsilon \approx u_x + (w_x^2/2) = 0,$$

so that

$$u(l) = \int_0^l u_x \, dx = -\tfrac{1}{2} \int_0^l w_x^2 \, dx, \qquad \delta u(l) = -\int_0^l w_x \, \delta w_x \, dx,$$

and therefore Eq. (2.3.1) becomes

$$\int_0^l \mu w_{tt} \, \delta w \, dx + \delta \Pi_i - (P_v + P_t) \int_0^l w_x \, \delta w_x \, dx + P_t w_x(l) \, \delta w \, (l) = 0.$$

We also have

$$(P_v + P_t) \int_0^l w_x \, \delta w_x \, dx = \delta \left[\tfrac{1}{2}(P_v + P_t) \int_0^l w_x^2 \, dx \right] = -\delta \Pi_a \,,$$

where Π_a is the potential of the external forces. Equation (2.3.1) can also be written in the form

$$\int_0^l \mu w_{tt} \, \delta w \, dx + \delta(\Pi_i + \Pi_a) + P_t w_x(l) \, \delta w \, (l) = 0. \tag{2.3.2}$$

Also we have

$$\Pi_i = \tfrac{1}{2} \int_0^l \alpha w_{xx}^2 \, dx,$$

where α is the bending stiffness of the column, assumed to be constant. After the variation is carried out and partial integration is used, we obtain

$$\delta(\Pi_i + \Pi_a) = \delta \left[\tfrac{1}{2} \int_0^l [\alpha w_{xx}^2 - (P_v + P_t) w_x^2] \, dx \right]$$

$$= \int_0^l [\alpha w_{xxxx} + (P_v + P_t) w_{xx}] \, \delta w \, dx$$

$$+ [-(P_v + P_t) w_x \, \delta w + \alpha w_{xx} \, \delta w_x - \alpha w_{xxx} \, \delta w]_0^l \,.$$

Substitution into (2.3.2) results in

$$\int_0^l \left[\mu w_{tt} + \alpha w_{xxxx} + (P_v + P_t)\, w_{xx} \right] \delta w \; dx$$

$$+ \left[-(P_v + P_t)\, w_x \, \delta w + \alpha w_{xx} \, \delta w_x - \alpha w_{xxx} \, \delta w \right]_0' + P_t w_x(l)\, \delta w\,(l) = 0.$$

Because of the prescribed geometric boundary conditions

$$w(0, t) = w_x(0, t) = 0,$$

the requirement that all boundary terms vanish, and because of the arbitrary nature of the variation δw, the differential equation

$$\mu w_{tt} + \alpha w_{xxxx} + (P_v + P_t)\, w_{xx} = 0 \qquad (2.3.3)$$

and the dynamic boundary conditions

$$w_{xx}(l, t) = 0, \qquad w_{xxx}(l, t) = -(P_v/\alpha)\, w_x(l, t) \qquad (2.3.4)$$

are obtained.

Equation (2.3.3) is the desired *variational equation*, which describes the perturbed motion around the original state.

The original state is called stable if the displacements $\mathbf{w}(x, y, z, t)$ remain sufficiently small for sufficiently small initial perturbations. In the example given above, this would have to be satisfied by the lateral displacements $w(x, t)$ if the column axis is subjected to a small transverse deflection from the vertical original state. This stability definition coincides with the definition of Lyapunov given in Section 1.1.2: If the initial perturbation of the original state is selected small enough, for stability the displacements $\mathbf{w}$ should remain within arbitrary limits at all times. The variational equations determine whether or not this is possible. In our example above, this is determined by Eq. (2.3.3) and their boundary conditions.

The following two *remarks* are in order.

In the example, we used the linearized equations, which we shall also do in the following. As stated above, this should be justified by showing that the displacements as well as their derivatives remain small when the original state is disturbed. Only when this is true should the classical theory of elasticity, which is based on linear constitutive equations, hold.

Also, in the example we did not vary all possible displacement components. Instead, only a special variation was carried out, the variation of the horizontal component w. This is certainly permissible in certain cases, because inspection of the problem will show which component and variation is really decisive for the behavior of the body. In principle, it is important to realize that, if not all variations are carried out, only *conditional* stability conditions with respect to special perturbations are obtained. This will not always be stated in the following. We will accept the restriction that our stability results are only conditional because of the special variations of the displacements considered. Usually this restriction is unimportant in practice.

2.3.3. *Conservative Stability Problems*

Conservative problems are especially important in particle mechanics and in the mechanics of rigid bodies, because there is an *energy integral* that can be used as the *Lyapunov function* to determine the stability of equilibrium positions. This is mentioned in Section 1.5.3 in connection with the theorem of *Lagrange–Dirichlet*. It is natural to proceed in the same way for conservative problems of elastomechanics. The total energy of the elastic body is given by

$$H = \int_V (\tfrac{1}{2}\mu\dot{\mathbf{w}}^2 + \varPi^*)\, dV = \int_V \mathfrak{H}\, dV, \qquad (2.3.5)$$

where V is the volume, μ the mass density, $\mathbf{w}$ the displacement vector, and $\varPi^*$ the specific potential of the internal and external forces of the body. It can be shown that the variational equations for the motion about the equilibrium position are given by

$$\delta H/\delta p_k = \dot{w}_k, \qquad \delta H/\delta w_k = -\dot{p}_k, \qquad p_k = \mu\dot{w}_k, \qquad (2.3.6)$$

where $\mathfrak{H} = \mu\dot{\mathbf{w}}^2/2 + \varPi^*$ is the Hamiltonian density appearing in Eq. (2.3.5).

Example: Let us consider a flat *plate* of thickness h which carries out small oscillations about the equilibrium position. If F is the area of the plate, we have

$$H = \int_F \left\{ \frac{1}{2\mu^*}\, p^2 + \frac{D}{2}\, [w_{xx}^2 + 2(1-\nu)\, w_{xy}^2 + 2\nu w_{xx}w_{yy} + w_{yy}^2] \right\} dF$$

$$= \int_F \mathfrak{H}\, dF,$$

where

$$p = \mu^*\dot{w}, \quad \mu^* = h\mu, \quad \varPi^* = w_{xx}^2 + 2(1-\nu)\, w_{xy}^2 + 2\nu w_{xx}w_{yy} + w_{yy}^2, \quad w(x,y,t)$$

is the vertical deflection and D is the plate stiffness. According to the first equation of (2.3.6)

$$\frac{\delta H}{\delta p} \equiv \frac{\partial \mathfrak{H}}{\partial p} = \frac{p}{\mu^*} = \dot{w},$$

and therefore

$$p = \mu^* \dot{w}.$$

The second equation of (2.3.6) yields

$$\frac{\delta H}{\delta w} = \frac{\partial \mathfrak{H}}{\partial w} - \frac{\partial}{\partial x}\left(\frac{\partial \mathfrak{H}}{\partial w_x}\right) - \frac{\partial}{\partial y}\left(\frac{\partial \mathfrak{H}}{\partial w_y}\right) + \frac{\partial^2}{\partial x^2}\left(\frac{d\mathfrak{H}}{\partial w_{xx}}\right)$$

$$+ \frac{\partial^2}{\partial x\, \partial y}\left(\frac{\partial \mathfrak{H}}{\partial w_{xy}}\right) + \frac{\partial^2}{\partial y^2}\left(\frac{\partial \mathfrak{H}}{\partial w_{yy}}\right) = -\dot{p}.$$

Differentiation of this expression and the relationship $\dot{p} = \mu^* \ddot{\mathbf{w}}$ results in

$$D(w_{xxxx} + 2w_{xxyy} + w_{yyyy}) = -\mu^* \ddot{w} \qquad \text{or} \qquad \Delta\, \Delta w = -(\mu^*/D)\, \ddot{w},$$

respectively, which is the equation of motion.

If $\partial \mathfrak{H}/\partial t = 0$, which is always assumed to hold, H is an integral of the variational equation because of (2.3.6). Making use of (2.3.6), we obtain

$$\frac{dH}{dt} = \int_V \left[\frac{\delta H}{\delta w_k}\frac{dw_k}{dt} + \frac{\delta H}{\delta p_k}\frac{dp_k}{dt}\right] dV$$

$$= \int_V \left[-\dot{p}_k \frac{dw_k}{dt} + \dot{w}_k \frac{dp_k}{dt}\right] dV \equiv 0,$$

which must be true for an integral.

Just as is done in particle mechanics and mechanics of rigid bodies, an attempt will be made to use the function H as the Lyapunov function, assuming that it is positive definite. This is always appropriate if the bounded nature of $\mathbf{w}$ were to follow from the fact that $\mathbf{w}$ always lies on the phase surface (and this is the case because $dH/dt = 0$). Unfortunately, this proof cannot always be given, and H cannot be used as the Lyapunov function in such a case. Therefore, there is *restricted validity of the energy criterion*. This can be shown as follows. Let H be positive definite. Equation (2.3.5) shows that this is the case if $\Pi = \int \Pi^* \, dV$ is positive definite, which means that the potential of the internal and external forces is positive definite. Now Π is the potential of the varied state. Therefore, we have

$$\Pi = \Pi_g + \delta\Pi_g + \tfrac{1}{2}\,\delta^2\Pi_g .$$

where the subscript g refers to the original state.

Since the original state is an equilibrium state, $\delta\Pi_g = 0$ must hold according to the principle of virtual displacements. Therefore

$$\Pi = \Pi_g + \tfrac{1}{2}\,\delta^2\Pi_g .$$

Since it is always possible to have $\Pi_g \geqslant 0$, it follows that for a positive definite H and consequently positive definite Π, the quantity $\delta^2\Pi_g$ is also positive definite, which is the condition for the energy stability criterion. Since for positive definite H, $\mathbf{w}$ is not always sufficiently bounded and therefore the equilibrium state is not stable according to our definition, H cannot be used as a Lyapunov function in all cases. Also, the energy stability criterion cannot be used in all cases.

Let us first assume the case in which the energy criterion does not fail for the selected definition of stability.

Example: Let us consider the elastic *buckling column with a conservative load* (Fig. 61). The origin of the coordinate system is located at a fixed support. Then $w(0, t) = 0$ holds for this point and all other "Euler cases" shown in Fig. 62. The internal and external forces have

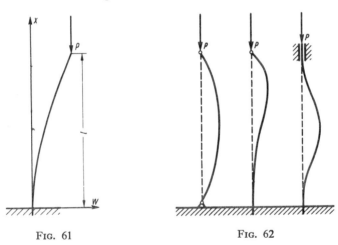

FIG. 61   FIG. 62

FIG. 61.   Buckling column with conservative load.
FIG. 62.   Euler cases of the buckling column.

the potential

$$\Pi = \tfrac{1}{2} \int_0^l (\alpha w_{xx}^2 - P w_x^2)\, dx. \tag{2.3.7}$$

Let us assume that it is positive definite and has a small magnitude, because the deflections w are the result of a small initial disturbance. Because $w(0, t) = 0$, and if

$$\| w_x \| = \left(\int_0^l w_x^2\, dx \right)^{1/2}$$

is the norm of w_x, using the Cauchy–Schwarz inequality results in the estimation

$$w^2 \leqslant x \int_0^x w_x{}^2 \, dx \leqslant l \int_0^l w_x{}^2 \, dx = l \, \| \, w_x \, \|^2. \qquad (2.3.8)$$

Furthermore, the same inequality gives

$$[w_x(x, t) - w_x(0, t)]^2 = \left[\int_0^x w_{xx} \, dx \right]^2 \leqslant x \int_0^x w_{xx}^2 \, dx$$

$$\leqslant x \int_0^l w_{xx}^2 \, dx \leqslant x \, \| \, w_{xx} \, \|^2,$$

where $\| \, w_{xx} \, \|$ is the norm of w_{xx}. Therefore

$$\int_0^l [w_x(x, t) - w_x(0, t)]^2 \, dx \leqslant \frac{l^2}{2} \, \| \, w_{xx} \, \|^2$$

is obtained. After the integration on the left side of the inequality is carried out, we obtain

$$\| \, w_x \, \|^2 - 2w_x(0, t) \, w(l, t) + w_x{}^2(0, t) \, l \leqslant (l^2/2) \| \, w_{xx} \, \|^2.$$

In all cases shown in Figs. 61 and 62 $w_x(0, t) = 0$ or $w(l, t) = 0$ holds. Therefore, we always have

$$\| \, w_x \, \|^2 \leqslant (l^2/2) \| \, w_{xx} \, \|^2, \qquad (2.3.9)$$

and finally, because of (2.3.8),

$$w^2 \leqslant (l^3/2) \| \, w_{xx} \, \|^2. \qquad (2.3.10)$$

Because the potential was assumed to be positive definite, we have

$$\Pi = (\alpha/2) \| \, w_{xx} \, \|^2 - (P/2) \| \, w_x \, \|^2 > 0. \qquad (2.3.11)$$

With (2.3.9), Π becomes

$$\Pi \geqslant \tfrac{1}{2} \| \, w_{xx} \, \|^2 (\alpha - (Pl^2/2)). \qquad (2.3.12)$$

We also impose the requirement

$$\tfrac{1}{2} \| \, w_{xx} \, \|^2 (\alpha - Pl^2/2) > 0. \qquad (2.3.13)$$

This is the case for

$$P < 2\alpha/l^2. \qquad (2.3.14)$$

P then remains below the critical Euler value P_E, which is equal to $2.46\alpha/l^2$ for the example shown in Fig. 61.

If (2.3.14) holds, then because of (2.3.12) we have

$$\| w_{xx} \|^2 \leqslant 4\Pi/(2\alpha - Pl^2),$$

and it follows from (2.3.10) that

$$w^2 \leqslant 2l^3\Pi/(2\alpha - Pl^2). \tag{2.3.15}$$

Because of the assumption (2.3.11) and (2.3.14) it is therefore possible to obtain an upper bound for the *magnitude* of the displacements w that guarantee that these magnitudes are small enough for stability to prevail. The smaller P, the more the above statement is valid, i.e., the more it will remain below the stability bound given by (2.3.14).

In the above it was shown that positive definite Π and sufficiently small load P guarantees stability of the elastic column according to the more severe Lyapunov stability definition. It is necessary to show that the derivative of w and therefore the *elongation* remains *small* for the column under these assumptions.

We have

$$[w_x(x, t) - w_x(0, t)]^2 \leqslant x \| w_{xx} \|^2 \leqslant l \| w_{xx} \|^2 \leqslant 4l\Pi/(2\alpha - Pl^2),$$

or

$$| w_x(x, t) - w_x(0, t)| \leqslant 2(l\Pi)^{1/2}/(2\alpha - Pl^2)^{1/2}.$$

We have the boundary condition $w_x(0, t) = 0$ for columns built in at the bottom. Therefore, the relationships

$$w_x^2(x, t) \leqslant 4l\Pi/(2\alpha - Pl^2) \qquad \text{or} \qquad | w_x(x, t)| \leqslant 2[l\Pi/(2\alpha - Pl^2)]^{1/2}$$

are immediately obtained. These are the desired bounds for w_x.

For columns hinged at the bottom, there exists a certain $x = \xi$ for which $w_x(\xi, t) = 0$ holds. Since the inequality given above holds for any x, we must therefore have

$$w_x^2(0, t) \leqslant 4l\Pi/(2\alpha - Pl^2) \qquad \text{or} \qquad | w_x(0, t)| \leqslant 2[l\Pi/(2\alpha - Pl^2)]^{1/2}.$$

Since

$$| w_x(x, t) - w_x(0, t)| \leqslant 2[l\Pi/(2\alpha - Pl^2)]^{1/2}$$

and owing to the upper bound for the magnitude of $w_x(0, t)$ just established, the extreme case is

$$| w_x(x, t)| = 4[l\Pi/(2\alpha - Pl^2)]^{1/2}.$$

This again shows the upper bound for w_x.

We have therefore shown that w and w_x are sufficiently small, which means it is possible to apply the linearized theory of elasticity.

As we already know, similar results do not hold for *multidimensional bodies*. In the case of the elastic plate we will see that for positive definite Π at least the *norm of w is bounded*. Stability would again exist according to a generalized and less rigorous stability definition.

Example: Let us consider the *plate* shown in Fig. 63, which is held

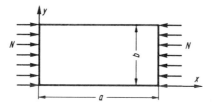

FIG. 63. Plate under compression loading.

along the edges. It is loaded by the forces N that lie in the plate plane and are applied along the edges $x = 0$, $x = a$. Their potential energy is given by

$$\Pi = \Pi_i - \int_F \tfrac{1}{2} N w_x^2 \, dF. \tag{2.3.16}$$

F is the area of the plate, Π_i the potential of the internal forces, and w the deflection perpendicular to the plane of the plate.

For $\nu > 0$ we have

$$\Pi_i = \frac{D}{2} \int_F [w_{xx}^2 + 2(1 - \nu)\, w_{xy}^2 + 2\nu w_{xx} w_{yy} + w_{yy}^2] \, dF$$

$$\geqslant \frac{D(1 - \nu)}{2} \int_F (w_{xx}^2 + 2w_{xy}^2 + w_{yy}^2) \, dF. \tag{2.3.17}$$

According to the Poincaré inequality, we have

$$\int_F w_x{}^2 \, dF \leqslant B \int_F (w_{xx}^2 + w_{xy}^2) \, dF + M \left(\int_F w_x \, dF \right)^2,$$

where $B > 0$, $M > 0$ are suitably selected constants. Introducing the angle α between the normal of the plate edge and the x axis, we have

$$\int_F w_x \, dF = \int_R w \cos \alpha \, ds,$$

which is equal to zero because w vanishes everywhere along the edge of the plate R which is held fixed. Therefore

$$\int_F w_x{}^2 \, dF \leqslant B \int_F (w_{xx}^2 + w_{xy}^2) \, dF$$

remains which, making use of (2.3.17), becomes

$$\int_F w_x{}^2 \, dF \leqslant \frac{2B}{D(1-\nu)} \Pi_i . \tag{2.3.18}$$

Similarly, we find

$$\int_F w_y{}^2 \, dF \leqslant \frac{2B}{D(1-\nu)} \Pi_i . \tag{2.3.19}$$

From (2.3.16) and (2.3.18) it follows that

$$\Pi \geqslant \Pi_i \{1 - BN/[D(1-\nu)]\}. \tag{2.3.20}$$

For sufficiently small N

$$N < D(1-\nu)/B,$$

and for positive definite Π, the following inequality is obtained:

$$\Pi_i \leqslant \frac{D(1-\nu)}{D(1-\nu) - BN} \Pi. \tag{2.3.21}$$

Since w was assumed zero along the edge R, it is also possible to use Friedrichs' inequality, which results in

$$\int_F w^2 \, dF \equiv \| w \|^2 < \kappa \int_F (w_x{}^2 + w_y{}^2) \, dF. \tag{2.3.22}$$

Introducing the relationships (2.3.18), (2.3.19), and (2.3.21) into (2.3.22), we obtain

$$\| w \|^2 < \{4\kappa B/[(1-\nu) D - BN]\} \Pi, \tag{2.3.23}$$

which is an upper bound for the norm of w.

This shows the pronounced difference with respect to the column problem. According to this calculation, and in spite of positive definite Π and sufficiently small N, it is only possible to guarantee that the *norm of w* is sufficiently small. The smallness of w itself cannot be guaranteed. In multidimensional problems it is therefore possible for the norm, but not the magnitude of the individual displacements at each point of the body to remain sufficiently small. An example of finite local displacements in spite of small initial perturbations is contained in the work by Shield and Green. Even though Π is positive definite, the rigorous Lyapunov stability definition is violated in such a case, and the energy criterion cannot be used. In order to save this criterion, W. T. Koiter suggests a looser stability definition where an elastic body is called stable if the norm $\| \mathbf{w} \| = (\int_v \mathbf{w}\mathbf{w} \, dV)^{1/2}$ of the displacement vector can be made as small as desired for sufficiently small initial perturbations. Since this condition can be met for a positive definite potential, the energy criterion holds again for this restricted definition. However, it should be remembered that locally there can be large displacements in the body.

The relationships (2.3.17)–(2.3.20) show that the statements regarding w also hold for the first derivatives of w with respect to x and y, and therefore also hold for the other displacements and elongations. It is only possible to estimate their norms, and only the quadratic mean of these quantities remains small. This is not sufficient to decide the stability condition according to our definition and to decide whether the linearized theory of elasticity can be used.

The linearized stability calculation and the energy criterion remain in question for multidimensional elastic bodies. The following theory is problematical for plates and shells but can be directly applied to one-dimensional bodies, such as elastic columns. If the theory is used for plates and shells anyway, as is always done in practice and as we shall do here, it should be realized that the stability results obtained can only have conditional validity.

2.3.4. Outline of an Algebraic Theory

The principle of virtual work gives the following equation for the perturbed motion of an elastic body about the original state:

$$\int_V \mu \mathbf{w}_{tt} \, \delta \mathbf{w} \, dV + \delta(\Pi_i + \Pi_a) - \int_O \mathbf{k}_a \, \delta \mathbf{w} \, dO - \sum_i \mathbf{K}_{ai} \, \delta \mathbf{w}_i = 0. \quad (2.3.24)$$

The following notation is used in equation (2.3.24): μ is the mass density, $\mathbf{w}$ the displacement vector, Π_i the potential of the internal forces, Π_a the potential of the external forces (if this is present), $\mathbf{k}_a$ are the nonconservative or polygenetic external surface forces, $\mathbf{K}_{ai}$ the nonconservative or polygenetic external concentrated forces, $\delta \mathbf{w}_i$ the displacement vectors of their points of application, V is the volume, O the surface area of the elastic body, and the subscripts to the variable $\mathbf{w}$ indicate partial differentiation.

Example: As is already known from Section 2.3.2, it is possible to obtain the differential equations and the boundary conditions of the stability problems under discussion from (2.3.24). For simplicity, let us demonstrate this for the *buckling column.* Let us consider the column shown in Fig. 64 with the conservative load P_v and the nonconservative loads P_t and q. Equation (2.3.24) gives

$$\int_0^l \mu \ddot{w}\, \delta w\, dx + \delta \left\{ \tfrac{1}{2} \int_0^l \left[\alpha w_{xx}^2 - \left(P_v + P_t + \int_x^l q(\xi)\, d\xi \right) w_x^2 \right] dx \right\}$$

$$+ \int_0^l q w_x\, \delta w\, dx + P_t w_x(l)\, \delta w\,(l) = 0. \tag{2.3.25}$$

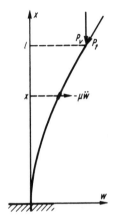

FIG. 64.   Buckling column with nonconservative load.

The second term of (2.3.25) results in the following expression after the variation is carried out:

$$\delta \left\{ \tfrac{1}{2} \int_0^l \left[\alpha w_{xx}^2 - \left(P_v + P_t + \int_x^l q(\xi)\, d\xi \right) w_x^2 \right] dx \right\}$$

$$= \int_0^l (\alpha w_{xxxx} + Q w_{xx} - q w_x)\, \delta w\, dx + [\alpha w_{xx}\, \delta w_x - \alpha w_{xxx}\, \delta w - Q w_x\, \delta w]_0^l, \tag{2.3.26}$$

with

$$Q = P_v + P_t + \int_x^l q(\xi)\, d\xi.$$

If this is substituted into (2.3.25), the result is

$$\int_0^l (\mu \ddot{w} + \alpha w_{xxxx} + Q w_{xx})\, \delta w\, dx + P_t w_x(l)\, \delta w(l)$$

$$+ \left[\alpha w_{xx}\, \delta w_x - \alpha w_{xxx}\, \delta w - P_t w_x\, \delta w - \left(P_v + \int_x^l q(\xi)\, d\xi \right) w_x\, \delta w \right]_0^l = 0,$$

or

$$\int_0^l [\mu\ddot{w} + L(w)]\,\delta w\,dx + R(w) = 0, \qquad (2.3.27)$$

where L is the linear differential operator $\alpha\,\partial^4/\partial x^4 + Q\,\partial^2/\partial x^2$ and $R(w)$ is the boundary value expression

$$R(w) = P_t w_x(0, t)\,\delta w(0, t)$$

$$+ \left\{ \alpha w_{xx}\,\delta w_x - \alpha w_{xxx}\,\delta w - \left[P_v + \int_x^l q(\xi)\,d\xi\right] w_x\,\delta w \right\}_{x=0}^{x=1}.$$

We require $R(w) = 0$, which results in the boundary values for the problem. The column shown in Fig. 64 is clamped at the bottom, which implies the geometric conditions $w_x(0, t) = w(0, t) = 0$. If $R(w) = 0$ holds, the dynamic boundary conditions

$$w_{xx}(l, t) = 0, \qquad w_{xxx}(l, t) = -(P_v/\alpha)\,w_x(l, t)$$

must be satisfied also. If the column is supported in a different way, the geometric boundary conditions are also different and the dynamical boundary conditions change as well. Because the variation δw is arbitrary, we must have

$$\mu\ddot{w} + L(w) = 0, \qquad (2.3.28)$$

which leads to the differential equations for the perturbed motion. Equation (2.3.27) therefore shows all the features of the column for the selected loading conditions.

The next step in solving the stability problem at hand consists of solving the partial differential equation obtained from the principle of virtual work. In our example this is Eq. (2.3.28). Since an exact solution is usually not possible, the question of a *sufficient mathematical approximation method* is raised which was already discussed at the end of Section 2.3.1. Since we are considering nonconservative or polygenetic loads, the differential expression $L(w)$ in (2.3.28) can be completely general and, in particular, it can be non-self-adjoint. This means that the Ritz method cannot be used. According to section 1.6.5, the Galerkin method can still be used and converges. We have thus obtained the following result: The *Galerkin method* becomes the basis of the entire stability theory of elastomechanics. This is why it is so important to determine its *convergence*. This was investigated in Section 1.6.5, and the conditions of the method can be obtained from there.

We will go further than the mathematical discussions of Section 1.6.5

by showing that the *Galerkin* method also converges for *nonconservative column problems*.

The discussion will be based on known theorems of functional analysis which can be found in the work of Michlin (*12*). The theorems imply the following.

The equation u + Tu = 0 is satisfied by the elements u of a Hilbert space H. The approximate solution obtained according to the Galerkin method converges, based on the norm of H, towards the exact solution u_0, if the equation has a unique solution and the operator T is completely continuous in H.

According to (2.3.28), the following differential equation holds for *column problems*:

$$\mu \ddot{w} + \alpha w_{xxxx} + Q w_{xx} = 0,$$

which is a special case of the general type

$$w_{xxxx} + a_1 w_{xx} + a_2 w_{tt} + a_3 w_t = 0. \tag{2.3.29}$$

The product trial solution $w = e^{\omega t} \varphi(x)$ applied to (2.3.29), and its boundary conditions results in the ordinary boundary value problem

$$\varphi^{\mathrm{IV}} + a_1 \varphi^{\mathrm{II}} + b\varphi = 0, \qquad U[\varphi]_R = 0, \qquad b = a_2 \omega^2 + a_3 \omega. \tag{2.3.30}$$

In operator notation, the differential equation (2.3.30) is

$$A\varphi + K\varphi = 0,$$

where

$$A = d^4/dx^4, \qquad K = a_1(d^2/dx^2) + b,$$

which can be transformed into

$$\varphi + A^{-1}K\varphi = 0.$$

According to the theorem above, in order to apply the Galerkin method it is necessary to show that the operator $T = A^{-1}K$ is *completely continuous*, because Eq. (2.3.30) will certainly have a unique solution for the prescribed boundary conditions.

As Michlin shows, $T = A^{-1}K$ is completely continuous if A^{-1} is completely continuous and K is bounded.

We have $A^{-1}\varphi(x) = \int_0^l G(x, \xi)\, \varphi(\xi)\, d\xi$, where l is the column length and $G(x, \xi)$ is the Green function corresponding to the operator $A = d^4/dx^4$ for the boundary conditions $U[\varphi]_R = 0$.

The following

$$
U[\varphi]_R = \begin{cases}
\varphi(0) = \varphi(l) \; = \varphi''(0) = \varphi''(l) \; = 0 & \text{(column hinged at both ends)} \\[2mm]
\varphi(0) = \varphi(l) \; = \varphi'(0) = \varphi'(l) \; = 0 & \text{(column clamped at both ends)} \\[2mm]
\varphi(0) = \varphi(l) \; = \varphi'(0) = \varphi''(l) \; = 0 & \text{(column clamped at one end and hinged at the other end)} \\[2mm]
\varphi(0) = \varphi'(0) = \varphi''(l) = 0, \varphi'''(l) = 0 & \text{(column clamped at one end and free at the other end without a conservative load } P_v \equiv 0)
\end{cases}
$$

are boundary conditions for which the Green's function[1] can easily be found. They are continuous and bounded, and it can be shown that A^{-1} must be completely continuous because $G(x, \xi)$ is quadratically integrable.

It remains to be shown that K is bounded. We have

$$
\| K\varphi \| \leqslant | a_1 |_{\max} \| \varphi'' \| + | b |_{\max} \| \varphi \|, \tag{2.3.31}
$$

where

$$
\| u \| = \left(\int_0^l u^2 \, dx \right)^{1/2}
$$

is the ordinary norm of an element u. The scalar product

$$
(A\varphi, \varphi) = \int_0^l (A\varphi) \cdot \varphi \, dx
$$

can be changed into the following form by partial integration and by consideration of the boundary conditions:

$$
(A\varphi, \varphi) = \int_0^l \varphi''^2 \, dx = \| \varphi'' \|^2. \tag{2.3.32}
$$

Applying the Cauchy–Schwarz inequality, we obtain

$$
\varphi^2 \leqslant x \int_0^x \varphi'^2 \, dx \leqslant x \int_0^l \varphi'^2 \, dx = x \| \varphi' \|^2
$$

---

[1] COLLATZ, L., "Eigenvalue Problems and their Numerical Treatment." Akademische Verlagsgesellschaft Geist und Portig K.-G., Leipzig, 1945.

and therefore

$$\| \varphi \|^2 \leqslant (l^2/2) \| \varphi' \|^2. \tag{2.3.33}$$

From (2.3.9) we have

$$\| \varphi' \|^2 \leqslant (l^2/2) \| \varphi'' \|^2, \tag{2.3.34}$$

and making use of (2.3.33), we obtain

$$\| \varphi \|^2 \leqslant (l^4/4) \| \varphi'' \|^2. \tag{2.3.35}$$

If the calculation is carried out in a function space that is bounded by

$$(A\varphi, \varphi) \leqslant N^2,$$

then, according to (2.3.32) and (2.3.35), we have

$$\| \varphi'' \| \leqslant N, \qquad \| \varphi \| \leqslant (l^2/2) \, N.$$

Substitution into (2.3.31) results in

$$\| K\varphi \| \leqslant CN$$

with the appropriately choosen constants C and N. Consequently, K is bounded. This means that all conditions for completely continuous T, and therefore for the convergence of the Galerkin method, are satisfied. It should be pointed out again that the convergence *is with respect to the norm of H*. This norm is defined as follows for an element φ of H:

$$\| \varphi \|_{(H)} = (A\varphi, \varphi)^{1/2}.$$

Now that the convergence of the Galerkin method has been established, the results of this method will now be examined.

Let us use the trial solution

$$w(x, t) = a_i w_i(x, t) = a_i f(t) \, \varphi_i(x) = f_i(t) \, \varphi_i(x). \tag{2.3.36}$$

A complete and orthonormal function system $\varphi_i(x)$ has been selected, which satisfies all the boundary conditions so that

$$R(w) = R[f_i(t) \, \varphi_i(x)] = 0.$$

As already discussed in Section 1.6.5, the *eigenfunctions $\varphi_i(x)$ of the "abbreviated" problem* are selected. They are obtained from (2.3.28) with the product trial solution $w = e^{\omega t} \varphi(x)$ if all terms that disturb the self-adjoint property are crossed out.

Substitution of (2.3.36) into (2.3.27) results in

$$\int_0^l \{\mu[\ddot{f}_k \varphi_k(x)] + L[f_k(t)\, \varphi_k(x)]\}[f(t)\, \varphi_r(x)\, \delta a_r]\, dx = 0. \qquad (2.3.37)$$

Since the variations δa_r are arbitrary, (2.3.37) becomes the following system of equations:

$$\int_0^l \mu f(t)\, \varphi_i(x)[\ddot{f}_k(t)\, \varphi_k(x)]\, dx + \int_0^l f(t)\, L[f_k(t)\, \varphi_k(x)]\, \varphi_i(x)\, dx = 0$$
$$i = 1, 2, 3,\dots. \qquad (2.3.38)$$

Let us now consider the orthonormal property of the coordinates $\varphi_k(x)$ and let us set $\mu = $ const. It is assumed that $f(t)$ is a function that does not vanish. Equation (2.3.38) then becomes

$$\mu \ddot{f}_i + f_k \int_0^l L(\varphi_k)\, \varphi_i\, dx = 0, \qquad i = 1, 2, 3,\dots. \qquad (2.3.39)$$

Making use of

$$a_{ik} = (1/\mu) \int_0^l L(\varphi_k)\, \varphi_i\, dx,$$

(2.3.39) can be written in the form of the linear, homogeneous system of differential equations

$$\ddot{f}_i + a_{ik} f_k = 0, \qquad i, k = 1, 2, 3,\dots. \qquad (2.3.40)$$

The trial solution $f_i = a_i e^{\omega t}$ is used so that $f(t) = e^{\omega t}$, which justifies the assumptions made above regarding $f(t)$.

From (2.3.40) it can be seen that the Galerkin method makes it possible to treat the stability problem of elastic-body equilibrium the same way as the problem of small oscillations that a discrete system of point masses with an infinite number of degrees of freedom carries out around its original position.

Introducing

$$f_k(t) = a_k e^{\omega t}, \qquad (2.3.41)$$

Eq. (2.3.40) becomes

$$(\mathbf{A} + \omega^2 \mathbf{E})\, \mathbf{X} = 0, \qquad (2.3.42)$$

where $\mathbf{A} = (a_{ik})$, $\mathbf{X} = (a_i)$ and $\mathbf{E}$ is the unit matrix. The stability problem is reduced to a discussion of the algebraic equation (2.3.42) for the components a_i of the vector $\mathbf{X}$ that are the coefficients of the trial solution (2.3.41). Since all further calculations are carried out using algebra and because stability is expressed in terms of algebraic theorems, we are justified in calling this theory the *algebraic stability theory*, which is the title of the following section.

ALGEBRAIC STABILITY THEORY: The following equation is obtained from (2.3.42):

$$\det(\mathbf{A} + \omega^2\mathbf{E}) = 0. \tag{2.3.43}$$

If

$$\omega = \pm(-\lambda_A)^{1/2} \tag{2.3.44}$$

and if λ_A is the eigenvalue of the matrix $\mathbf{A}$, Eq. (2.3.43) is equivalent to the eigenvalue problem

$$\det(\mathbf{A} - \lambda_A\mathbf{E}) = 0. \tag{2.3.45}$$

According to (2.3.36) and (2.3.41) $w(x, t) = a_k e^{\omega t}\varphi_k(x)$, so that the behavior of the disturbed motion as well as stability is decided by the term $e^{\omega t}$. If ω is *imaginary*, there is an oscillation motion that can be called *stable*. Whether or not this is the case depends, according to (2.3.44), on the behavior of the eigenvalues λ_A. In the final analysis, stability is decided by λ_A and the structure of the matrix $\mathbf{A}$. The stability investigation is carried out by tracing the path of λ_A in the complex plane (Fig. 65). The positive real axis is the stability region (only if λ_A is positive and real can ω be imaginary according to (2.3.44), which is sufficient for stability). The path of λ_A results because λ_A is function of a parameter p (for example, the load). If p is varied, the image point of λ_A changes position in the complex plane. Stability is lost when the image point leaves the positive real axis. This can occur if either λ_A passes through the origin or if λ_A becomes complex. The parameter value for which this occurs is the *critical value* p_{crit}, which is the buckling value. We shall restrict ourselves to considering only the *smallest* critical value, which corresponds to the case where one of the eigenvalues λ_A leaves the stability region for the first time.

STRUCTURE INVESTIGATIONS: The behavior of λ_A depends on the *structure of the matrix A*. Two cases must be distinguished: $\mathbf{A}$ is symmetric or Hermitian or $\mathbf{A}$ is asymmetric or non-Hermitian.

Let us consider the first case and assume that $\mathbf{A}$ is *symmetric or Hermitian*. Then $\mathbf{A}$ has only *real eigenvalues* and the image point λ_A must always move along the real axis. As shown in Fig. 65a, stability can only be lost if λ_A passes through the origin and becomes negative. ω then has a positive real part according to (2.3.44), and instability occurs according to Theorem 2 of section 1.3.1. The stability boundary is given by $\lambda_A = 0$.

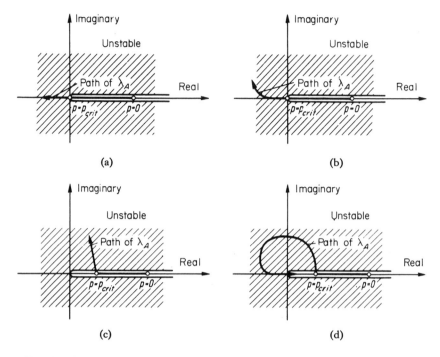

FIG. 65. Path of the smallest eigenvalue in the complex plane. (a) and (b) Behavior of λ_A for a symmetric matrix $\mathbf{A}$, for a matrix $\mathbf{A}$ which can be made symmetric, and for a slightly asymmetric matrix $\mathbf{A}$. (c) and (d) Behavior of λ_A for a strongly asymmetric matrix $\mathbf{A}$.

λ_A is a root of the characteristic equation (2.3.45) and can only be zero if the absolute term of this equation,—that is, det $\mathbf{A}$—vanishes. In the case of a symmetric matrix $\mathbf{A}$, instead of $\lambda_A = 0$ the criterion

$$\det \mathbf{A} = 0 \qquad\qquad (2.3.46)$$

can also be used which is nothing else than the *static stability criterion,* as we shall now see. If a static calculation is carried out under the assumption that a nontrivial equilibrium position exists, then according to the principle of virtual work the following is obtained in place of (2.3.24):

$$\delta(\Pi_i + \Pi_a) - \int_0 \mathbf{k}_a \,\delta\mathbf{w}\, dO - \sum_i \mathbf{K}_{ai}\,\delta\mathbf{w}_i = 0. \qquad (2.3.24a)$$

Instead of (2.3.27), we obtain

$$\int_0^l L(w)\,\delta w\, dx + R(w) = 0. \qquad\qquad (2.3.27a)$$

Instead of using (2.3.36), we now use the trial solution which is independent of time t:

$$w(x) = \rho_i \varphi_i(x), \qquad (2.3.36a)$$

and (2.3.27a) becomes

$$\int_0^l L(\rho_k \varphi_k)(\varphi_r \delta \rho_r)\, dx = 0. \qquad (2.3.37a)$$

From (2.3.37a) it finally follows that

$$a_{ik}\rho_k = 0, \qquad a_{ik} = \int_0^l L(\varphi_k)\, \varphi_i\, dx, \qquad i = 1, 2, 3,..., \qquad (2.3.40a)$$

and a nontrivial equilibrium position exists if the homogeneous system of equations (2.3.40a) results in a nontrivial solution for the coefficients ρ_k. The necessary condition for this is det $\mathbf{A} = 0$. But this is exactly (2.3.46), which shows that this condition is indeed the static stability criterion expressed in an algebraic way, because, if nontrivial equilibrium positions exist, the stability disappears.

Based on these results, we may now attempt to answer the second question of Section 2.3.1, which is concerned with the validity of the static stability criterion. It can at least be used if the matrix $\mathbf{A}$ is symmetric!

It will be shown below that $\mathbf{A}$ is always *symmetric* for *conservative or monogenetic* problems, so that the static criterion is acceptable for this class of problem.

In the conservative or monogenetic case, Eq. (2.3.24a) reduces to the following: because of $\mathbf{k}_a = 0$, $\mathbf{K}_{ai} = 0$

$$\delta(\Pi_i + \Pi_a) = \delta\Pi = 0. \qquad (2.3.24b)$$

According to the discussion in Section 2.3.3,

$$\Pi = \Pi_g + \tfrac{1}{2}\delta^2\Pi_g.$$

Since $\delta\Pi_g = 0$ holds for the original state according to the definition, it follows that

$$\delta\Pi = \tfrac{1}{2}\delta(\delta^2\Pi_g)$$

so that (2.3.24b) becomes $\tfrac{1}{2}\delta(\delta^2\Pi_g) = 0$.

The following relationship can be found from (2.3.26):

$$\tfrac{1}{2}\delta(\delta^2\Pi_g) = \int_0^l L(w)\,\delta w\, dx + R(w),$$

and in addition this must now be made zero. With the trial solution $w = \rho_i\varphi_i(x)$, which satisfies the boundary conditions, we find $R(w) = 0$, and the relationship

$$\tfrac{1}{2}\,\delta(\delta^2 \Pi_g) = \int_0^l L(w)\,\delta w\,dx = 0$$

is obtained. Instead of using (2.3.27a), we may substitute (2.3.36a) into the completely equivalent relationship $\tfrac{1}{2}\,\delta(\delta^2 \Pi_g) = 0$. $\delta^2\Pi_g$ consists only of second-order terms, and, in particular, for the example of a column problem, it consists of *quadratic terms* only. For the elastic column, it has the form

$$\delta^2\Pi_g = \int_0^l \sum_i c_i D_i^2(w)\,dx, \tag{2.3.47}$$

where c_i are coefficients and $D_i(w)$ are differential quotients. Because of (2.3.47) we have

$$\delta(\delta^2\Pi_g) = 2\int_0^l \sum_i c_i D_i(w)\,\delta D_i(w)\,dx = 0,$$

which, according to (2.3.36a), becomes

$$\int_0^l \sum_i (c_i D_i \left[\sum_k \rho_k\varphi_k\right] D_i \left[\sum_k \varphi_k\delta\rho_k\right] dx = 0.$$

Since the variations are independent, the system of equations

$$\sum_k \rho_k \int_0^l \sum_i \{c_i D_i(\varphi_k)\,D_i(\varphi_j)\}\,dx = 0, \qquad j = 1, 2, 3,...,$$

follows. In can be written in the abbreviated form

$$a_{jk}\rho_k = 0, \qquad j = 1, 2, 3,...,$$

where

$$a_{jk} = \int_0^l \sum_i c_i D_i(\varphi_k)\,D_i(\varphi_j)\,dx. \tag{2.3.48}$$

The static stability criterion $\det \mathbf{A} = \det a_{jk} = 0$ makes use of a symmetric matrix $\mathbf{A} = (a_{jk})$, because $a_{jk} = a_{kj}$ follows from (2.3.48). It follows from this discussion that the matrix $\mathbf{A}$ is always symmetric for conservative or monogenetic problems and the static stability criterion can be used.

The *self-adjoint property* of the differential expression $L(\varphi)$ follows from the symmetry of the a_{jk}. This differential expression plays a role in the differential equation of the problem. If a_{jk} (2.3.48) is symmetric, the element a_{ik} (2.3.40a) that is equivalent to it must also be symmetric. This means that

$$\int_0^l L(\varphi_k)\, \varphi_i \, dx = \int_0^l L(\varphi_i)\, \varphi_k \, dx,$$

which means that $L(\varphi)$ is self-adjoint. Therefore, conservative or monogenetic stability problems have self-adjoint differential equations so that the *Ritz method* can be applied to them, as was shown in Section 1.6.5.

Example: Let us use the buckling column with conservative load as an example. We have $P_t = 0$, $q = 0$, $P_v \neq 0$

$$\delta^2 \Pi_g = \tfrac{1}{2} \int_0^l (\alpha w_{xx}^2 - P_v w_x^2) \, dx.$$

$\delta^2 \Pi_g$ has quadratic terms as was postulated. From $\delta(\delta^2 \Pi_g) = 0$ it follows that

$$\int_0^l (\alpha w_{xx} \, \delta w_{xx} - P_v w_x \, \delta w_x) \, dx = 0.$$

Therefore, $c_1 = \alpha$, $c_2 = -P_v$, $D_1 = \partial^2/\partial x^2$, $D_2 = \partial/\partial x$ and the trial solution $w = \rho_i \varphi_i(x)$ results in

$$a_{jk} = \int_0^l (\alpha \varphi_k'' \varphi_j'' - P_v \varphi_k' \varphi_j') \, dx,$$

which is symmetric. The differential equation of the problem is

$$L(w) = \alpha w_{xxxx} + P_v w_{xx} = 0,$$

or, after making use of $w = \rho_i \varphi_i(x)$,

$$L(\varphi_i) = \alpha \varphi_i^{IV} + P_v \varphi_i'' = 0.$$

For the boundary conditions

$$U[\varphi]_R = \begin{cases} \varphi(0) = \varphi(l) = \varphi''(0) = \varphi''(l) = 0, & \text{(column hinged at both ends)} \\ \varphi(0) = \varphi(l) = \varphi'(0) = \varphi'(l) = 0, & \text{(column clamped at both ends)} \\ \varphi(0) = \varphi(l) = \varphi'(0) = \varphi''(l) = 0, & \text{(column clamped at one end and hinged at the other end)} \\ \varphi(0) = \varphi'(0) = \varphi''(l) = 0,\ \varphi'''(l) = -(P_v/\alpha)\,\varphi'(l), & \text{(column clamped at one end and free at the other end)} \end{cases}$$

The differential expression $L(\varphi)$ is self-adjoint, as can easily be established. This is because for two functions u, v, which satisfy these boundary conditions, the relationship

$$\int_0^l L(v)\, u\, dx = \int_0^l L(u)\, v\, dx.$$

always holds.

The discussion above can be summarized as follows.

THEOREM: *For conservative or monogenetic stability problems the matrix* **A** *is symmetric, the differential equation is self-adjoint, and the static stability criterion is applicable.*

It is interesting to note that the static criterion is not restricted to conservative or monogenetic problems. It can remain valid for *nonconservative* problems, as will now be shown.

The nature of a stability problem can be inferred from the load frequency curve (eigenvalue curve) of the buckling column hinged at both ends with a concentrated load having constant direction. The buckling value P_{crit} and the stability boundary is the location where this curve intersects the load axis (P axis). Since $\omega \equiv 0$ for the intersection point, P_{crit} is the load for which a nontrivial equilibrium position is possible. It may therefore be concluded that only when the eigenvalue curves intersect the load axis can nontrivial equilibrium positions exist. The eigenvalue curves of other conservative problems have a similar course.

Example: For the column clamped at one end and for a concentrated load with fixed direction at the free end, if

$$a = \mu/\alpha, \qquad p = P/\alpha,$$

$$\lambda_{1,2} = [\tfrac{1}{2}(4a\omega^2 + p^2)^{1/2} \mp p]^{1/2}$$

(where μ is the specific mass, α the bending stiffness, P the load on the column), the eigenvalue equation is

$$2a\omega^2 - a^{1/2}p\omega \sinh(\lambda_1 l)\sin(\lambda_2 l) + (2a\omega^2 + p^2)\cosh(\lambda_1 l)\cos(\lambda_2 l) = 0.$$

The important branch that results in the buckling load is shown in Fig. 66.

It is possible for nonconservative problems to have completely similar eigenvalue curves. The approximate eigenvalue equation for a

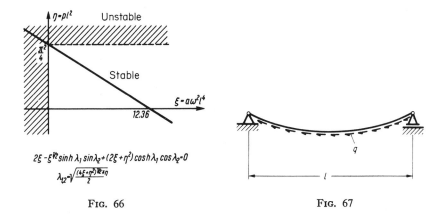

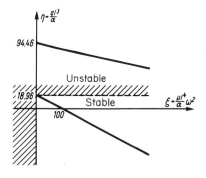

FIG. 66

FIG. 67

FIG. 66. Eigenvalue curve.

FIG. 67. Buckling column with nonconservative load.

beam pinned at both ends with a uniformly distributed tangential load (Fig. 67) is given by

$$q = 56.7 \frac{\alpha}{l^3} - 0.15\mu l\omega^2 \pm \left(1424.6 \frac{\alpha^2}{l^6} + 2.5 \frac{\alpha\mu\omega^2}{l^2} + 0.011\mu^2 l^2 \omega^4\right)^{1/2}.$$

Figure 68 shows two of the branches. The eigenvalue curve intersects the load axis in this case as well. Therefore, nontrivial equilibrium positions exist. Even though the problem is nonconservative, the stability can only disappear, as can be seen from the course of the eigenvalue curve, because of the fact that these intersection points exist and because ω^2 as a function of the load parameter passes through zero. The stability limit is therefore given by $\omega = 0$. According to (2.3.44), it is given by $\lambda_A = 0$, i.e., the static criterion det $\mathbf{A} = 0$ is decisive in this nonconservative case as well.

FIG. 68. Eigenvalue curve.

It can be seen that in such cases the matrix $\mathbf{A}$ is of the special form $\mathbf{A} = \mathbf{HD}$, where $\mathbf{H}$ is real and symmetric or Hermitian and $\mathbf{D}$ is a positive diagonal matrix. Therefore, it belongs to the class of symmetrizable matrices that have only real eigenvalues just like the symmetric matrices. If $\mathbf{A}$ can be made symmetric, it can be represented in the form $\mathbf{A} = \mathbf{T^{-1}HT}$, where $\mathbf{H}$ is real symmetric or Hermitian and $\mathbf{T}$ is non-singular. The eigenvalues $\lambda_A = -\omega^2$ of $\mathbf{A}$ are the roots of the characteristic equation $\det(\mathbf{A} - \lambda_A \mathbf{E}) = 0$. It can be transformed into

$$\det(\mathbf{A} - \lambda_A \mathbf{E}) = \det(\mathbf{T^{-1}HT} - \lambda_A \mathbf{E}) = \det[\mathbf{T^{-1}(H} - \lambda_A \mathbf{E})\,\mathbf{T}]$$
$$= \det \mathbf{T^{-1}} \det \mathbf{T} \det(\mathbf{H} - \lambda_A \mathbf{E}) = 0,$$

or

$$\det(\mathbf{A} - \lambda_A \mathbf{E}) = \det(\mathbf{H} - \lambda_A \mathbf{E}) = 0.$$

From this it is seen that λ_A is also an eigenvalue of the symmetric or Hermitian matrix $\mathbf{H}$ and therefore is always real. This means that the image point of $\lambda_A = -\omega^2$ can only move along the real axis when the load parameter is varied, just as was the case in Fig. 65a for conservative problems. Stability can only be lost when λ_A passes through the origin, and this means that the static criterion is applicable and decisive.

The "Pflüger" buckling column shown in Fig. 67 is an example of *nonconservative problems with a symmetrizable matrix*. Other examples are the beam hinged at both ends with the axial torsion T (Fig. 69) and the beam clamped at both ends with a uniformly distributed tangential longitudinal load q (Fig. 70). Apparently the symmetric boundary

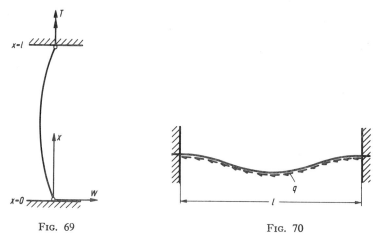

FIG. 69   FIG. 70

FIG. 69.   Column with axial torsion moment.

FIG. 70.   Buckling column with nonconservative load.

conditions have a decisive influence on the structure of the matrix $\mathbf{A}$ and can make it symmetrizable.

The final result can be expressed as follows.

THEOREM: *The static stability criterion is applicable and decisive in the case of nonconservative or polygenetic stability problems if the matrix $\mathbf{A}$ of the problem can be made symmetrical, i.e., if $\mathbf{A}$ is symmetrizable.*

Even if the problem is nonconservative or polygenetic and the matrix cannot be made symmetrical, i.e., there is a *truly asymmetrical* (or non-Hermitian) matrix, the static criterion sometimes can be applicable and decisive.

Example: Let us assume that the structure of the matrix $\mathbf{A}$ is governed by a parameter. The column discussed in Section 1.6.5 with a subtangential compressive load (Fig. 34) is an example of this. The parameter α, which specifies the inclination of the column, determines the structure of the matrix $\mathbf{A}$. The reader is referred to the calculation presented in Section 1.6.5. According to this calculation and to the degree of approximation considered, the eigenvalue problem is given by

$$\det[\mathbf{A} + \omega^2\mathbf{C}] = 0,$$

where

$$\mathbf{A} = \begin{pmatrix} c_1 & d_1 \\ d_2 & c_2 \end{pmatrix}, \qquad \mathbf{C} = \mu \begin{pmatrix} a_1 & b \\ b & a_2 \end{pmatrix},$$

according to (1.6.76). The meaning of the elements a_1, a_2, b, c_1, c_2, d_1, d_2 can be found in Section 1.6.5. Apparently the matrix $\mathbf{C}$ is a function of the parameter α, because it is a function of c_2 and d_1. The matrix $\mathbf{C}$ takes the place of the matrix $\mathbf{E}$ of the special eigenvalue problem $\det[\mathbf{A} + \omega^2\mathbf{E}] = 0$. Since $\mathbf{C}$ is symmetric and always positive definite, the nature of the problem is not changed.

It is assumed that the elements of the matrix $\mathbf{A}$ are whole algebraic and continuous functions of the structure parameter α and the load parameter p, as is the case in the example (in the example $p \equiv P$, where P is the concentrated load applied to the column). Let us assume that $\mathbf{A}$ becomes symmetric for $\alpha = 0$ and is also positive definite for $p = 0$. In other words: The nonconservative problem becomes a conservative one for $\alpha = 0$, and for $p = 0$, it becomes the problem of free oscillations of the elastic body.

It is easily be seen that this is the case for the example mentioned above. Assuming that only positive values of p and the smallest critical load is of interest, we must still determine whether and how long the static criterion remains in effect. For this purpose, let us consider the *stability diagram* plotted in the p, α plane. It is shown in Fig. 35 for example, and Fig. 71 shows a more general case. The boundaries of the instability regions are partly given by the static criterion det $\mathbf{A} = 0$ and partly by the kinetic criterion, which will now be discussed. Which of the criterions is applicable depends on the different α intervals.

It is certain that at least *one* interval must exist, the interval $0 \leqslant \alpha \leqslant \alpha_{g,\min}$, in which the static criterion holds in spite of the non-conservative nature of the problem. This means that the static criterion remains in force for sufficiently small α, even though for $\alpha > 0$ the matrix $\mathbf{A}$ is nonsymmetric. This can be expressed in another way. The static criterion remains in force for a "weakly asymmetric" matrix $\mathbf{A}$!

Proof: $\mathbf{A}$ is symmetric for $\alpha = 0$. Consequently, it only has real eigenvalues λ_A. This property of $\mathbf{A}$ will remain in spite of the asymmetry of $\mathbf{A}$ which occurs for $\alpha > 0$, because of the continuous dependence of the matrix elements on α in the region $0 \leqslant \alpha \leqslant \alpha_{g,\min}$. In this region the stability can only disappear for increasing p if one of the real eigenvalues λ_A changes from positive to negative. The stability boundary is given by $\lambda_A = 0$, which implies that the static criterion holds in this α interval.

Let us summarize these facts as follows.

THEOREM: *In nonconservative or polygenetic stability problems, in which the matrix $\mathbf{A}$ depends on a structure parameter α, the static stability criterion applies and is decisive if $\mathbf{A}$ is slightly asymmetric (or slightly non-Hermitian) for $0 \leqslant \alpha \leqslant \alpha_{g,\min}$. $\alpha_{g,\min}$ is a limiting value which depends on the problem under discussion.*

We have now completely answered the question of the validity of the static stability criterion. The criterion is applicable and decisive for all conservative or monogenetic stability problems, as well as for those nonconservative or polygenetic stability problems having a matrix $\mathbf{A}$ that can be made symmetrical or are slightly asymmetric. Let us give a broader definition of the term *slightly asymmetric* matrix. As Fig. 65b shows, the static criterion remains applicable and decisive if the image point of the eigenvalue λ_A of an asymmetric matrix $\mathbf{A}$ only enters the complex plane for sufficiently large load parameter p, if it has *first* passed through the origin, which is the first and lowest stability boundary. Therefore, matrices for which λ_A has this property can also be called slightly asymmetric, even though their eigenvalues do not always remain real, as is the case for symmetric matrices or matrices which can be made symmetrical.

Let us consider the second case where $\mathbf{A}$ is asymmetric (or non-

Hermitian). Just as in the first case, for stability λ_A must be real and $\lambda_A \geqslant 0$, because then according to (2.3.44) ω is imaginary and the perturbed motion consists of small oscillations around the equilibrium position. Let us assume that **A** is *"strongly asymmetric"* (or "strongly non-Hermitian"), which means that none of the special cases discussed above is being considered such as the case of matrices **A** which can be made symmetrical or slightly asymmetrical matrices. This means that the matrix can have complex eigenvalues for sufficiently large values of the parameter p without any of the eigenvalues first becoming zero. As soon as λ_A becomes complex, one of the ω values must have a positive real part according to (2.3.44), which means that stability is lost. This process can be visualized as follows. The fact that one eigenvalue λ_A becomes complex for p_{crit} means that the image point of λ_A moves from the real positive axis into the complex plane, the region of instability, as Figs. 65c and 65d shows, *before* it has passed through the origin, which is a stability boundary. This first (or only) stability boundary is now the edge of the groove in the complex plane outside of the origin.

Under these conditions it can be said that the equilibrium position of the body at rest has become *kinetically unstable,* because instability occurs owing to induced oscillations. It is not caused by the fact that the body passes through nontrivial equilibrium positions, as would be the case in the classical conservative case. These relationships can be followed by means of the corresponding eigenvalue curve.

Example: Let us consider the *"Beck" buckling column* shown in Fig. 72a. It is clamped at one end, and there is a tangential compression force P_t at the free end. Figure 72b shows the corresponding eigenvalue curve, which is the dependence of P_t on $\omega^2 \equiv -\lambda_A$. The eigenvalue curve never intersects the load axis. No nontrivial equilibrium positions exist, and the static stability criterion cannot be used. Instability occurs because there are no real but only complex values of ω^2 and therefore of λ_A for $P_t > P_{t,\mathrm{crit}}$.

A new set of circumstances has arisen that cannot be treated with the classical criteria, the static and energy criteria. This type of instability can only be discussed in terms of the third criterion, the *kinetic* criterion.

Kinetic instability cannot exist if it is specified that all roots λ_A of the characteristic equation $\det(\mathbf{A} - \lambda_A \mathbf{E}) = 0$ remain real. For this, the matrix

$$\mathbf{B}_0 = \begin{pmatrix} S_0 & S_1 & \cdots & S_{n-1} \\ S_1 & S_2 & \cdots & S_n \\ \cdot \cdot \cdot \cdot \cdot \cdot \cdot \cdot \cdot \cdot \\ S_{n-1} & S_n & \cdots & S_{2n-2} \end{pmatrix}, \qquad (2.3.49)$$

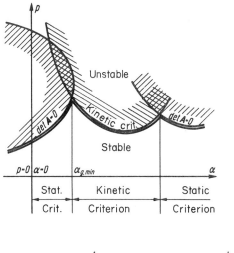

FIG. 71. Stability diagram.

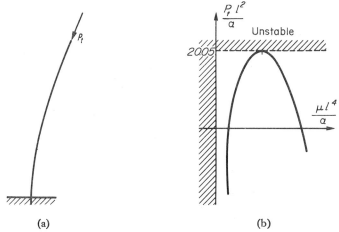

(a)   (b)

FIG. 72. (a) The "Beck" buckling column. (b) Eigenvalue curve of the "Beck" buckling column.

which will be called the "Borchardt matrix,"[1] must be *positive definite.* The elements S_k of this matrix are sums of powers of the roots of the characteristic equation.

The criterion can therefore be formulated as follows.

Kinetic stability criterion: If the stability problem is nonconservative and if its matrix is strongly asymmetric or strongly non-Hermitian, the

---

[1] BORCHARDT, A., *J. Math.* pp. 50–67 (1847).

stability disappears if one of the eigenvalues λ_A becomes complex. This happens when the Borchardt matrix is no longer positive definite. Therefore, the kinetic criterion will be called the "Borchardt" criterion in the future.

It should be pointed out that if the calculation is restricted to matrices **A** of second and third order, which depend on the degree of approximation used, the kinetic stability criterion is greatly simplified. This is because in this case it is sufficient to set the *discriminant* of the characteristic equation equal to zero in order to calculate the stability boundary.

Example: Let us consider the column shown in Fig. 72a. It is a special case of the column shown in Fig. 64 ($P_v = 0$, $q = 0$, $P_t \neq 0$). We must set $Q \equiv P_t$ and

$$L(w) = \alpha\, \partial^4 w/\partial x^4 + P_t\, \partial^2 w/\partial x^2,$$

$$w(0, t) = w_x(0, t) = w_{xx}(l, t) = w_{xxx}(l, t) = 0.$$

In order to form the matrix $\mathbf{A} = (a_{ik})$, let us form

$$a_{ik} = (1/\mu) \int_0^l L(\varphi_k)\, \varphi_i\, dx. \tag{2.3.50}$$

The φ_k are the eigenfunctions of the "abbreviated" problem

$$\mu\omega^2\varphi(x) + \alpha\varphi^{IV}(x) = 0,$$

$$\varphi(0) = \varphi'(0) = \varphi''(l) = \varphi'''(l) = 0, \tag{2.3.51}$$

which is obtained from $\mu\ddot{w} + L(w) = 0$ and the corresponding boundary conditions after $w = e^{\omega t}\varphi(x)$ is substituted. The term containing P_t in $L(w)$ is crossed out. The boundary value problem (2.3.51) is self-adjoint. Two of its eigenfunctions are used, i.e., the orthonormal functions

$$\varphi_k(x) = \kappa_k \left[c_{1k} \left(\cos \frac{m_k x}{l} - \cosh \frac{m_k x}{l} \right) - c_{2k} \left(\sin \frac{m_k x}{l} - \sinh \frac{m_k x}{l} \right) \right], \quad k = 1, 2$$

$$m_1 = 1.875, \qquad m_2 = 4.694,$$

$$c_{11} = 4.148, \quad c_{21} = 3.037, \qquad c_{12} = 53.640, \qquad c_{22} = 54.631,$$

$$\kappa_1 = (17.259l)^{-1/2}, \quad \kappa_2 = (2868.283l)^{-1/2}.$$

The elements of $\mathbf{A}$ are obtained from (2.3.50) as follows:

$$a_{11} = \frac{1}{17.259\mu l} \left(\frac{\alpha}{l^3} \cdot 1.875^4 \cdot 17.259 + \frac{P_t}{l} 1.875^2 \cdot 4.277 \right),$$

$$a_{12} = \frac{P_t}{l} \frac{1.875^2 \cdot 118.707}{(17.259 \cdot 2868.283)^{1/2} \mu l},$$

$$a_{21} = -\frac{P_t}{l} \frac{4.694^2 \cdot 118.664}{(17.259 \cdot 2868.283)^{1/2} \mu l},$$

(2.3.52)

$$a_{22} = \frac{1}{2868.283\mu l} \left(\frac{\alpha}{l^3} \cdot 4.694^4 \cdot 2868.283 - \frac{P_t}{l} \cdot 4.694^2 \cdot 1733.631 \right).$$

The characteristic equation $\det(\mathbf{A} - \lambda_A \mathbf{E}) = 0$ results in

$$\lambda_A{}^2 - (a_{11} + a_{22}) \lambda_A + a_{11}a_{22} - a_{12}a_{21} = 0.$$

The kinetic stability criterion consists of setting the discriminant D of this quadratic equation equal to zero. For $D < 0$, λ_A can take on complex values, and the stability can become questionable. Therefore, the stability boundary follows from $D = 0$. If the expressions (2.3.52) are substituted into

$$D = (a_{11} - a_{22})^2 + 4a_{12}a_{21} = 0,$$

the following equation is obtained for the critical value P_t:

$$P_{t,\text{crit}} = 20.07(\alpha/l^2).$$

The exact value calculated by *Beck* is $P_{t,\text{crit}} = 20.05\alpha/l^2$. The accuracy of our approximate calculation is remarkable considering that we only used two eigenfunctions. We have therefore shown how effective the Galerkin method is.

The realization that nonconservative or polygenetic problems exist for which only the kinetic and not the static stability criterion hold—thus making it necessary to carry out the kinetic calculation—leads us to the third question posed in Section 2.3.1.

INFLUENCE OF THE MASS DISTRIBUTION: When it is permissible to use the static criterion, there are no quantities which contain the mass distribution, because of $\det \mathbf{A} = \det(a_{ik}) = 0$. The mass distribution is of no consequence in this case. This is not the case if it is necessary to carry out the kinetic calculation, as mentioned above. Then the mass distribution is important and must be taken into account. A paper by Pflüger[1] is

---

[1] PFLÜGER, A., The stability of a tangentially loaded compression column. *Z. Angew. Math. Mech.* **35**, 191 (1955).

devoted to the subject that considers a concentrated mass M at the end of the column (Fig. 73). If ml is the column mass, there is a pronounced dependence of the load parameter $p = P_t l^2/\alpha$ on the mass distribution parameters $\mu_1 = M/ml$ or $\mu_2 = ml/M$ shown in Fig. 74, according to this calculation.

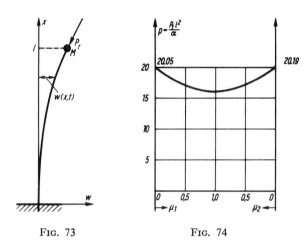

FIG. 73   FIG. 74

FIG. 73.   Influence of mass distribution for nonconservative stability problems.

FIG. 74.   Influence of mass distribution on the magnitude of the "follower" buckling force.

THE INFLUENCE OF DAMPING: Consideration of this problem also touches on the fourth question of justifying the linearization. A detailed answer to these questions will be given in the following.

Theorem 5 of Section 1.5.3, given by Lyapunov, states that it is possible to carry out the stability calculation with linearized equations as long as all exponents of the characteristic equation have negative real parts. If only one characteristic exponent has a vanishing real part, the critical case occurs and the calculation with the linearized equations become questionable.

If we assume that there is no damping and if we use the system (2.3.40), we will use the theory of small oscillations and we will find that the equilibrium position of the elastic body is stable because the roots $\omega = \pm(-\lambda_A)^{1/2}$ of the characteristic equation are all purely imaginary. This is exactly the critical case according to the definition of Lyapunov, so that the stability results obtained from (2.3.40) will appear uncertain at first.

In order to satisfy Theorem 5 of Lyapunov, it would be better to consider damping. Every time the damping would make the roots ω of the characteristic equation have negative real parts, there would be asymptotic stability and the stability investigation could be carried out with the linearized equations.

In nonconservative stability problems it can also happen that the stability becomes worse, in contrast to conservative systems, if damping is taken into account. In the transition from the damped to the undamped system, there can be a jump in the stability condition. H. Ziegler was probably the first to point this out.

All the following reasons make it necessary to investigate the influence of damping: Occurrence of the critical case according to the Lyapunov definition when damping is absent, possible degradation of stability by damping, possible jumps in stability conditions in the transition from the damped to the undamped process.

When damping is considered, the term

$$\int_0^l \beta w_t \, \delta w \, dx$$

must be added to the relationship (2.3.25) for the buckling column, for example. β is the damping coefficient. Instead of (2.3.27), one obtains

$$\int_0^l [\mu \ddot{w} + L^*(w)] \, \delta w \, dx + R(w) = 0, \tag{2.3.53}$$

where

$$L^*(w) = \alpha w_{xxxx} + Q w_{xx} + \beta w_t. \tag{2.3.54}$$

The expression $R(w)$ remains unchanged. On the other hand, (2.3.37) now becomes

$$\int_0^l \{\mu[\ddot{f}_k \varphi_k(x)] + \beta[\dot{f}_k \varphi_k(x)] + L[f_k(t) \, \varphi_k(x)]\}[f(t) \, \varphi_r(x) \, \delta a_r] \, dx = 0,$$

and in place of (2.3.39),

$$\mu \ddot{f}_i + f_k \int_0^l L(\varphi_k) \, \varphi_i \, dx + \beta \dot{f}_i = 0.$$

In place of (2.3.40), we have

$$\ddot{f}_i + a_{ik} f_k + b \dot{f}_i = 0, \qquad i, k = 1, 2, 3,\ldots. \tag{2.3.55}$$

Equation (2.3.55) is equivalent to the equations of motion of a system of discrete point masses with an infinite number of degrees of freedom, which carry out *damped* small oscillations about its equilibrium position.

Until now, β and therefore $b = \beta/\mu$ were assumed to be constant. For purposes of greater generality, we will not exclude the case where the b's can be different, so that instead of (2.3.55) we have

$$\ddot{f}_i + a_{ik}f_k + b_{(i)}\dot{f}_i = 0 \qquad \text{(not to be summed over } i). \qquad (2.3.55a)$$

The following eigenvalue equation is obtained from (2.3.55a) with $f_k = a_k e^{\omega t}$:

$$(\mathbf{A} + \omega\mathbf{B} + \omega^2\mathbf{E})\,\mathbf{X} = 0, \qquad (2.3.56)$$

in which there is the *diagonal* damping matrix $\mathbf{B} = (b_i)$, in contrast to (2.3.42).

Let $\mathbf{A}$ be *symmetric* and $\mathbf{B}$ be *an arbitrary positive diagonal matrix* so that the damping coefficients b_i, which are the elements of $\mathbf{B}$, are different from each other. This is the case of an initially conservative system that is disturbed by purely dissipative forces. To investigate this further, the eigenvalue problem

$$(\mathbf{A} + \omega\mathbf{B} + \omega^2\mathbf{E})\,\mathbf{X} = 0,$$

which contains the arbitrary vector $\mathbf{X}$, is multiplied on the left by the conjugate transposed vector $\mathbf{X}'$. The result is

$$A + \omega^*B + \omega^{*2} = 0,$$

where A, B are numbers that involve values of the matrices $\mathbf{A}$, $\mathbf{B}$. According to our assumption, A is always real, since $\mathbf{A}$ has only real elements because it is a symmetric matrix. B is real and positive, because $\mathbf{B}$ is symmetric and positive definite. The relationship

$$\omega^* = -\tfrac{1}{2}B \pm (\tfrac{1}{4}B^2 - A)^{1/2}$$

specifies the region of values in which the quantity ω must be located so that it corresponds to the original eigenvalue problem.

As long as $A > 0$, any possible ω^* and therefore the ω, which is of interest to us, has a negative real part and the equilibrium of the elastic body is asymptotically stable. This means that the linearized calculation is justified. The stability boundary is given by $A = 0$. Since we have $\lambda_{A,\min} \leqslant A$, $\lambda_{A,\min} = 0$ is sufficient for stability, which means that the static stability criterion is sufficient. This fact, established for symmetric $\mathbf{A}$—that is, for conservative problems—agress with the well-known theorem of Lord Kelvin stating that the stability of a conservative system is not disturbed by dissipative forces and that the originally stable equilibrium of the system even becomes asymptotically stable because of dissipative forces.

Let us assume that the problem is nonconservative but the matrix $\mathbf{A}$ *is symmetrizable or is weakly asymmetrical* and the damping matrix is positive and diagonal. All the elements are assumed to be the same, so that $\mathbf{B} = b\mathbf{E}$, i.e., all the damping coefficients are equal. Then from (2.3.56) we have

$$[\mathbf{A} + (\omega b + \omega^2)\,\mathbf{E}]\,\mathbf{X} = 0,$$

which, because $\omega^2 + \omega b = -\lambda_A$, becomes

$$(\mathbf{A} - \lambda_A\mathbf{E})\,\mathbf{X} = 0.$$

Because of the assumption that $\mathbf{A}$ can be made symmetrical or is weakly asymmetrical, all the λ_A are real or only become complex if the λ_A, which was previously real, first passes through zero.

From the relationship $\omega^2 + \omega b = -\lambda_A$, it follows that

$$\omega = -\tfrac{1}{2}b \pm (\tfrac{1}{4}b^2 - \lambda_A)^{1/2} \tag{2.3.57}$$

will have a negative real part *only* when $\lambda_A > 0$, which means that the equilibrium position of the elastic body is asymptotically stable and the linearized calculation can be used. The stability boundary is given by $\lambda_A = 0$ because of the properties of λ_A. This means that it is also given by det $\mathbf{A} = 0$, which means that the static stability criterion is satisfied.

Let us assume that the stability problem is nonconservative, the matrix $\mathbf{A}$ is asymmetric and that $\mathbf{B} = b\mathbf{E}$, i.e., all *damping coefficients are equal.* The eigenvalue problem can again be written in the form

$$[\mathbf{A} + (\omega b + \omega^2)\,\mathbf{E}]\,\mathbf{X} = 0,$$

which leads to the condition $\det[\mathbf{A} + (\omega b + \omega^2)\,\mathbf{E}] = 0$. If this is compared with the characteristic equation $\det[\mathbf{A} - \lambda_A\mathbf{E}] = 0$ corresponding to $\mathbf{A}$, it can be seen that $\omega b + \omega^2 = -\lambda_A$ so that

$$\omega^2 + \omega b + \lambda_A = 0. \tag{2.3.58}$$

As before, we have

$$\omega = -\tfrac{1}{2}b \pm (\tfrac{1}{4}b^2 - \lambda_A)^{1/2}.$$

As long as the load parameter p remains sufficiently small, the λ_A are real, positive and move to the left along the real positive axis of the λ_A plane (Fig. 75) as p is increased. At a certain value p the image point of

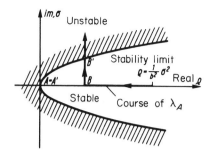

FIG. 75. Stability diagram taking damping into account.

an eigenvalue λ_A passes through the point B into the complex plane and $\lambda_A = \rho + i\sigma$, so that it becomes complex. It follows from

$$\omega = \tfrac{1}{2}b \pm (\tfrac{1}{4}b^2 - \rho - i\sigma)^{1/2} \qquad (2.3.59)$$

that ω has a negative real part if

$$\rho > \sigma^2/b^2, \qquad (2.3.60)$$

which means that the equilibrium of the elastic body is asymptotically stable and the linearization can be used. The stability boundary is given by $\rho = (\sigma/b)^2$. A comparison of Fig. 75 with Figs. 65c and 65d shows that the instability region has become smaller because of damping. As b is increased, i.e., for stronger damping, the stability region becomes inflated and the damping has a favorable effect. The critical case of loss of stability occurs at B' where the image of λ_A crosses the stability boundary, and not at B. In particular it can be seen that the kinetic Borchardt criterion for the calculation without damping, which is represented by point B in Fig. 75, is sufficient for stability. In retrospect it can be seen that in a nonconservative case the corresponding criterion of the undamped case (static or kinetic criterion, according to Borchardt) remains at least sufficient in the case with damping, regardless of whether **A** is symmetrizable or is asymmetric, if the damping matrix satisfies **B** $= b$**E**, i.e., all damping coefficients are equal. The damping cannot neutralize the stability. In addition, damping makes the stability become asymptotic and makes it possible to use the linearized calculation. Therefore, all presently known results regarding the stability of buckling columns for nonconservative loading would remain in effect, even though it was determined without damping, provided the damping is introduced in the same way as in (2.3.54). This means that $b_i = b$ for all i in (2.3.55), i.e., the same damping coefficients throughout. The formulation (2.3.54) contains all the cases discussed in the literature, if special values for $Q(x)$ and special boundary conditions $R(w)$ are assumed.

It is necessary to proceed with caution if **A** is asymmetric and the damping coefficients are different. As H. Ziegler pointed out, it can happen that the damping may have a *neutralizing* effect. The independent work of O. Bottema and V. V. Bolotin for second-order systems has shown that in the nonconservative case and for different damping coefficients the stability condition is discontinuous with respect to the undamped case.

This statement can be generalized as follows.

In the undamped case, the behavior of the roots ω of the characteristic equation $\det(\mathbf{A} + \omega^2 \mathbf{E}) = 0$ must be established. Since $\omega = \pm(-\lambda_A)^{1/2}$ we must have $\lambda_A > 0$ and real for stability. This means that the roots of the algebraic equation $\det(\mathbf{A} - \lambda \mathbf{E}) = 0$ must remain real. A necessary and sufficient condition for this is that the matrix (2.3.49) is positive definite. This is the kinetic *Borchardt criterion for the undamped* case.

On the other hand, for the damping case the behavior of the roots of the characteristic equation $\det(\mathbf{A} + \omega \mathbf{B} + \omega^2 \mathbf{E}) = 0$ must be followed. If **A** is of order n, this leads to an algebraic equation of the type

$$\omega^{2n} + a_1 \omega^{2n-1} + \cdots + a_{2n-1}\omega + a_{2n} = 0, \qquad (2.3.61)$$

where the coefficients a_j ($j = 1, 2,..., 2n$) are all different from zero and are formed in certain ways from the elements a_{ik}, b_i of the matrices **A** and **B**.

Stability occurs when all roots ω have negative real parts. The necessary and sufficient conditions for this are $a_j > 0$ and the Hurwitz matrix must be positive definite. The Hurwitz matrix is not necessarily identical to (2.3.49), which is the Borchardt matrix. Therefore there will be *stability conditions* which are *different* from the undamped case.

In the special case $\mathbf{B} \equiv b\mathbf{E}$, the following discussion will show that for zero damping, the stability conditions which can follow from the Hurwitz criterion coincide exactly with the stability conditions which apply in the undamped case and are based on the Borchardt criterion. It is very fortunate that in this special case there is no jump in the stability conditions in the transition from the damped to the undamped system.

If, on the other hand, **B** is a general positive diagonal matrix, the coefficients of (2.3.61) will change as the damping is decreased. Even though the equation coefficients change, the difference in the b_i will cause a difference between the stability conditions applicable for (2.3.61) that follow from the Hurwitz criterion and the stability conditions for the undamped case that follow from the Borchardt criterion. This is true no matter how small the elements b_i of **B** are. In the limiting case, where there is still very small damping, there will be a jump in the stability conditions, which can be dangerous. This means that for nonconservative systems, for which the diagonal damping matrix has elements that are all different, it is not possible to use the calculation method for the undamped system.

Let us give the method with the following *third-order system:*

$$\begin{vmatrix} \omega^2 + \omega b_1 + a_{11} & a_{12} & a_{13} \\ a_{21} & \omega^2 + \omega b_2 + a_{22} & a_{23} \\ a_{31} & a_{32} & \omega^2 + \omega b_3 + a_{33} \end{vmatrix} = 0.$$

The characteristic equation for the damped system is

$$\omega^6 + a_1\omega^5 + a_2\omega^4 + a_3\omega^3 + a_4\omega^2 + a_5\omega + a_6 = 0, \qquad (2.3.62)$$

where

$$a_1 = b_1 + b_2 + b_3 ,$$

$$a_2 = a_{11} + a_{22} + a_{33} + b_1b_2 + b_2b_3 + b_3b_1 ,$$

$$a_3 = a_{11}(b_2 + b_3) + a_{22}(b_3 + b_1) + a_{33}(b_1 + b_2) + b_1b_2b_3 ,$$

$$a_4 = (a_{11}a_{22} - a_{12}a_{21}) + (a_{22}a_{33} - a_{23}a_{32}) + (a_{33}a_{11} - a_{31}a_{13})$$
$$\qquad + a_{11}b_2b_3 + a_{22}b_3b_1 + a_{33}b_1b_2 , \qquad (2.3.63)$$

$$a_5 = b_1(a_{22}a_{33} - a_{23}a_{32}) + b_2(a_{33}a_{11} - a_{31}a_{13}) + b_3(a_{11}a_{22} - a_{12}a_{21}),$$

$$a_6 = a_{11}(a_{22}a_{33} - a_{23}a_{32}) + a_{12}(a_{23}a_{31} - a_{21}a_{33})$$
$$\qquad + a_{13}(a_{21}a_{32} - a_{31}a_{22}) = \det \mathbf{A}.$$

In the undamped case, the characteristic equation is

$$\omega^6 + a_2{}^*\omega^4 + a_4{}^*\omega^2 + a_6{}^* = 0, \qquad (2.3.64)$$

where

$$a_2{}^* = a_{11} + a_{22} + a_{33} ,$$

$$a_4{}^* = (a_{11}a_{22} - a_{12}a_{21}) + (a_{22}a_{33} - a_{23}a_{32}) + (a_{33}a_{11} - a_{31}a_{13}), \quad (2.3.65)$$

$$a_6{}^* = \det \mathbf{A}.$$

The stability of the *undamped* mechanical system exists if (2.3.64) has only pure imaginary roots ω. For this purpose, it is necessary that the auxiliary equation

$$(\omega^2)^3 + a_2{}^*(\omega^2)^2 + a_4{}^*(\omega^2)^1 + a_6{}^* = 0,$$

considered as a cubic equation, have no complex but only real roots ω^2. This occurs when the discriminant D of this cubic equation is greater than zero. This condition results in the necessary stability condition in the form of the following inequality:

$$D = a_2^{*2}a_4^{*2} - 4a_2^{*3}a_6^* - 4a_4^{*3} + 18a_2^*a_4^*a_6^* - 27a_6^{*2} > 0. \qquad (I)$$

Because of

$$D \equiv -4(a_4^* - \tfrac{1}{3}a_2^{*2})^3 - 27(a_6^* - \tfrac{1}{3}a_2^*a_4^* + (2/27)\,a_2^{*3})^2,$$

the condition $D > 0$ can only be satisfied if

$$a_4^* - \tfrac{1}{3}a_2^{*2} < 0,$$

or

$$a_2^{*2} > 3a_4^* . \qquad (II)$$

This is another condition for stability of the undamped system. In the case *with damping*, let the b_i first all be equal to b. Since we are investigating the behavior for small damping, let us also specify $|b| \ll 1$. This means that the higher powers of b can be ignored, and the following new equations are obtained for the coefficients a_i of the characteristic equations (2.3.62), in place of Eqs. (2.3.63):

$$a_1^{**} = 3b, \; a_2^{**} = a_2^*, \; a_3^{**} = 2ba_2^*, \; a_4^{**} = a_4^*, \; a_5^{**} = ba_4^*, \; a_6^{**} = a_6^*.$$

$$\qquad (2.3.66a)$$

The characteristic equation is

$$\omega^6 + a_1^{**}\omega^5 + a_2^{**}\omega^4 + a_3^{**}\omega^3 + a_4^{**}\omega^2 + a_5^{**}\omega + a_6^{**} = 0.$$

The Hurwitz criterion is used to investigate stability of the corresponding mechanical system. It follows from this that the damped mechanical system is stable if the two inequalities

$$(a_1^{**}a_2^{**} - a_3^{**})[a_5^{**}(a_4^{**}a_3^{**} - a_2^{**}a_5^{**}) + a_6^{**}(2a_1^{**}a_5^{**} - a_3^{**2})]$$

$$+ (a_1^{**}a_4^{**} - a_5^{**})[a_1^{**}a_3^{**}a_6^{**} - a_5^{**}(a_1^{**}a_4^{**} - a_5^{**})] - a_1^{**3}a_6^{**2} > 0$$

and

$$a_3^{**}(a_1^{**}a_2^{**} - a_3^{**}) - a_1^{**}(a_1^{**}a_4^{**} - a_5^{**}) > 0$$

are satisfied. If relationships (2.3.66a) are substituted for a_i^{**} $(i = 1, 2,..., 5)$ in these inequalities, the result is

$$(3ba_2^* - 2ba_2^*)[ba_4^*(a_4^*2ba_2^* - a_2^*ba_4^*) + a_6^*(6b^2a_4^* - 4b^2a_2^{*2})]$$

$$+ (3ba_4^* - ba_4^*)[6b^2a_2^*a_6^* - ba_4^*(3ba_4^* - ba_4^*)] - 27b^3a_6^{*2} > 0$$

and

$$2ba_2^*(3ba_2^* - 2ba_2^*) - 3b(3ba_4^* - ba_4^*) > 0.$$

By cancellation of b^3 in the upper inequality and b^2 in the lower inequality, and after a simple calculation, the new inequalities

$$a_2^{*2}a_4^{*2} + 18a_2^*a_4^*a_6^* - 4a_2^{*3}a_6^* - 4a_4^{*3} - 27a_6^{*2} > 0 \quad \text{and} \quad 2a_2^{*2} - 6a_4^* > 0$$

are obtained. These are identical to the conditions (I) and (II) of the undamped case. It has been shown that the transition between the stability conditions for the undamped and damped case for zero damping occurs without a jump *only if* the damping coefficients $b_i = b$ are the *same*. On the other hand, if the $|b_i|$ differ for the damping case, and if they are again assumed to be small, $|b_i| \ll 1$, in the limit where the damping strives to zero, it is possible to ignore the products and powers of the b_i. If we set

$$b_2 = b_1(1 + e_2), \qquad b_3 = b_1(1 + e_3),$$

where e_2 and e_3 are suitable factors and $b_1 = b$, the following relationships are obtained for the coefficients a_i of (2.3.62) instead of (2.3.63) for these assumptions:

$$\bar{a}_1 = 3b + b(e_2 + e_3) = 3b + b\alpha_1,$$

$$\bar{a}_2 = a_2^*,$$

$$\bar{a}_3 = 2ba_2^* + b[a_{11}(e_2 + e_3) + a_{22}e_3 + a_{33}e_2] = 2ba_2^* + b\alpha_3,$$

$$\bar{a}_4 = a_4^*, \hspace{5cm} (2.3.66b)$$

$$\bar{a}_5 = ba_4^* + b[e_2(a_{33}a_{11} - a_{13}a_{31}) + e_3(a_{11}a_{22} - a_{12}a_{21})] = ba_4^* + b\alpha_5,$$

$$\bar{a}_6 = a_6^*.$$

The meaning of the quantities α_1, α_3, α_5 used as abbreviations is obvious. The characteristic equation is

$$\omega^6 + \bar{a}_1\omega^5 + \bar{a}_2\omega^4 + \bar{a}_3\omega^3 + \bar{a}_4\omega^2 + \bar{a}_5\omega + \bar{a}_6 = 0.$$

The Hurwitz stability criterion must again be used, which leads to the inequalities

$$(\bar{a}_1\bar{a}_2 - \bar{a}_3)[\bar{a}_5(\bar{a}_4\bar{a}_3 - \bar{a}_2\bar{a}_5) + \bar{a}_6(2\bar{a}_1\bar{a}_5 - \bar{a}_3^2)]$$
$$+ (\bar{a}_1\bar{a}_4 - \bar{a}_5)[\bar{a}_1\bar{a}_3\bar{a}_6 - \bar{a}_5(\bar{a}_1\bar{a}_4 - \bar{a}_5)] - \bar{a}_1^3\bar{a}_6^2 > 0$$

and

$$\bar{a}_3(\bar{a}_1\bar{a}_2 - \bar{a}_3) - \bar{a}_1(\bar{a}_1\bar{a}_4 - \bar{a}_5) > 0.$$

If (2.3.66b) is substituted, the result is

$$a_2^{*2}a_4^{*2} + 18a_2^*a_6^*a_4^* - 4a_2^{*3}a_6^* - 4a_4^{*3} - 27a_6^{*2}$$

$$> (\alpha_3 - \alpha_1 a_2^*)\{(a_4^* + \alpha_5)[a_4^*(2a_2^* + \alpha_3) - a_2^*(a_4^* + \alpha_5)]$$

$$+ a_6^*[2(3 + \alpha_1)(a_4^* + \alpha_5) - (2a_2^* + \alpha_3)^2]\} - a_2^*\{a_4^*\alpha_3(a_4^* + \alpha_5)$$

$$- \alpha_5^2 a_2^* + a_6^*[\alpha_5(6 + 2\alpha_1) + 2a_4^*\alpha_1 - \alpha_3(4a_2^* + \alpha_3)]\}$$

$$+ (\alpha_5 - \alpha_1 a_4^*)\{(3 + \alpha_1)(2a_2^* + \alpha_3) a_6^* - (a_4^* + \alpha_5)[(3 + \alpha_1) a_4^*$$

$$- (a_4^* + \alpha_5)]\} - 2a_4^*\{\alpha_1[a_6^*(2a_2^* + \alpha_3) - a_4^{*2}] + 3\alpha_3 a_6^*$$

$$- \alpha_5[a_4^*(1 + \alpha_1) - \alpha_5]\} + (27\alpha_1 + 9\alpha_1^2 + \alpha_1^2) a_6^{*2}$$

and

$$2a_2^{*2} - 6a_4^* > a_2^*[\alpha_3 - \alpha_1(2a_2^* + \alpha_3)] + \alpha_3^2 + \alpha_1 a_4^*(5 + \alpha_1) - \alpha_5(3 + \alpha_1),$$

which differs from inequalities (I) and (II) of the undamped case, because the right hand side of the inequalities are not zero, as was the case there. There is a *jump* in the stability conditions with respect to the undamped case. This jump disappears only when the quantities e_2 and e_3, which measure the difference in the damping coefficients, become zero. This makes the two conditions equal, because for α_1, α_3, α_5 we have

$$\alpha_1 = e_2 + e_3 ,$$

$$\alpha_3 = a_{11}(e_2 + e_3) + a_{22}e_3 + a_{33}e_2 ,$$

$$\alpha_5 = e_2(a_{33}a_{11} - a_{13}a_{31}) + e_3(a_{11}a_{22} - a_{12}a_{21}),$$

and the right side of the inequalities becomes zero for $e_2 = e_3 = 0$, as is easily established.

Let us now generalize our assumptions regarding the structure of the matrix **B** in another direction. The elements of **B** outside of the diagonal are assumed to be *skew symmetric*. This means that *gyroscopic forces* also effect the mechanical system. The restriction is made that the matrix **A** is *symmetric*. Problems of this type occur often in practice when the buckling of *pipes containing flows* is investigated.

The differential equation for a pipe hinged at both ends and having finite length is

$$\alpha w_{xxxx} + S w_{xx} + Q w_{xt} + \mu w_{tt} + \beta w_t + K w = 0,$$

where S and Q are load parameters and K is the modulus of foundation of the elastic foundation of the pipe. The boundary conditions for hinged supports are

$$w(0, t) = w(l, t) = w''(0, t) = w''(l, t) = 0,$$

so that

$$w = f_1 \sin(\pi/l)\, x + f_2 \sin(2\pi/l)\, x, \qquad f_i = a_i e^{\omega t} \qquad (i = 1, 2),$$

is a permissible, two-termed approximate trial solution. This leads to the system

$$\ddot{f}_1 + a_{11} f_1 + b_1 \dot{f}_1 - b_{12} \dot{f}_2 = 0,$$

$$\ddot{f}_2 + a_{22} f_2 + b_{12} \dot{f}_1 + b_2 \dot{f}_2 = 0,$$

(2.3.67)

where

$$b_1 = b_2 = b = \beta/\mu, \qquad b_{12} = 8Q/3\mu l.$$

This is the case mentioned above:

$$\mathbf{A} = \begin{pmatrix} a_{11} & 0 \\ 0 & a_{22} \end{pmatrix}$$

is symmetric, whereas

$$\mathbf{B} = \begin{pmatrix} b_1 & -b_{12} \\ b_{12} & b_2 \end{pmatrix} \equiv \begin{pmatrix} b & -b_{12} \\ b_{12} & b \end{pmatrix}$$

satisfies the relationship $\mathbf{B} = b\mathbf{E} + \mathbf{G}$, where

$$\mathbf{G} = \begin{pmatrix} 0 & -b_{12} \\ b_{12} & 0 \end{pmatrix}$$

is skew symmetric.

After differentiation with respect to time, the system of Eqs. (2.3.67) can be written as follows:

$$(\mathbf{A} + \mathbf{B}\omega + \mathbf{E}\omega^2)\,\mathbf{X} \equiv (\mathbf{A} + b\mathbf{E}\omega + \mathbf{G}\omega + \mathbf{E}\omega^2)\,\mathbf{X} = 0, \qquad (2.3.68)$$

where $\mathbf{A}$ and $\mathbf{E}$ are symmetric, $\mathbf{G}$ is skew symmetric, and $\mathbf{E}$ is positive definite. In this case, Eq. (2.3.68) results in the relationship

$$A + b\omega^* + G\omega^* + \omega^{*2} = 0, \qquad (2.3.69)$$

where the numbers A and G are taken from values in the matrices $\mathbf{A}$ and $\mathbf{G}$. For $\omega^* = -i\sigma$, Eq. (2.3.69) becomes

$$\sigma^2 + \sigma(ib + \tilde{G}) - A = 0,$$

where A and $\tilde{G} = iG$ are real. We have

$$\omega^* = -\tfrac{1}{2}b + \tfrac{1}{2}i\tilde{G} \pm \tfrac{1}{2}[(b^2 - \tilde{G}^2 - 4A) - i2b\tilde{G}]^{1/2}.$$

It is easy to see that for $A < 0$, the quantity ω^* has a positive real part which causes the mechanical system to become unstable. Therefore, $A \geqslant 0$ must be required for stability. Since $\lambda_{A,\min}$ is the lower bound for the values that A can take on, the condition

$$\lambda_{A,\min} \geqslant 0$$

is also sufficient for stability. However, λ_A is the eigenvalue of the symmetric matrix $\mathbf{A}$, which is always real. The condition $\lambda_{A,\min} \geqslant 0$ means that the static stability criterion obtained with the buckling determinant $\mathbf{A}$ is sufficient for the stability of the kinetic system.

SUMMARY: The Galerkin method makes it possible to reduce the treatment of nonconservative or polygenetic stability problems to the discussion of a discrete system with an infinite number of degrees of freedom. The latter can approximately be calculated by a reduction method using a finite number of degrees of freedom (with any degree of accuracy).

The presence of damping and asymptotic stability is required to justify the linearized calculation. This condition leads to the specification of the buckling load.

The static and kinetic criterion can be used to determine the buckling load. In the case *without damping*, the static criterion holds for conservative or monogenetic problems with symmetric matrix $\mathbf{A}$ and for nonconservative or polygenetic problems with matrices $\mathbf{A}$ which can be made symmetrical or which are slightly asymmetric. On the other hand, the kinetic Borchardt criterion holds for nonconservative or polygenetic problems with a strongly asymmetrical matrix $\mathbf{A}$.

When *damping is considered*, the static criterion remains sufficient if $\mathbf{A}$ is symmetric and the damping matrix $\mathbf{B}$ is an arbitrary positive diagonal matrix or a skew-symmetric matrix. It also remains in force if $\mathbf{A}$ is symmetrizable or is slightly asymmetric and $\mathbf{B} = B\mathbf{E}$, $b > 0$.

If $\mathbf{A}$ is strongly asymmetric, the kinetic Borchardt criterion remains sufficient for the damping case, if $\mathbf{B} = b\mathbf{E}$, $b > 0$ again holds. On the other hand, if $\mathbf{B}$ is a positive diagonal matrix with elements *that are different from each other*, there is a jump in the stability condition for asymmetric $\mathbf{A}$ with respect to the corresponding undamped case. Such a nonconservative or polygenetic problem should therefore be calculated with the Hurwitz criterion and the calculation should not be made without damping.

The accompanying table summarizes the results given above.

Damping	Type of matrix	Sufficient stability criterion
Without	**A** symmetrical, symmetrizable, or slightly asymmetric	Static
	A strongly asymmetric	Kinetic (Borchardt)
With	**A** symmetric; **B** positive, diagonal with different elements or skew symmetric	Static
	A symmetrizable or slightly asymmetric; $\mathbf{B} = b\mathbf{E}, b > 0$	Static
	A strongly asymmetric; $\mathbf{B} = b\mathbf{E}, b > 0$	Kinetic (Borchardt)
	A asymmetric; **B** positive diagonal with different elements	Kinetic (Hurwitz)

For **A**, in the class of asymmetric matrices, a distinction was made between the symmetrizable and the true asymmetric matrices (matrices which cannot be made symmetrical). The true asymmetric matrices were divided into slightly asymmetric and strongly asymmetric matrices. The slightly asymmetric matrices are matrices whose eigenvalues always remain real when the load parameter p is varied, even though they are asymmetric, or at least they only become complex *after* the image point has passed through the origin as it moves along the real axis. The strongly asymmetric matrices have eigenvalues that, as the load parameter is varied, already have become complex *before* the image point has passed through the origin as it moves along the real axis.

The above discussion has shown the possible matrix structures and their influence on the type of stability problem. It is now possible to find the appropriate stability criterion and determine the method of calculation. The structure of the matrix **A** follows naturally from the nature of the given problem. On the other hand, it is especially difficult to make realistic assumptions for the damping matrix **B**. It would be necessary to use exact results from experiments. Since there are not yet enough such results available, it is very important that experiments be carried out to obtain the required data. The structure of **B** is discussed in the books by Bolotin (*1.7*, *1.8*) already mentioned. According to this author, it is quite possible for the damping matrix to have elements which are different from each other. It must always be assumed that the internal resistance of the elastic column is proportional to the time variation of the deformation of the longitudinal fibers. Then Eqs. (2.3.53) and (2.3.54) must be supplemented as follows:

$$\int_0^l [\mu w_{tt} + \alpha w_{xxxx} + Q w_{xx} + \beta_1 w_{txxxx} + \beta_2 w_t] \, \delta w \, dx + R(w) = 0,$$

where β_1 is the coefficient of the internal damping and β_2 is the coefficient of the external damping.

Let us use the trial solution $w(x, t) = a_i f(t) \varphi_i(x) = f_i(t) \varphi_i(x)$, where $\varphi_i(x)$ are the eigenfunctions of the free eigenoscillation of the column.

Then

$$R(w) = 0, \varphi_i^{IV} = k_{(i)} \varphi_i , \quad \int_0^l \varphi_k^{IV} \varphi_i \, dx = k_{(i)} \, \delta_{ik} , \quad \int_0^l \varphi_k \varphi_i \, dx = \delta_{ik} ,$$

where $k_{(i)}$ is a constant which contains the square of the eigenfrequencies of the column as a factor. This trial solution then leads to

$$\int_0^l \{\mu(\ddot{f}_k \varphi_k(x)) + [f_k(\alpha \varphi_k^{IV}(x) + Q \varphi_k''(x)]$$
$$+ [\dot{f}_k(\beta_1 \varphi_k^{IV}(x) + \beta_2 \varphi_k(x))]\}(f(t) \varphi_r(x) \, \delta a_r) \, dx = 0.$$

Since the variations δa_r are independent, $f(t) \neq 0$ and due to the special properties of the φ_i, the following system of equations is obtained:

$$\mu \ddot{f}_i + f_i \alpha k_{(i)} + f_k \int_0^l Q \varphi_k'' \varphi_i \, dx + \dot{f}_i (\beta_1 k_{(i)} + \beta_2) = 0, \qquad i, k = 1, 2, 3, \ldots .$$

If

$$a_{ik} = (1/\mu)\left(\alpha k_{(i)} \, \delta_{ik} + \int_0^l Q \varphi_k'' \varphi_i \, dx\right) , \qquad (1/\mu)(\beta_1 k_i + \beta_2) = b_i ,$$

this can also be written in the form $\ddot{f}_i + a_{ik} f_k + b_{(i)} \dot{f}_i = 0, i, k = 1, 2, 3.$ Therefore, $f_k = a_k e^{\omega t}$ leads to the eigenvalue equation

$$(\mathbf{A} + \omega \mathbf{B} + \omega^2 \mathbf{E}) \, \mathbf{X} = 0,$$

where $\mathbf{A} = (a_{ik})$, $\mathbf{X} = (a_i)$ and the $\mathbf{B} = (b_i)$ is now a positive diagonal matrix with elements that are all different!

Because of the assumptions made above regarding the internal damping of the elastic column, it would be necessary to reject all the results of the stability calculation without damping for nonconservative problems because the destabilizing effect of damping must be considered, which is due to the structure of $\mathbf{B}$.

CONVERGENCE DISCUSSION: In order to determine the limits of the algebraic theory discussed above, it is necessary to determine the conditions under which the bounded nature of the approximate solution

$$w_{(n)} = e^{\omega t} \varphi_{(n)}(x), \qquad \varphi_{(n)}(x) = \sum_{i=l}^{n} a_i \varphi_i(x)$$

and its derivative guarantee that the true solution

$$w = e^{\omega t} \varphi(x)$$

and its derivative will be bounded. The Galerkin method converges for the norm of the Hilbert space employed. For sufficiently large n we have

$$\| w - w_{(n)} \| = \epsilon^{1/2},$$

where ϵ is arbitrarily small. Let us consider the *column problem* as an example. We have

$$\| w - w_{(n)} \|_{(H)}^2 = \int_0^l \{ A(w - w_{(n)}), \quad (w - w_{(n)}) \} \, dx = \epsilon,$$

with $A = d^4/dx^4$. Therefore

$$\int_0^l (w^{IV} - w_{(n)}^{IV})(w - w_{(n)}) \, dx$$
$$= \int_0^l (w^{IV} w - w^{IV} w_{(n)} - w_{(n)}^{IV} w + w_{(n)}^{IV} w_{(n)}) \, dx = \epsilon.$$

For two functions u and v, which satisfy the possible boundary conditions of the column, integration by parts yields

$$\int_0^l u^{IV} v \, dx = \int_0^l u'' v'' \, dx.$$

w as well as $w_{(n)}$ satisfy the boundary conditions and therefore satisfy this condition. The result is

$$\int_0^l (w^{IV} - w_{(n)}^{IV})(w - w_{(n)}) \, dx = \int_0^l (w'' - w_{(n)}'')^2 \, dx = \epsilon.$$

Since for the column problem we always have $w(0, t) = w_{(n)}(0, t) = 0$ because of the boundary conditions, we have

$$w - w_{(n)} = \int_0^x (w' - w_{(n)}') \, dx.$$

The Cauchy–Schwarz inequality results in

$$(w - w_{(n)})^2 \leqslant x \int_0^x (w' - w'_{(n)})^2 \, dx \leqslant x \int_0^l (w' - w'_{(n)})^2 \, dx \leqslant l \int_0^l (w' - w'_{(n)})^2 \, dx.$$

Furthermore

$$w'(x, t) - w'(0, t) - w'_{(n)}(x, t) + w'_{(n)}(0, t) = \int_0^x (w'' - w''_{(n)}) \, dx.$$

Then, according to the Cauchy–Schwarz inequality

$$\{[w'(x, t) - w'_{(n)}(x, t)] - [w'(0, t) - w'_{(n)}(0, t)]\}^2 \leqslant x \int_0^l (w'' - w''_{(n)})^2 \, dx$$

and therefore

$$\int_0^l \{[w'(x, t) - w'_{(n)}(x, t)] - [w'(0, t) - w'_{(n)}(0, t)]\}^2 \, dx$$

$$\leqslant (l^2/2) \int_0^l (w'' - w''_{(n)})^2 \, dx = l^2 \epsilon / 2.$$

We have

$$\int_0^l \{[w'(x, t) - w'_{(n)}(x, t)] - [w'(0, t) - w'_{(n)}(0, t)]\}^2 \, dx$$

$$= \int_0^l [w'(x, t) - w'_{(n)}(x, t)]^2 \, dx - 2[w'(0, t) - w'_{(n)}(0, t)] \int_0^l [w'(x, t)$$

$$- w'_{(n)}(x, t)] \, dx + [w'(0, t) - w'_{(n)}(0, t)]^2 \, l$$

or, after carrying out the integration in the second term on the right side and with $w(0, t) = w_{(n)}(0, t) = 0$, we obtain

$$\int_0^l \{[w'(x, t) - w'_{(n)}(x, t)] - [w'(0, t) - w'_{(n)}(0, t)]\}^2 \, dx$$

$$= \int_0^l [w'(x, t) - w'_{(n)}(x, t)]^2 \, dx - 2[w'(0, t) - w'_{(n)}(0, t)][w(l, t)$$

$$- w_{(n)}(l, t)] + [w'(0, t) - w'_{(n)}(0, t)]^2 \, l.$$

The possible boundary conditions of the column are either

$$w'(0, t) = w'_{(n)}(0, t) = 0 \quad \text{or} \quad w(l, t) = w_{(n)}(l, t) = 0,$$

so that

$$\int_0^l \{[w'(x, t) - w'_{(n)}(x, t)] - [w'(0, t) - w'_{(n)}(0, t)]\}^2 \, dx$$

$$= \int_0^l [w'(x, t) - w'_{(n)}(x, t)]^2 \, dx + [w'(0, t) - w'_{(n)}(0, t)]^2 \, l \leqslant l^2 \epsilon/2.$$

This results in

$$\int_0^l [w'(x, t) - w'_{(n)}(x, t)]^2 \, dx \leqslant l^2 \epsilon/2,$$

which finally leads to

$$(w - w_{(n)})^2 \leqslant l \int_0^l (w' - w'_{(n)})^2 \, dx \leqslant l^3 \epsilon/2.$$

Therefore

$$|w - w_{(n)}| \leqslant (l^3 \epsilon/2)^{1/2}$$

holds. *For elastic columns, convergence with respect to the norm of the Hilbert space means the same as convergence with respect to the magnitude.* This also means that the bounded nature of the approximate solution $w_{(n)}$ calculated for the stability case also implies that the true solution w will be bounded. Therefore, in the case of stability problems of elastic columns, the requirement of the Lyapunov stability definition, which requires w to be bounded everywhere, is satisfied.

Since we have based our calculation method on linear elasticity theory, which assumes that the strains are also small, it is necessary to show that not only w but also its *derivative* remain sufficiently bounded.

For this, we may use the relationship

$$\{[w'(x, t) - w'_{(n)}(x, t)] - [w'(0, t) - w'_{(n)}(0, t)]\}^2 \leqslant x \int_0^l (w'' - w''_{(n)})^2 \, dx.$$

It follows that

$$\{[w'(x, t) - w'_{(n)}(x, t)] - [w'(0, t) - w'_{(n)}(0, t)]\}^2$$

$$\leqslant l \int_0^l (w'' - w''_{(n)})^2 \, dx \leqslant l\epsilon.$$

If the lower end of the column is damped,

$$w'(0, t) = w'_{(u)}(0, t) = 0.$$

Hence

$$[w'(x, t) - w'_{(u)}(x, t)]^2 \leqslant l\epsilon,$$

$$[w'(x, t) - w'_{(u)}(x, t)] \leqslant \sqrt{l\epsilon},$$

respectively, i. e., w' converges with respect to the magnitude, like w.

If the ends of the column are hinged, we have

$$\int_0^l w'(x, t)\, dx = \int_0^l w'_{(n)}(x, t)\, dx = 0,$$

because the boundary conditions

$$w(l, t) = w_{(n)}(l, t) = w(0, t) = w_{(n)}(0, t) = 0$$

are satisfied. We may also set

$$\int_0^l [w'(x, t) - w'_{(n)}(x, t)]\, dx = 0.$$

The integrand is continuous, and therefore—according to the mean-value theorem—$w'(\xi, t) - w'_{(n)}(\xi, t) = 0$ must hold for a certain $0 < x = \xi < l$. The above inequality holds for any $x = \xi$, which means that

$$[w'(0, t) - w'_{(n)}(0, t)]^2 \leqslant l\epsilon, \qquad |\, w'(0, t) - w'_{(n)}(0, t)| \leqslant (l\epsilon)^{1/2}.$$

This means that

$$|\, w'(x, t) - w'_{(n)}(x, t)| \leqslant 2(l\epsilon)^{1/2},$$

and this again means that the absolute value of w' is convergent. The smallness of $w'_{(n)}$, which is assumed here, therefore implies that w' is small. We are therefore justified in the application of linear elasticity theory.

For *multidimensional* elastic bodies, such as plates, it is *not* directly possible to make estimates. For them, the convergence of the Galerkin method with respect to the norm of H only makes it possible to conclude that w is bounded with respect to the quadratic mean. Therefore, for multidimensional bodies, the kinetic criterion, together with the algebraic method, is only conditionally valid, as was the energy criterion. In this case it may be necessary to introduce a new stability definition, as we did previously. This new condition would require that the displacements and their derivatives are bounded only in the mean. Under certain conditions, this may raise questions regarding the validity of the linear elasticity theory.

2.3.5. *Examples for the Calculation of Elastostatic Stability Problems*

Let us first consider *conservative or mongenetic* problems. We already know that the *static stability criterion* holds for them. It can be applied in many ways, which will now be discussed.

One way is to assume that a *nontrivial equilibrium position* exists, which is justified for conservative problems, and then to apply the intersection method to the deflected position of the elastic body. The equilibrium conditions are then written down for the parts that have been cut out. This leads to the differential equation for the problem. The corresponding boundary conditions lead to the formulation of a boundary value problem. Since the equations contain the load parameter, which is decisive for the existence of the solution and of a nontrivial equilibrium position, the problem becomes an *eigenvalue problem*. The buckling load, which is the critical value of the load parameter, is the eigenvalue of the problem. This is the classical stability calculation used in elastostatics, which has been shown to have restricted validity according to new research. It can only be applied for conservative problems and some non-conservative problems.

Example 1: Let us consider the *buckling column* shown in Fig. 76. It is loaded by a unidirectional, vertical, concentrated load P. A column element having length dx is cut out. Let us formulate the equilibrium conditions for it, taking into account that $\cos \varphi \approx 1$, $\sin \varphi \approx \varphi$.

The equilibrium condition for all forces in the direction of the column axis results in $N = P$, $dN = 0$. Equilibrium of all forces perpendicular to the column axis results in $Q + dQ = P(\varphi + d\varphi)$, $Q = P\varphi$ or $dQ = P\,d\varphi$. The equilibrium of all moments results in $dM = Q\,dx$.

In addition, $\varphi = dw/dx$, $d\varphi = (d^2w/dx^2)\,dx$.

Since $M = -\alpha w''$, the relationship $dM/dx = Q$, $d^2M/dx^2 = dQ/dx$ results in

$$d^2M/dx^2 = dQ/dx = -\alpha\,d^4w/dx^4.$$

On the other hand,

$$dQ/dx = P\,d\varphi/dx = P\,d^2w/dx^2,$$

so that the differential equation of the problem is

$$\alpha\,d^4w/dx^4 + P\,d^2w/dx^2 = 0. \tag{2.3.70}$$

It has been assumed that the bending stiffness α of the column is constant.

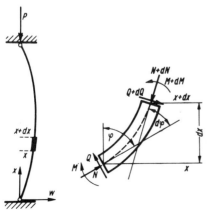

FIG. 76.   Euler buckling column.

The boundary value problem is completely formulated, once the corresponding conditions have been specified. They are

$$w(0) = w''(0) = w(l) = w''(l) = 0$$

for the column hinged at both ends shown in Fig. 76. For the other possible support possibilities, the boundary conditions are

$$w(0) = w(l) = w'(0) = w'(l) = 0 \qquad \text{for the column clamped at both ends}$$

$$w(0) = w(l) = w'(0) = w''(l) = 0 \qquad \text{for the column clamped at the bottom and hinged at the top,}$$

$$w(0) = w'(0) = w''(l) = 0, \; w'''(l) = -(P/\alpha)\, w'(l) \qquad \text{for the column clamped at the bottom free at the top.}$$

The general solution of Eq. (2.3.70) is

$$w = C_1 f_1(p, x) + C_2 f_2(p, x) + C_3 x + C_4 , \qquad (2.3.71)$$

where $p = P/\alpha$ is the load parameter and is an *eigenvalue* of the problem. The functions

$$f_1(p, x) = \cos p^{1/2} x, \qquad f_2(p, x) = \sin p^{1/2} x$$

have been introduced.

When (2.3.71) is substituted into a set of boundary conditions, we obtain

$$a_{ik} C_k = 0, \qquad i, k = 1, 2, 3, 4,$$

which is an algebraic system for the determination of the integration constants C_k. Since it is homogeneous, it will only have nontrivial solutions for the C_k and consequently there will only be a nontrivial equilibrium position for the column if $\det a_{ik} = 0$. The determinant $\det a_{ik}$ is called the *buckling determinant*. The condition

$$\det a_{ik} = 0 \qquad (2.3.72)$$

is the equation for determining the eigenvalue p_{crit}. It therefore determines the buckling load (Euler load) $P_E = \alpha p_{\text{crit}}$.

For example, let us consider the column clamped at the bottom and hinged at the top. The corresponding boundary conditions result in

$$w(0) = C_1 + C_4 = 0,$$

$$w(l) = C_1 \cos p^{1/2}l + C_2 \sin p^{1/2}l + C_3 l + C_4 = 0,$$

$$w'(0) = C_2 p^{1/2} + C_3 = 0,$$

$$w''(l) = -C_1 p \cos p^{1/2}l - C_2 p \sin p^{1/2}l = 0,$$

and Eq. (2.3.72) is given by

$$\begin{vmatrix} 1 & 0 & 0 & 1 \\ \cos p^{1/2}l & \sin p^{1/2}l & l & 1 \\ 0 & p^{1/2} & 1 & 0 \\ -p \cos p^{1/2}l & -p \sin p^{1/2}l & 0 & 0 \end{vmatrix} = 0,$$

which results in the following equation for determining the eigenvalue p:

$$\tan(p^{1/2}l) - p^{1/2}l = 0.$$

The smallest value that satisfies this transcendental equation is $p^{1/2}l = 4.493$. The smallest buckling load is

$$P_E = \alpha p_{\text{crit}} = 4.493^2 \, \alpha/l^2.$$

The method of applying the equilibrium method used here assumes that the differential equation of the problem can be solved exactly. If this is not possible, which is often the case, it is necessary to use the "algebraic" method in the stability calculation, i.e., the Galerkin method. Since we have assumed a conservative problem, the Galerkin method could also be used in its special form, the *Ritz method*.

Example 2: Let us consider the buckling column that carries uniformly distributed *longitudinal forces* which have a fixed direction q (for example, column weight). This case can be exactly solved by means of Bessel functions. The advantage of this example is that the accuracy of the algebraic method can be compared to the exact result.

Let us first determine the differential equation and the boundary conditions. This can either be done by consideration of the deflected column and the intersection principle, or we can employ the principle of virtual work. In this case, it has the simple form $\delta\Pi = 0$. We have

$$\Pi = \tfrac{1}{2} \int_0^l \left[\alpha w''^2 - q(l - x) \, w'^2 \right] dx.$$

Therefore

$$\delta\Pi = \int_0^l [\alpha w'' \, \delta w'' - q(l - x) \, w' \, \delta w'] \, dx = 0. \qquad (2.3.73)$$

Partial integration results in

$$\int_0^l [\alpha w^{\mathrm{IV}} + q(l - x) \, w'' - qw'] \, \delta w \, dx + [\alpha w'' \, \delta w'$$
$$- \alpha w''' \, \delta w - q(l - x) \, w' \, \delta w]_0^l = 0.$$

The condition that the boundary value expression

$$R(w) = [\alpha w'' \, \delta w' - \alpha w''' \, \delta w - q(l - x) \, w' \, \delta w]_0^l$$

be zero and the fact that the geometric boundary conditions

$$w(0) = w(l) = 0$$

hold for the column pinned at both ends, leads to the dynamic boundary conditions

$$w''(0) = w''(l) = 0,$$

and the differential equation

$$L(w) = \alpha w^{\mathrm{IV}} + q(l - x) \, w'' - qw' = 0.$$

This follows from the relationship (2.3.73) obtained by partial integration.

Let us set $w = \rho_1 \varphi_1(x) + \rho_2 \varphi_2(x)$ and $\varphi_1 = \sin(\pi/l) \, x$, $\varphi_2 = \sin 2(\pi/l) \, x$. These are coordinate functions that satisfy all the boundary conditions. According to (2.3.50), the elements of the matrix $\mathbf{A}$ are

$$a_{ik} = \frac{1}{\mu} \int_0^l L(\varphi_k) \, \varphi_i \, dx.$$

In this case

$$a_{11} = \frac{1}{\mu} \int_0^l \left[\alpha \frac{\pi^4}{l^4} \sin \frac{\pi}{l} x - q(l - x) \frac{\pi^2}{l^2} \sin \frac{\pi}{l} x - \frac{\pi}{l} q \cos \frac{\pi}{l} x \right] \sin \frac{\pi}{l} x \, dx$$

$$= \frac{1}{\mu} \left(\frac{\alpha \pi^4}{2l^3} - q \frac{\pi^2}{4} \right),$$

$$a_{12} = \frac{1}{\mu} \int_0^l \left[\alpha \frac{\pi^4}{l^4} \sin \frac{\pi}{l} x - q(l-x) \frac{\pi^2}{l^2} \sin \frac{\pi}{l} x - \frac{\pi}{l} q \cos \frac{\pi}{l} x \right] \sin \frac{2\pi}{l} x \, dx$$

$$= \frac{1}{\mu} \left(-\frac{8}{9} q - \frac{4}{3} q \right) = -\frac{1}{\mu} \frac{20}{9} q,$$

$$a_{21} = \frac{1}{\mu} \int_0^l \left[\alpha \frac{16\pi^4}{l^4} \sin \frac{2\pi}{l} x - q(l-x) \frac{4\pi^2}{l^2} \sin \frac{2\pi}{l} x \right.$$

$$\left. - \frac{2\pi}{l} q \cos \frac{2\pi}{l} x \right] \sin \frac{\pi}{l} x \, dx = \frac{1}{\mu} \left(-\frac{32}{9} q + \frac{4}{3} q \right) = -\frac{1}{\mu} \frac{20}{9} q,$$

$$a_{22} = \frac{1}{\mu} \int_0^l \left[\alpha \frac{16\pi^4}{\mu} \sin \frac{2\pi}{l} x - q(l-x) \frac{4\pi^2}{l^2} \sin \frac{2\pi}{l} x \right.$$

$$\left. - \frac{2\pi}{l} q \cos \frac{2\pi}{l} x \right] \sin \frac{2\pi}{l} x \, dx = \frac{1}{\mu} \left(\frac{8\alpha\pi^4}{l^3} - q\pi^2 \right).$$

The static criterion requires that

$$\det(a_{ik}) = \frac{1}{\mu} \begin{vmatrix} \dfrac{\alpha\pi^4}{2l^3} - \dfrac{\pi^2}{4} q & -\dfrac{20}{9} q \\[2ex] -\dfrac{20}{9} q & \dfrac{8\alpha\pi^4}{l^3} - q\pi^2 \end{vmatrix} = 0.$$

This leads to the quardatic equation $p^2 - 123.8p + 1955 = 0$ in which $p = ql^3/\alpha$ is the load parameter. The smallest critical value is the root of this equation, which is $p_{\text{crit}} = 18.58$. Therefore, the smallest buckling load is

$$q_E = 18.58 \, \alpha/l^3.$$

This value is very accurate compared to the exact value first calculated by F. Willers, which is $q_E = 18.57 l^3/\alpha$.

If the boundary conditions are known some other way, so that the coordinate functions can be immediately selected, it is possible to directly use (2.3.73). The elements of the matrix **A** are then obtained from (2.3.48) as follows:

$$a_{jk} = \int_0^l \sum_i c_i D_i(\varphi_k) \, D_i(\varphi_j) \, dx,$$

where

$$i = 1.2, \quad c_1 = \alpha, \quad c_2 = -q(l-x), \quad D_1 = d^2/dx^2, \quad D_2 = d/dx.$$

Therefore, for

$$\varphi_1 = \sin(\pi/l) x, \quad \varphi_2 = \sin(2\pi/l) x,$$

we have

$$a_{11} = \int_0^l \left[\alpha \frac{\pi^4}{l^4} \sin^2 \frac{\pi}{l} x - q(l-x) \frac{\pi^2}{l^2} \cos^2 \frac{\pi}{l} x \right] dx$$

$$= -q \frac{\pi^2}{4} + \alpha \frac{\pi^4}{l^4} \frac{l}{2},$$

$$a_{12} = a_{21} = \int_0^l \left[\alpha \frac{4\pi^4}{l^4} \sin \frac{\pi}{l} x \sin \frac{2\pi}{l} x \right.$$

$$\left. - q(l-x) \frac{2\pi^2}{l^2} \cos \frac{\pi}{l} x \cos \frac{2\pi}{l} x \right] dx = -\frac{20}{9} q,$$

$$a_{22} = \int_0^l \left[\alpha \frac{16\pi^4}{l^4} \sin^2 \frac{2\pi}{l} x - q(l-x) 4 \frac{\pi^2}{l^2} \cos^2 \frac{2\pi}{l} x \right] dx$$

$$= -q\pi^2 + \alpha \frac{\pi^4}{l^4} 8l,$$

so that the static criterion again is given by

$$\det(a_{jk}) = \begin{vmatrix} -q \dfrac{\pi^2}{4} + \alpha \dfrac{\pi^4}{2l^3} & -\dfrac{20}{9} q \\[2ex] -\dfrac{20}{9} q & -q\pi^2 + \alpha \dfrac{8\pi^4}{l^3} \end{vmatrix} = 0.$$

For $p = ql^3/\alpha$, the same quadratic equation

$$p^2 - 123.8p + 1955 = 0$$

is again obtained so that the smallest root is $p_{\text{crit}} = 18.58$ and the same buckling load $q_E = 18.58\alpha/l^3$ is obtained.

The third method is to use the *energy criterion*. As was shown in Sections 2.3.1 and 2.3.3, caution must be used when this criterion is applied to multidimensional problems. In spite of this, we will calculate with the energy method in the following example, in order to give an impression of the method normally used to calculate the stability boundary. The theoretical bases for multidimensional problems will be the subject of further research. This is especially necessary for shells, where there are discrepancies between theoretical predictions and experimental results. In any case, the condition $\delta(\delta^2 \Pi_g) > 0$ of the energy criterion, which is equivalent to saying that $\delta^2 \Pi_g$ is positive definite for stability, or has a minimum, is at least necessary. Instability occurs if it is violated.

Example 3: Let us again consider the *plate* shown in Fig. 63. It is loaded by the edge forces N in the x direction. It is assumed that the plate is thin and that the Kirchhoff plate theory holds. The potential of the internal and external forces for the deflected plate, which is in a perturbed state with respect to the trivial equilibrium position, is given by

$$\Pi = \tfrac{1}{2}\,\delta^2\Pi_g$$
$$= (D/2) \int_F [w_{xx}^2 + w_{yy}^2 + 2\nu w_{xx}w_{yy} + 2(1 - \nu)\,w_{xy}^2 - (N/D)\,w_x{}^2]\,dF.$$

$$(2.3.74)$$

This also implies that the potential Π of the deflected plate is equal to one-half of the second variation of the potential Π_g of the plate in the original state.

The stability is given by $\delta\Pi \equiv \delta(\tfrac{1}{2}\,\delta^2\Pi_g) = 0$, because it is exactly when this occurs that $\delta^2\Pi_g$ is no longer definite and the stability is lost. It is not impossible for Lyapunov stability to be lost before this, because the energy criterion is only necessary and not sufficient for the Lyapunov stability theory.

The condition $\delta\Pi = 0$ is identical to the condition that (2.3.74) has an extremum. The *Ritz method* will be used to find this extreme value. This is permissible in conservative problems like that with which we are dealing.

The trial solution

$$w(x, y) = \sum_m \sum_n \rho_{mn} \sin\frac{m\pi x}{a} \sin\frac{n\pi y}{b} \qquad (2.3.75)$$

can be used for the freely supported plate. It satisfies all the geometric boundary conditions, because w is zero for $x = 0$, $x = a$, $y = 0$, $y = b$, at the edges of the plate.

The expression (2.3.74) can be rewritten as

$$\tfrac{1}{2}\,\delta^2\Pi_g = D/2 \int_F [(w_{xx} + w_{yy})^2 - 2(1 - \nu)(w_{xx}w_{yy} - w_{xy}^2) - (N/D)\,w_x{}^2]\,dF.$$

$$(2.3.76)$$

When (2.3.75) is substituted into (2.3.76), it can be seen that the term multiplied by $2(1 - \nu)$ is zero. The remainder of this expression is

$$\frac{1}{2}\delta^2\Pi_g = \frac{D}{2} \int \left[\sum_m \sum_n \rho_{mn} \left(\frac{m^2\pi^2}{a^2} + \frac{n^2\pi^2}{b^2}\right) \sin\frac{m\pi x}{a} \sin\frac{n\pi y}{b} \right]^2 dx\,dy$$
$$- \frac{N}{2} \int \left[\sum_m \sum_n \rho_{mn}^2 \frac{m^2\pi^2}{a^2} \cos^2\frac{m\pi x}{a} \sin^2\frac{n\pi y}{b} \right] dx\,dy.$$

After the integrations have been carried out, we obtain

$$\frac{1}{2}\delta^2\Pi_g = \frac{ab}{8}D\sum_m\sum_n\rho_{mn}^2\left(\frac{m^2\pi^2}{a^2}+\frac{n^2\pi^2}{b^2}\right)^2 - \frac{ab}{8}N\sum_m\sum_n\rho_{mn}^2\frac{m^2\pi^2}{a^2}.$$

(2.3.77)

The conditions that require that $\delta^2\Pi_g$ takes on an extreme value are

$$(\partial/\partial\rho_{mn})(\delta^2\Pi_g) = 0.$$

Applied to (2.3.77), this results in

$$\frac{ab}{4}D\rho_{mn}\left(\frac{m^2\pi^2}{a^2}+\frac{n^2\pi^2}{b^2}\right)^2 - \frac{ab}{4}N\rho_{mn}\frac{m^2\pi^2}{a^2} = 0.$$

Since $\rho_{mn}\neq 0$, this is satisfied for

$$N_{\text{crit}} = D\frac{\pi^2a^2}{m^2}\left(\frac{m^2}{a^2}+\frac{n^2}{b^2}\right)^2 = \frac{\pi^2D}{b^2}\left(\frac{mb}{a}+\frac{n^2a}{mb}\right)^2. \qquad (2.3.78)$$

The smallest critical value is obtained for $n = 1$ and the desired buckling load is

$$N_{\text{crit,min}} = \frac{k\pi^2D}{b^2}, \qquad k = \left(\frac{mb}{a}+\frac{a}{mb}\right)^2 = (\xi + \xi^{-1})^2, \qquad \xi = \frac{mb}{a}.$$

From this it can be seen that the plate can have several half-wave dents in the load direction, and their number depends on m. There is only *one* half-wave perpendicular to the load direction (because of $n = 1$).

The smallest buckling load is obtained for the smallest possible value of k. It is obtained from

$$dk/d\xi = 2(\xi + \xi^{-1})(1 - \xi^{-2}) = 0,$$

where $\xi = 1$. It is $(N_{\text{crit,min}})_{\text{min}} = 4\pi^2D/b^2$.

For fixed values of m, k is a function of the side ratio a/b of the plate. Figure 77 shows the dependence of k for different values of m. The diagram is used in the following way. For example, let $a/b = 2.7$. The ordinate along the lower boundary curve (solid line) is found which corresponds to this value. The values $m = 3$ and $k = 4.04$ are read off. This means that the plate will deform in the load direction in the form of three half-waves, and the buckling load is $N_{\text{crit,min}} = 4.04\pi^2D/b^2$.

The result (2.3.78) is also obtained if another formulation of the energy criterion is used, stating that $\delta^2\Pi_g$ is positive definite. This can only be the case if each term of the double series (2.3.77) is itself positive.

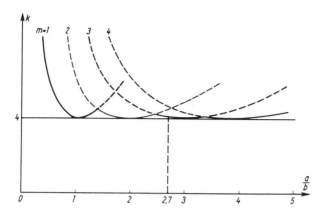

FIG. 77. Denting diagram of a rectangular plate.

Instability will certainly occur if each term of the series is zero. This occurs for

$$N_{\text{crit}} = D \frac{\pi^2 a^2}{m^2} \left(\frac{m^2}{a^2} + \frac{n^2}{b^2} \right)^2.$$

This is again the relationship (2.3.78), and everything else proceeds as before.

Let us now consider nonconservative or polygenetic problems. Let us first consider the case where there is a *strongly asymmetric* matrix, which means that the *kinetic* criterion applies.

Example 4: Let us consider the column clamped at one end with a uniformly distributed, *tangential load* (Fig. 78a). Equation (2.3.27) holds,

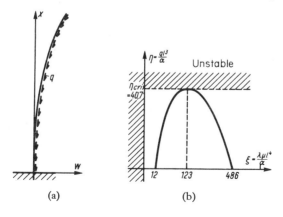

(a)   (b)

FIG. 78. (a) Buckling column with nonconservative load. (b) Eigenvalue curve.

where

$$Q = q(l - x),$$

$$R(w) = [\alpha w_{xx}\, \delta w_x - \alpha w_{xxx}\, \delta w - q(l - x)\, w_x\, \delta w]_0^l ,$$

so that

$$L(w) = \alpha w_{xxxx} + q(l - x)\, w_{xx} .$$

The condition $R(w) = 0$ results in the geometrical boundary conditions $w(0, t) = w_x(0, t) = 0$ and the dynamic conditions

$$w_{xx}(l, t) = w_{xxxx}(l, t) = 0.$$

The *"abbreviated"* problem is given by

$$\mu\omega^2\varphi(x) + \alpha\varphi^{IV}(x) = 0,$$

$$\varphi(0) = \varphi'(0) = \varphi''(l) = \varphi'''(l) = 0,$$

which is obtained from $\mu\ddot{w} + L(w) = 0$ and the boundary conditions by substituting $w = e^{\omega t}\varphi(x)$ and crossing out the term $q(1 - x)\, w_{xx}$ because it disturbs the self-adjoint property. The orthonormal eigenfunctions are

$$\varphi_k(x) = \kappa_k \left[c_{1k} \left(\cos\frac{m_k x}{l} - \cosh\frac{m_k x}{l} \right) - c_{2k} \left(\sin\frac{m_k x}{l} - \sinh\frac{m_k x}{l} \right) \right],$$

$$k = 1.2, \quad m_1 = 1.875, \quad m_2 = 4.694, \quad c_{11} = 4.148, \quad c_{21} = 3.037,$$

$$c_{12} = 53.640, \quad c_{22} = 54.631, \quad \kappa_1 = (17.259l)^{-1/2}, \quad \kappa_2 = (2868.283l)^{-1/2}.$$

The elements a_{ik} of the matrix $\mathbf{A}$ can be obtained from the equation

$$a_{ik} = (1/\mu) \int_0^l L(\varphi_k)\, \varphi_i\, dx$$

which are

$$a_{11} = \frac{1}{17.259\mu l} \left(213.356\,\frac{\alpha}{l^3} + 7.5q \right),$$

$$a_{12} = \frac{264.435}{(17.259 \cdot 2868.283)^{1/2}}\,\frac{q}{\mu l}, \quad a_{21} = -\frac{972.229}{(17.259 \cdot 2868.283)^{1/2}}\,\frac{q}{\mu l},$$

$$a_{22} = \frac{1}{2868.283\mu l} \left(1{,}392{,}542.792\,\frac{\alpha}{l^3} - 19{,}077.477q \right).$$

The characteristic equation $\det(\mathbf{A} - \lambda\mathbf{E}) = 0$ or

$$\begin{vmatrix} a_{11} - \lambda & a_{12} \\ a_{21} & a_{22} - \lambda \end{vmatrix} = 0$$

results in

$$\lambda^2 - \lambda(a_{11} + a_{22}) + a_{11}a_{22} - a_{12}a_{21} = 0.$$

When the expressions for a_{ik} are substituted, the following *eigenvalue curve* equation is obtained:

$$F(\xi, \eta) = \eta^2 + \eta(55.9 + 2.7\xi) - 216.2\xi + 0.43\xi^2 + 2606.0 = 0.$$

It is shown in Fig. 78b. The notation

$$\eta = ql^3/\alpha, \qquad \xi = \lambda\mu l^4/\alpha$$

was used. It can be seen that this is the same type of eigenvalue curve as was obtained for the Beck problem (Fig. 72). It does not intersect the η axis (load axis). Consequently, there are no nontrivial equilibrium positions, which means that the static stability criterion cannot be used. Only kinetic instability of the column is possible. This occurs for $q > 40.7\alpha/l^3$, as can be seen from Fig. 78b. The critical load value

$$q_{\text{crit,min}} = 40.7\ \alpha/l^3$$

can either be found by determining the maximum of the eigenvalue curve or by applying the algebraic kinetic stability criterion.

In the first case, the conditions

$$F(\xi, \eta) = 0, \qquad \partial F/\partial \xi = 0$$

must be used. The condition $\partial F/\partial \xi = 0$ results in $\eta^* = 80.1 - 0.32\xi$, so that $F(\xi, \eta^*) = 0$ becomes $\xi^2 + 210\xi - 40{,}908 = 0$. The positive solution $\xi^* = 123$ is obtained from which $\eta_{\text{crit}} = 80.1 - 0.32\xi^*$ or $\eta_{\text{crit}} = 40.7$ is calculated. The buckling load obtained is the one mentioned above $q_{\text{crit}} = 40.7\alpha/l^3$ and follows from the definition of η.

In the second case, the discriminant D of the characteristic equation must be used, where

$$D = (a_{11} - a_{22})^2 + 4a_{12}a_{21} = 0.$$

In round numbers, the relationships for the a_{ik} result in

$$\left(7\frac{q}{\mu l} - 473\frac{\alpha}{\mu l^4}\right)^2 - 21\frac{q^2}{\mu^2 l^2} = 0,$$

so that we are again led to the following quadratic equation with $\eta = ql^3/\alpha$:

$$\eta^2 - 237\eta + 7990 = 0.$$

The smallest solution is $\eta = 40.7$, which is the already known critical value.

Let us now consider a nonconservative problem for which the matrix *is symmetrizable*.

Example 5: Let us consider the column clamped at both ends with a continuously distributed *tangential load* (Fig. 79a). W. Hauger provided this example.

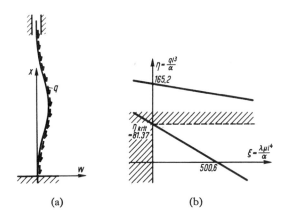

(a)   (b)

FIG. 79. (a) Buckling column with nonconservative load. (b) Eigenvalue curve.

Just as in the preceding case, we have $L(w) = \alpha w_{xxxx} + q(l - x)\, w_{xx}$. The boundary conditions are now

$$w(0, t) = w_x(0, t) = w(l, t) = w_x(l, t) = 0,$$

so that the following orthonormal eigenfunctions of the "abbreviated" problem

$$\mu\omega^2 \varphi(x) + \alpha \varphi^{IV}(x) = 0,$$

$$\varphi(0) = \varphi'(0) = \varphi(l) = \varphi'(l) = 0$$

must be used. The two functions

$$\varphi_k(x) = \kappa_k \left[c_{1k} \left(\cos \frac{m_k x}{l} - \cosh \frac{m_k x}{l} \right) - c_{2k} \left(\sin \frac{m_k x}{l} - \sinh \frac{m_k x}{l} \right) \right],$$

$$k = 1.2, \quad m_1 = 4.730, \quad m_2 = 7.853, \quad c_{11} = 57.646, \quad c_{21} = 56.637,$$

$$c_{12} = 1\,285.985, \quad c_{22} = 1286.984, \quad \kappa_1 = (3.323 \times 10^3 l)^{-1/2}, \quad \kappa_2 = (1.654 \times 10^6 l)^{-1/2}$$

are used in the calculation.

The elements

$$a_{ik} = (1/\mu) \int_0^l L(\varphi_k)\,\varphi_i\,dx$$

of the matrix **A** are obtained from them,

$$a_{11} = \frac{1}{3.323 \cdot 10^3 \mu l}\left(1.663 \cdot 10^6\,\frac{\alpha}{l^3} - 2.044 \cdot 10^4 q\right),$$

$$a_{12} = -\frac{4.488 \cdot 10}{(3.323 \cdot 1.654 \cdot 10^9)^{1/2}}\,\frac{q}{\mu l}, \qquad a_{21} = -\frac{4.955 \cdot 10^5}{(3.323 \cdot 1.654 \cdot 10^9)^{1/2}}\,\frac{q}{\mu l},$$

$$a_{22} = \frac{1}{1.654 \cdot 10^6 \mu l}\left(6.290 \cdot 10^9\,\frac{\alpha}{l^3} - 3.808 \cdot 10^7 q\right).$$

From $\det(\mathbf{A} - \lambda \mathbf{E}) = 0$ and the notations $\eta = ql^3/\alpha$, $\xi = \lambda \mu l^4/\alpha$, the following *eigenvalue curve* equation is obtained:

$$F(\xi, \eta) = \xi^2 + 2 \cdot 15.110\xi\eta + 141.631\eta^2$$
$$- 2 \cdot 2152.05\xi - 2 \cdot 17{,}461.1\eta + 1{,}903{,}920 = 0.$$

The two branches of this curve are shown in Fig. 79b. Owing to the symmetry in the loads and the boundary conditions, the situation is essentially the same as is found for conservative problems (Fig. 66). The eigenvalue curve intersects the load axis (η axis) and the lowest inter-section point results in the critical load value

$$q_{\text{crit,min}} = 81.37\,\alpha/l^3,$$

which is the buckling load.

Since there are intersection points with the load axis and because the eigenvalue curve shows that in this case there cannot be kinetic instability, the critical load value can be determined by the static stability criterion $\det \mathbf{A} = 0$. It would have been possible to determine it directly from

$$\det \mathbf{A} = a_{11}a_{22} - a_{12}a_{21} = 0$$

without determining the eigenvalue curve. When the values of the a_{ik} given above are substituted into this equation, and if $\eta = ql^3/\alpha$, the following equation is obtained, in round numbers:

$$\eta^2 - 248\eta + 13{,}570 = 0.$$

The smallest root is

$$\eta_{\text{crit}} = 81.5$$

TABLE I[a]

Support conditions	Load	Tangential concentrated force	Tangential distributed load; $q = $ const	Tangential distributed load; $q = q_0(l - x)$
	Buckling load:	$P_k = 20.05\alpha/l^2$	$q_k = 40.7\alpha/l^3$	$q_{0k} = 158.2\alpha/l^4$
	Eigenvalue curve:			
	Stability criterion:	Kinetic	Kinetic	Kinetic
	Author:	M. Beck	H. Leipholz	W. Hauger
	Load:	Tangential concentrated load	$q = $ const	$q = q_0(l - x)$
	Buckling load:	$P_k = P_E = 9.87\alpha/l^2$	$q_k = 18.96\alpha/l^3$	$q_{0k} = 62.28\alpha/l^4$
	Eigenvalue curve:			
	Stability criterion:	Static	Static	Static
	Authors:	—	E. Abody and A. Petur, A. Pflüger, H. Leipholz	W. Hauger

[a] l is the length, and α the bending stiffness of the column.

which compares quite well with the more exact value $\eta_{\text{crit}} = 81.37$ mentioned before.

Finally, Table I summarizes the nonconservative load cases of the elastic column treated up to the present.

Let us now consider a nonconservative system where a *slightly asymmetric* matrix occurs.

TABLE I (continued)

Support conditions	Load	Tangential concentrated force	Tangential distributed load; q = const	Tangential distributed load; $q = q_0(l - x)$
	Load:	Tangential concentrated load	q = const	$q = q_0(l - x)$
	Buckling load:	$P_k = P_E = 20.2\alpha/l^2$	$q_k = 57.95\alpha/l^3$	$q_{0k} = 402.3\alpha/l^4$
	Eigenvalue curve:			
	Stability criterion:	Static	Static	Kinetic
	Author:	—	W. Hauger	W. Hauger
	Load:	Tangential concentrated force	q = const	$q = q_0(l - x)$
	Buckling load:	$P_k = P_E = 39.5\alpha/l^2$	$q = 81.37\,\alpha/l^3$	$q = 328.0\,\alpha/l^4$
	Eigenvalue Curve:			
	Stability criterion:	Static	Static	Static
	Author:	—	W. Hauger	W. Hauger

Example 6: This example was given by G. Herrmann and R. W. Bungay. It is a *pinned column* (Fig. 80) with a *subtangential or supertangential buckling load P* (subtangential for $\alpha < 1$, supertangential for $\alpha > 1$).

The kinetic energy T of the system is

$$T = \tfrac{1}{2}ml^2(3\dot{\varphi}_1^2 + 2\dot{\varphi}_1\dot{\varphi}_2 + \dot{\varphi}_2^2).$$

The potential energy of the restoring moments is

$$U = \tfrac{1}{2}c(2\varphi_1^2 - 2\varphi_1\varphi_2 + \varphi_2^2),$$

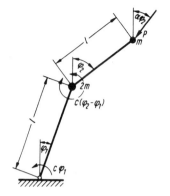

FIG. 80. Jointed rod with sub-tangential or supertangential buckling load.

and the load results in the following generalized forces:

$$Q_1 = Pl(\varphi_1 - \alpha\varphi_2), \qquad Q_2 = Pl(1 - \alpha)\,\varphi_2.$$

The Lagrange equations of the second kind are

$$\frac{d}{dt}\left(\frac{\partial T}{\partial \dot{\varphi}_k}\right) + \frac{\partial U}{\partial \varphi_k} = Q_k$$

and result in the equations of motion. They are

$$3ml^2\ddot{\varphi}_1 + ml^2\ddot{\varphi}_2 + (2c - Pl)\,\varphi_1 + (\alpha Pl - c)\,\varphi_2 = 0,$$

$$ml^2\ddot{\varphi}_1 + ml^2\ddot{\varphi}_2 - c\varphi_1 + [c - (1 - \alpha)\,Pl]\,\varphi_2 = 0.$$

The trial solution $\varphi_1 = A_1 e^{\omega t}$, $\varphi_2 = A_2 e^{\omega t}$ results in the following algebraic system of equations:

$$(2c - Pl + 3ml^2\omega^2)\,A_1 + (\alpha Pl - c + ml^2\omega^2)\,A_2 = 0,$$

$$(-c + ml^2\omega^2)\,A_1 + [c - (1 - \alpha)\,Pl + ml^2\omega^2]\,A_2 = 0,$$

which can be written in the form $(\mathbf{A} + \Omega^2\mathbf{C})\,\mathbf{X} = 0$, where

$$p = \frac{Pl}{c}, \qquad \Omega^2 = \frac{ml^2\omega^2}{c},$$

$$\mathbf{A} = \begin{pmatrix} 2 - p & \alpha p - 1 \\ -1 & 1 - (1 - \alpha)\,p \end{pmatrix}, \qquad \mathbf{C} = \begin{pmatrix} 3 & 1 \\ 1 & 1 \end{pmatrix}, \qquad \mathbf{X} = \begin{pmatrix} A_1 \\ A_2 \end{pmatrix}.$$

In order to determine the stability of the nontrivial equilibrium position of the jointed system, the eigenvalue problem

$$\det(\mathbf{A} - \lambda\mathbf{C}) = 0, \qquad \lambda = -\Omega^2 \tag{2.3.79}$$

must be discussed. The problem is slightly asymmetric, because all the assumptions regarding this case described in Section 2.3.4 are satisfied, as we will now show. However, it must be realized that instead of $\mathbf{E}$ in (2.3.79) there is the more general, positive definite matrix $\mathbf{C}$. It is now necessary to formulate all results for the eigenvalue $\lambda_{A,C}$ of the pair of matrices $\mathbf{A}$, $\mathbf{C}$.

It can be seen that $\mathbf{A}$ is symmetric for $\alpha = 0$ and that it is even positive definite for $p = 0$. In addition, (2.3.79) continues to have only real eigenvalues $\lambda_{A,C}$ for sufficiently small α. It may be shown that this is at least the case for $0 \leqslant \alpha \leqslant \alpha_{g,\min} = 0.345$, so that the static stability criterion must remain in force at least in this α interval. This means that the conditions for weak asymmetry are satisfied.

The proof is carried out as follows. Equation (2.3.79) is multiplied on the left by

$$\mathbf{C}^{-1} = \begin{pmatrix} \tfrac{1}{2} & -\tfrac{1}{2} \\ -\tfrac{1}{2} & \tfrac{3}{2} \end{pmatrix}$$

so that

$$(\mathbf{C}^{-1}\mathbf{A} - \lambda\mathbf{E})\,\mathbf{X} = \begin{pmatrix} \tfrac{1}{2}(3 - p) & \tfrac{1}{2}(p - 2) \\ \tfrac{1}{2}(p - 5) & \tfrac{1}{2}[4 + (2\alpha - 3)\,p] \end{pmatrix}\begin{pmatrix} A_1 \\ A_2 \end{pmatrix} = 0 \qquad (2.3.80)$$

is obtained. The eigenvalue of interest $\lambda_{A,C}$ is obtained as the root of the characteristic equation $\det(\mathbf{C}^{-1}\mathbf{A} - \lambda\mathbf{E}) = 0$. If $\mathbf{B} = \mathbf{C}^{-1}A = (b_{ik})$ and

$$b_{11} = \tfrac{1}{2}(3 - p), \quad b_{12} = \tfrac{1}{2}(p - 2), \quad b_{21} = \tfrac{1}{2}(p - 5), \quad b_{22} = \tfrac{1}{2}[4 + (2\alpha - 3)\,p],$$
$$(2.3.81)$$

the characteristic equation is

$$\lambda_{A,C}^2 - \lambda_{A,C}(b_{11} + b_{22}) + b_{11}b_{22} - b_{12}b_{21} = 0. \qquad (2.3.82)$$

Complex eigenvalues $\lambda_{A,C}$ are only possible if the discriminant of (2.3.82) is smaller than zero. This would be the case for

$$D = (b_{11} - b_{22})^2 + 4b_{12}b_{21} < 0.$$

The limiting case is given by $D = 0$, which is also the *kinetic stability criterion*. When $D = 0$ is calculated from (2.3.81), the result is

$$p = \frac{8 - \alpha \pm \{(8 - \alpha)^2 - 41[1 + (1 - \alpha)^2]\}^{1/2}}{2[1 + (1 - \alpha)^2]}. \qquad (2.3.83)$$

The relationship (2.3.83) is the connection between the α, p values at the kinetic instability boundary. Figure 81 shows the corresponding boundary curve on the stability diagram based on this law. In the follow-

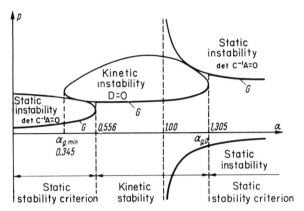

FIG. 81. Stability diagram.

ing discussion, we will only consider the positive quadrant of the p, α plane. The diagram shows that there are α values for which *no* real p values can be found that would satisfy (2.3.83). This is the case for

$$(8 - \alpha)^2 - 41[1 + (1 - \alpha)^2] < 0. \qquad (2.3.84)$$

It is easily seen that (2.3.84) is satisfied for $\alpha < \alpha_{g,\min} = 0.345$ and for $\alpha > \alpha_{g,0} = 1.305$. It immediately follows that there are no complex $\lambda_{A,C}$ for $0 \leqslant \alpha \leqslant 0.345$ according to our assumption.

The *static stability criterion* is given by det $\mathbf{C}^{-1}\mathbf{A} = 0$ and therefore by $b_{11}b_{22} - b_{12}b_{21} = 0$. Making use of (2.3.81), the following equation is obtained:

$$p^2(1 - \alpha) - 3p(1 - \alpha) + 1 = 0.$$

This is the boundary curve for static instability shown in Fig. 81.

The curve **G** is drawn as a thick curve and delimits the important lower region of stability. The stability diagram shows that for $\alpha = 0.556$ and $\alpha = 1.305$ there is a *jump* in the stability conditions. It can also be seen that the regions of static and kinetic stability alternate as the parameter α is varied.

It should also be noted that in the intervals $0.345 < \alpha < 0.556$ and $1.00 < \alpha < 1.305$, it is possible for both criteria to hold simultaneously. The *decisive* one will always be the one that results in a portion of the heavy boundary curve G. Therefore, the static criterion holds for $0.345 < \alpha < 0.556$ and the kinetic criterion holds for $1.00 < \alpha < 1.305$. This becomes apparent, when, for example, the eigenvalue curve

$$p^2 - (3 + 18\lambda_{A,C})\,p + 70\lambda_{A,C} - 20\lambda_{A,C}^2 - 10 = 0,$$

which follows from (2.3.82) is plotted for $\alpha = 1.1$ (Fig. 82). The course of the $\lambda_{A,C}$ curve in the complex plane is marked with the corresponding p values (Fig. 83). If it is assumed that negative p values are not possible,

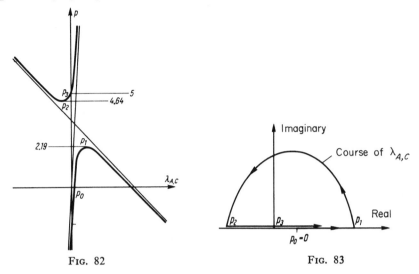

FIG. 82.   FIG. 83.

FIG. 82.   Eigenvalue curve.
FIG. 83.   Path of the eigenvalue $\lambda_{A,C}$ in the complex plane.

the curve begins at $p_0 = 0$ on the real axis and stays along this axis as p increases. At $p_1 = 2.19$, it turns into the complex plane. The kinetic instability begins exactly at this point. At $p_2 = 4.64$, the image point of $\lambda_{A,C}$ again reaches the real axis and follows it until it goes through the origin for $p_3 = 5$. At this point static instability occurs. Since kinetic instability occurs first in this case, because the image point turns into the complex plane before passing through the origin, the kinetic stability criterion holds for $\alpha = 1.1$ as well as for all α values of the interval $1 < \alpha < 1.305$.

Let us show the *effect of mass distribution* according to the method of W. Hauger.

Example 7:   Let us consider the *column*, shown in Fig. 84, which is clamped at the bottom. There is a *concentrated mass m* at the point $x = l_1$ whose weight is $G = mg$. There is a *tangential concentrated load P_t* at this point. At the free end, $x = l$, there is a *vertical concentrated load P_v*. The total mass of the column is $M = \mu l$. Its constant bending stiffness is α.

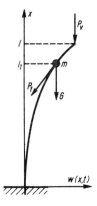

FIG. 84.   Influence of mass distribution.

The principle of virtual work results in the equation

$$\mu \left[\int_0^{l_1} \ddot{w}_1 \, \delta w_1 \, dx + \int_{l_1}^l \ddot{w}_2 \, \delta w_2 \, dx \right]$$

$$+ \delta(V_i + V_a) + [m\ddot{w}(l_1, t) + P_i w_x(l_1, t)] \, \delta w(l_1, t) = 0, \quad (2.3.85)$$

where

$$V_i = \tfrac{1}{2} \left[\int_0^{l_1} \alpha w_{1xx}^2 \, dx + \int_{l_1}^l \alpha w_{2xx}^2 \, dx \right],$$

$$V_a = -\tfrac{1}{2} \left[(P_v + P_t + G) \int_0^{l_1} w_{1x}^2 \, dx + P_v \int_{l_1}^l w_{2x}^2 \, dx \right],$$

and the boundary conditions are

$$w_1(0, t) = w_{1x}(0, t) = w_{2xx}(l, t) = 0, \quad\quad\quad (2.3.86)$$

$$w_{2xxx}(l, t) = - (P_v/\alpha) \, w_{2x}(l, t). \quad\quad\quad (2.3.87)$$

The transition conditions are

$$w_1(l_1, t) = w_2(l_1, t) = w(l_1, t), \; \ddot{w}_1(l_1, t) = \ddot{w}_2(l_1, t) = \ddot{w}(l_1, t),$$

$$\quad\quad\quad\quad\quad\quad\quad\quad\quad\quad\quad\quad\quad\quad\quad\quad (2.3.88)$$

$$w_{1x}(l_1, t) = w_{2x}(l_1, t) = w_x(l_1, t), \; w_{1xx}(l_1, t) = w_{2xx}(l_1, t) = w_{xx}(l_1, t).$$

The deflection of the column axis is given by $w_1(x, t)$ in the interval $0 \leqslant x \leqslant l_1$ and by $w_2(x, t)$ in the interval $l_1 \leqslant x \leqslant l$.

Taking the boundary conditions (2.3.86) into account, the equation (2.3.85) becomes

$$\int_0^{l_1} [\mu \ddot{w}_1 + \alpha w_{1xxxx} + (P_v + P_t + G) w_{1xx}] \, \delta w_1(x, t) \, dx + \int_{l_1}^l [\mu \ddot{w}_2$$

$$+ \alpha w_{2xxxx} + P_v w_{2xx}] \, \delta w_2(x, t) \, dx + [m \ddot{w}(l_1, t) + P_t w_x(l_1, t)$$

$$- (P_t + G) w_x(l_1, t) - \alpha w_{1xxx}(l_1, t) + \alpha w_{2xxx}(l_1, t)] \, \delta w_1(l_1, t)$$

$$- [\alpha w_{2xxx}(l, t) + P_v w_{2x}(l, t)] \, \delta w_2(l, t) = 0 \qquad (2.3.89)$$

after the variation is carried out.

In order to apply the *Galerkin method*, the coordinate functions must be selected so that (2.3.86) is indeed satisfied. This is easy to do because the eigenfunctions for free rod oscillations satisfy them. The functions are also orthogonal, which is useful in the calculations. It can be shown that the two-term trial solution

$$w(x, t) \equiv w_1(x, t) \equiv w_2(x, t) = e^{i\omega t}[\rho_1 \varphi_1(x) + \rho_2 \varphi_2(x)] \qquad (2.3.90)$$

can be used, where φ_1, φ_2 are the mentioned eigenfunctions. By means of (2.3.90), Eq. (2.3.89) can first be transformed into

$$\int_0^{l_1} [\alpha(\rho_1 \varphi_1^{IV} + \rho_2 \varphi_2^{IV}) - \mu\omega^2(\rho_1 \varphi_1 + \rho_2 \varphi_2) + (P_v + P_t + G)(\rho_1 \varphi_1'' + \rho_2 \varphi_2'')]$$

$$\times (\delta\rho_1 \varphi_1 + \delta\rho_2 \varphi_2) \, dx + \int_{l_1}^l [\alpha(\rho_1 \varphi_1^{IV} + \rho_2 \varphi_2^{IV}) - \mu\omega^2(\rho_1 \varphi_1 + \rho_2 \varphi_2)$$

$$+ P_v(\rho_1 \varphi_1'' + \rho_2 \varphi_2'')](\delta\rho_1 \varphi_1 + \delta\rho_2 \varphi_2) \, dx$$

$$- \{G[\rho_1 \varphi_1'(l_1) + \rho_2 \varphi_2'(l_1)] + m\omega^2[\rho_1 \varphi_1(l_1) + \rho_2 \varphi_2(l_1)]\}$$

$$\times [\delta\rho_1 \varphi_1(l_1) + \delta\rho_2 \varphi_2(l_1)] - \{\alpha[\rho_1 \varphi_1'''(l) + \rho_2 \varphi_2'''(l)]$$

$$+ P_v[\rho_1 \varphi_1'(l) + \rho_2 \varphi_2'(l)]\}[\delta\rho_1 \varphi_1(l) + \delta\rho_2 \varphi_2(l)] = 0.$$

After multiplication and noting that the variations $\delta\rho_1$, $\delta\rho_2$ are independent, the following algebraic system of equations $\alpha_{ik}\rho_k = 0$ is obtained, whose coefficients are

$$\alpha_{ik} = \int_0^l [\alpha\varphi_k^{IV}\varphi_i + P_v\varphi_k''\varphi_i - \mu\omega^2\varphi_k\varphi_i] \, dx + \int_0^{l_1} (P_t + G) \, \varphi_k''\varphi_i \, dx$$

$$- [G\varphi_k'(l_1) \, \varphi_i(l_1) + m\omega^2\varphi_k(l_1) \, \varphi_i(l_1) + \alpha\varphi_k'''(l) \, \varphi_i(l) + P_v\varphi_k'(l) \, \varphi_i(l)].$$

$$(2.3.91)$$

The eigenvalue equation is obtained from the condition $\det(\alpha_{ik}) = 0$. It is given by

$$a\xi^2 + 2b\xi\eta + c\eta^2 + 2d\xi + 2e\eta + f = 0, \qquad (2.3.92)$$

where

$$\xi = \mu l^4 \omega^2 / \alpha, \qquad \eta = P_t l^2 / \alpha.$$

The coefficients of (2.3.92) have the following meaning:

$a = \det(c_{ij})$, $\qquad\qquad\qquad\qquad b = \tfrac{1}{2}(b_{11}c_{22} + b_{22}c_{11} - b_{12}c_{21} - b_{21}c_{12})$,

$c = \det(b_{ij})$, $\qquad\qquad\qquad\qquad d = \tfrac{1}{2}(c_{11}e_{22} + c_{22}e_{11} - c_{12}e_{21} - c_{21}e_{12})$,

$e = \tfrac{1}{2}(b_{11}e_{22} + b_{22}e_{11} - b_{12}e_{21} - b_{21}e_{12})$, $\quad f = \det(e_{ij})$,

$\qquad a_{ij} = k_i^4 A_{ij} + t(B_{ij} - F_{ij})$, $\qquad b_{ij} = C_{ij}$,

$\qquad c_{ij} = -(A_{ij} + RD_{ij})$, $\qquad\qquad d_{ij} = R(C_{ij} - E_{ij}),\, e_{ij} = a_{ij} + d_{ij}\zeta$,

$\qquad A_{ij} = (1/l) \int_0^l \varphi_i \varphi_j\, dx$, $\qquad B_{ij} = l \int_0^l \varphi_i' \varphi_j\, dx$, $\qquad C_{ij} = l \int_0^l \varphi_i'' \varphi_j\, dx$,

$\qquad D_{ij} = \varphi_i(l_1)\, \varphi_j(l_1)$, $\qquad\qquad E_{ij} = l\varphi_i'(l_1)\, \varphi_j(l_1)$, $\qquad F_{ij} = l\varphi_i'(l)\, \varphi_j(l)$.

In addition, the following data are required:

$t = P_v l^2 / \alpha$, $R = m/\mu l \equiv G/\mu lg$, $\zeta = \mu l^3 g / \alpha$, and $k_1 = 1.875$, $k_2 = 4.694$.

The critical load parameters may be found from the kinetic stability criterion. It is the value for which (2.3.92) has a horizontal tangent (see, for example, Fig. 78b):

$$\eta_{\text{crit}} = -\frac{1}{ac - b^2}\,[(ae - bd) \pm \{(ae - bd)^2 - (ac - b^2)(af - d^2)\}^{1/2}]. \qquad (2.3.93)$$

From (2.3.93) it follows that the buckling load is $P_{t,\text{crit}} = \eta_{\text{crit}}\, \alpha/l^2$.

W. Hauger evaluated this for $l_1 = l$, $t = l$, $\zeta = 0.01$. This ζ value corresponds approximately to a steel column 2.2 m long having a circular cross section with radius $r = 4$ cm. When (2.3.93) was evaluated, he found that there is an optimum value of R and therefore an optimum ratio of concentrated mass m to total mass $M = \mu l$ of the column, for which the buckling load P_t has a maximum value. Figure 85 shows this relationship.

A large number of nonconservative stability problems occur in *aeroelasticity*. There are a number of structures which are loaded by aerodynamic forces ("follower" forces). Let us consider one example

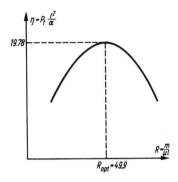

FIG. 85. Calculation of the optimum mass ratio R.

and also refer to the book by Bisplinghoff and Ashley (*13*) which contains this example as well as a number of others.

Example 8: Let us consider the "flutter" of a plane, thin plate, which is infinitely long in the y direction and has length l in the x direction. We will show that this flutter is a kinetic instability due to the effect of aerodynamic forces at supersonic speeds. The original state is the plane, unstressed plate.

If it is assumed that the perturbed state is essentially given by deflections $W(x, t)$ perpendicular to the plane of the plate, the variational equations are given by

$$\frac{\partial^4 W}{\partial \xi^4} + p \frac{\partial W}{\partial \xi} + \kappa \frac{\partial^2 W}{\partial t^2} = 0. \qquad (2.3.94)$$

$\xi = x/l$ is a dimensionless variable, $\kappa = \mu l^4/D$, $p = 2ql^3/D(M^2 - 1)^{1/2}$, μ is the mass density, D the stiffness of the plate, M the mach number, $q = \rho_\infty U^2/2$ the stagnation pressure, ρ_∞ the density of the undisturbed stream, U the stream velocity. The derivation of (2.3.94) is given by Bisplinghoff and Ashley (*13*). The influence of damping by the aerodynamic forces has been ignored for purposes of simplicity.

If we consider a perturbation of the original state in only one direction, perpendicular to the plane of the plate, we are only dealing with one case of conditional stability. However, this case certainly includes the important features of the process.

The trial solution $W(\xi, t) = w(\xi) e^{\omega t}$ is made for (2.3.94), so that stability will prevail as long as ω is imaginary. Equation (2.3.94) becomes

$$w^{IV} + pw' + \kappa \omega^2 w = 0. \qquad (2.3.95)$$

The trial solution $w = \rho_i w_i$ is used for (2.3.95), and it is assumed that the coordinate functions satisfy *all* boundary conditions. Then the

Galerkin method in the restricted sense can be carried out and we have the following according to (1.6.67):

$$\sum_k \rho_k \int_0^l L^*(w_k)\, w_j\, d\xi = 0, \qquad L^*(y) = \frac{d^4y}{d\xi^4} + p\,\frac{dy}{d\xi} + \kappa\omega^2 y. \quad (2.3.96)$$

We will assume that the plate is freely supported at the edges. The boundary conditions are then $w(0) = w''(0) = w(l) = w''(l) = 0$. We also restrict ourselves to a two-term trial solution w. We have

$$w_1(\xi) = \sqrt{2}\sin \pi\xi, \qquad w_2(\xi) = \sqrt{2}\sin 2\pi\xi. \quad (2.3.97)$$

As usual, $(\mathbf{A} - \lambda\mathbf{E})\,\mathbf{X} = 0$ is written for (2.3.96). The relationships $\mathbf{X} = (\rho_k)$ and $\mathbf{A} = (\alpha_{ik})$ hold, where

$$a_{ik} = \int_0^l L(w_k)\, w_i\, d\xi, \qquad L(y) = \frac{d^4y}{d\xi^4} + p\,\frac{dy}{d\xi}. \quad (2.3.98)$$

Furthermore, $\lambda = -\kappa\omega^2$, so that λ must be positive and real for stability.

The following is calculated from (2.3.98):

$$a_{11} = \pi^4, \qquad a_{12} = -(8/3)\,p, \qquad a_{21} = +(8/3)\,p, \qquad a_{22} = 16\pi^4.$$

The static stability criterion det $\mathbf{A} = 0$ results in

$$16\pi^2 + (64/9)\,p^2 = 0, \qquad p = \pm i(3/2)\,\pi^4,$$

so that there is apparently no real critical load parameter. Therefore, the criterion fails, and the kinetic calculation must be carried out.

The characteristic equation $\det(\mathbf{A} - \lambda\mathbf{E}) = 0$ is

$$\lambda^2 - 17\pi^4\lambda + 16\pi^8 + (64/9)\,p^2 = 0. \quad (2.3.99)$$

The eigenvalue curve

$$p = \pm\left(-\frac{9}{64}\lambda^2 + \frac{153}{64}\pi^4\lambda - \frac{9}{4}\pi^8\right)^{1/2} \quad (2.3.100)$$

is derived and is shown in Fig. 86. The course of this curve shows that the static criterion cannot be satisfied, because there is no intersection point with the load axis (p axis).

The kinetic stability criterion requires that the discriminant of (2.3.99) is zero. This is identical to the condition $(a_{11} = a_{22})^2 + 4a_{12}a_{21} = 0$, so that, because of the meaning of a_{ik}, we have

$$15^2\pi^8 - (4 \cdot 64/9)\,p^2 = 0$$

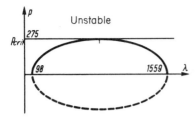

FIG. 86.   Eigenvalue curve.

and therefore

$$p_{\text{crit}} = (3 \cdot 15\pi^4)/(2 \cdot 8) = 275.$$

The critical flow velocity U_{crit} is obtained from p_{crit}, which follows from the relationships between p and q and q and U. Flutter of the plate and kinetic instability must begin at U_{crit}.

2.3.6. Problems of Elastokinetics

The previously discussed elastostatic problems were obtained by considering the perturbed motion around an equilibrium position. We refer to "stability problems of elastostatics" because the unperturbed original state was an equilibrium position.

Let us generalize and consider a certain (if possible stationary) *motion* as the original state of the elastic body. We then investigate motions that are adjacent to the undisturbed motion, which leads to the concept of stability problems of elastokinetics. This is a restricted definition, because any stability problem is kinetic in the broadest sense. An equilibrium position can always be considered as a special case of an undisturbed original motion. There will always be disturbed motions around the original state. This "kinetic" content of any stability theory was already apparent in stability problems of elastostatics, because there we consider small oscillations and use the kinetic stability criterion.

Therefore, we will now use the motion of the elastic body as the undisturbed state. It is assumed that the points of the body have the displacements $u_i^{(0)}$, $i = 1, 2, 3$ for this original motion. The disturbance of the original state results in adjacent motions with the displacements $u_i^{(0)} + u_i$. The u_i are the "variations." The most important thing is to establish the differential equations for the u_i within the framework of the general stability theory—that is, the variational equations. The solution of these equations determines the nature of the motions adjacent to the original state. If the u_i, which are obtained from the variational equations, can be made arbitrarily small at all times for sufficiently small disturbances, the undisturbed original motion is called "stable" according to Lyapunov's definition. Strictly speaking, it would again be necessary to determine that the strains remain sufficiently small, so that we are certain that none of the foundations of the theory of elasticity have been violated. However, for purposes of simplicity, we will not do this and shall simply assume that the strains are small enough.

The variational equations can again be obtained from the principle of virtual work. There are two ways in which this can be done, just as was the case for elastostatic problems, which result from the distinction between conservative or monogenetic problems and nonconservative or polygenetic problems.

Let us first consider *conservative or monogenetic* problems. The theory was advanced by Mettler.[1]

The principle of virtual work results in

$$\int_V \mu \ddot{\mathbf{u}} \, \delta\mathbf{u} \, dV + \delta(\Pi_i + \Pi_a) = 0. \qquad (2.3.101)$$

Let

$$T = \tfrac{1}{2} \int_V \mu \dot{\mathbf{u}}^2 \, dV$$

be the kinetic energy. Then $\delta T = \int_V \mu \dot{\mathbf{u}} \, \delta\dot{\mathbf{u}} \, dV$. If δT is added and subtracted from the left side of (2.3.101), the following result is obtained:

$$\delta(T - \Pi_i - \Pi_a) = (\mu \ddot{\mathbf{u}} \, \delta\mathbf{u} + \mu \dot{\mathbf{u}} \, \delta\dot{\mathbf{u}}) \, dV.$$

Making use of the Lagrangian $L = T - (\Pi_i + \Pi_a)$, the following relationship

$$\delta L = \frac{d}{dt} \int_V \mu \dot{\mathbf{u}} \, \delta\mathbf{u} \, dV \qquad (2.3.102)$$

is obtained. If both sides of (2.3.102) are integrated with respect to time t between the limits t_1, t_2, the result is

$$\int_{t_1}^{t_2} \delta L \, dt = \int_V [\mu \dot{\mathbf{u}} \, \delta\mathbf{u}]_{t_1}^{t_2} \, dV,$$

and the final result is

$$\delta \int_{t_1}^{t_2} L \, dt = 0, \qquad (2.3.103)$$

---

[1] METTLER, E., A theory of stability of elastic motion, *Ing. Arch.* **16**, 135–146 (1947).

which is the Hamilton principle if it is assumed that the variations satisfy $\delta u(t_1) = \delta u(t_2) = 0$.

The Lagrangian L of the adjacent motions can be expanded with respect to the Lagrangian L_g of the undisturbed original motion, so that

$$L = L_g + \delta L_g + \tfrac{1}{2}\delta^2 L_g . \tag{2.3.104}$$

When (2.3.104) is substituted into (2.3.103), we obtain

$$\delta \int_{t_1}^{t_2} L_g \, dt + \delta \left[\delta \int_{t_1}^{t_2} L_g \, dt \right] + \tfrac{1}{2}\delta \int_{t_1}^{t_2} \delta^2 L_g \, dt = 0. \tag{2.3.105}$$

The first two terms in (2.3.105) are zero, because the condition (2.3.103) also holds for the undisturbed original motion. Therefore

$$\delta \int_{t_1}^{t_2} L_g \, dt = 0.$$

The following term from (2.3.105) remains for the derivation of the variational equations for the u_i:

$$\delta \int_{t_1}^{t_2} \delta^2 L_g \, dt = 0. \tag{2.3.106}$$

It is analogous to the equation

$$\delta(\delta^2 \Pi_g) = 0,$$

which holds in elastostatics and is given in Sections 2.3.4 and 2.3.5.

If it is assumed that only small strains are present, we have

$$\delta \int_{t_1}^{t_2} \delta^2 L_g \, dt = - \int_V \left[\frac{\partial \Delta\sigma_{ik}}{\partial x_k} + \frac{\partial}{\partial x_k}\left(\sigma_{jk} \frac{\partial u_i}{\partial x_j} \right) \right] \delta u_i \, dV$$

$$- \int_0 \left[\Delta\sigma_{ik} + \frac{\partial u_i}{\partial x_j} \sigma_{jk} \right] n_k \, \delta u_i \, dO + \int_V \mu \frac{\partial^2 u_i}{\partial t^2} \delta u_i \, dV = 0, \tag{2.3.107}$$

which can be found in Mettler's paper. V is the volume, O the surface area, μ the density of the elastic body. σ_{ij} is the stress tensor, $\Delta\sigma_{ij}$ the variation of the stress tensor for the disturbed motion due to the perturbation, u_i the vector of displacement variations, and n_i the unit vector in the direction of the surface normal.

Since the variations δu_i are independent, Eq. (2.3.107) results in the system of differential equations

$$\frac{\partial \Delta\sigma_{ik}}{\partial x_k} + \frac{\partial}{\partial x_k}\left(\sigma_{jk} \frac{\partial u_i}{\partial x_j} \right) = \mu \frac{\partial^2 u_i}{\partial t^2} , \tag{2.3.108}$$

as well as the boundary conditions

$$[\Delta\sigma_{ik} + (\partial u_i/\partial x_j)\,\sigma_{jk}]\,n_k = 0. \tag{2.3.109}$$

It should be noted that $\sigma_{jk}n_k = p_j$ is the prescribed surface load. In addition, the relationship

$$\Delta\sigma_{ik} = \lambda^*\,\delta_{ik}\,\frac{\partial u_j}{\partial x_j} + \mu^*\left(\frac{\partial u_i}{\partial x_k} + \frac{\partial u_k}{\partial x_i}\right) \tag{2.3.110}$$

must be used which follows from Hooke's law (in which λ^*, μ^* are the Lamé constants and δ_{ik} is the Kronecker delta). The final differential equations of the displacement "variations" u_i are given by

$$\mu^*\ \nabla^2 u_i + (\lambda^* + \mu^*)\,\frac{\partial^2 u_j}{\partial x_i\,\partial x_j} + \frac{\partial}{\partial x_k}\left(\sigma_{jk}\,\frac{\partial u_i}{\partial x_j}\right) = \mu\,\frac{\partial^2 u_i}{\partial t^2}, \tag{2.3.111}$$

which follows from (2.3.108) and holds for small displacements of the undisturbed state.

As already mentioned, the solutions of (2.3.111) determine the stability of the original state. This must be investigated in each individual case. The application of this theory to a practical example can be found in the paper by Mettler, which also contains additional references.

Additional terms must be added for *nonconservative, polygenetic* problems. For example, it must be remembered that when the original motion is disturbed, there can also be changes in the surface and body forces. This is the case for "follower" forces, i.e., when they change dependent of the variations in the displacements.

This time, let us use the equations of motion

$$\partial S_{ik}/\partial\xi_k + X_i = \mu\,d^2v_i/dt^2. \tag{2.3.112}$$

S_{ik} is the stress tensor and ξ_k are the Euler coordinates, X_i the specific volume forces, and v_i the displacements. Equation (2.3.112) is transformed into Lagrangian coordinates as follows, assuming small strains:

$$\frac{\partial}{\partial x_k}\left[\left(\delta_{ij} + \frac{\partial v_i}{\partial x_j}\right)S_{jk}\right] + X_i = \mu\,\frac{\partial^2 v_i}{\partial t^2}. \tag{2.3.113}$$

The following equilibrium conditions also hold for any point of the loaded surface of the elastic body:

$$(\delta_{ij} + \partial v_i/\partial x_j)\,S_{jk}n_k = p_i. \tag{2.3.114}$$

It is assumed that the parameters appearing in (2.3.113), (2.3.114) are obtained as the result of a disturbance to the original state, so that

$$S_{ij} = \sigma_{ij} + \Delta\sigma_{ij}, \qquad p_i = p_{i0} + \Delta p_i,$$
$$X_i = X_{i0} + \Delta X_i, \qquad v_i = u_{i0} + u_i.$$

$$(2.3.115)$$

In (2.3.115), σ_{ij}, p_{i0}, X_{i0}, u_{i0} are parameters for the undisturbed state and $\Delta\sigma_{ij}$, Δp_i, ΔX_i, u_i are the variations that arise because of a disturbance of the original state.

When (2.3.115) is substituted into (2.3.113), the result is

$$\frac{\partial}{\partial x_k}\left[\left(\delta_{ij} + \frac{\partial u_{i0}}{\partial x_j} + \frac{\partial u_i}{\partial x_j}\right)(\sigma_{jk} + \Delta\sigma_{jk})\right] + X_{i0} + \Delta X_i$$
$$= \mu\,\frac{\partial^2 u_{i0}}{\partial t^2} + \mu\,\frac{\partial^2 u_i}{\partial t^2}.$$

$$(2.3.116)$$

Since relationship (2.3.113) also holds for the original state, we have

$$\frac{\partial}{\partial x_k}\left[\left(\delta_{ij} + \frac{\partial u_{i0}}{\partial x_j}\right)\sigma_{jk}\right] + X_{i0} = \mu\,\frac{\partial^2 u_{i0}}{\partial t^2}.$$

Let us also assume that the products

$$(\partial u_i/\partial x_j)\,\Delta\sigma_{jk} \qquad \text{and} \qquad (\partial u_{i0}/\partial x_j)\,\Delta\sigma_{jk} \qquad\qquad (2.3.117)$$

can be ignored because they are small. Then Eq. (2.3.116) becomes

$$\frac{\partial}{\partial x_k}\Delta\sigma_{ik} + \frac{\partial}{\partial x_k}\left(\sigma_{jk}\,\frac{\partial u_i}{\partial x_j}\right) + \Delta X_i = \mu\,\frac{\partial^2 u_i}{\partial t^2}, \qquad\qquad (2.3.118)$$

which coincides with Eq. (2.3.108) except for an additional term.

Again assuming that the products (2.3.117) are small and can be ignored, the following boundary conditions are obtained from (2.3.114) and (2.3.115):

$$[\Delta\sigma_{ik} + (\partial u_i/\partial x_j)\,\sigma_{jk}]\,n_k = \Delta p_i, \qquad\qquad (2.3.119)$$

which corresponds to the conditions (2.3.109).

The difference between (2.3.118), (2.3.119) and (2.3.108), (2.3.109) is that "follower" loads are assumed to be present in the first case, so that $\Delta X_i \neq 0$, $\Delta p_i \neq 0$. The loads are "unidirectional" in the conservative case and $\Delta X_i = \Delta p_i = 0$. In this case, the more general equations (2.3.118), (2.3.119) immediately become the special equations (2.3.108), (2.3.109).

Finally, (2.3.118) can be rewritten as follows using the Hooke's law:

$$\mu^* \nabla^2 u_i + (\lambda^* + \mu^*) \frac{\partial^2 u_j}{\partial x_i \, \partial x_j} + \frac{\partial}{\partial x_k} \left(\sigma_{jk} \frac{\partial u_i}{\partial x_j} \right) + \varDelta X_i = \mu \frac{\partial^2 u_i}{\partial t^2}, \quad (2.3.120)$$

which corresponds to (2.3.111).

In the mathematical discussion of this problem, we will introduce the following simplifications:

(1) $\varDelta X_i = \varDelta p_i = 0$, i.e., there are no "follower" loads, and Eqs. (2.3.108), (2.3.109) are used.

(2) The undisturbed state is identified with the equilibrium state. This means that the quantities $s_{ik}^{(0)}$ and $s_{ik}^{(t)}$ which determine the original state of stress

$$\sigma_{ik} = -\alpha s_{ik}^{(0)} - \beta \Phi(t) s_{ik}^{(t)}, \quad (2.3.121)$$

can be determined in a quasistatic way, i.e., the $s_{ik}^{(0)}$ can be determined from the static loads $p_i^{(0)}$, $X_i^{(0)}$, which depends on a parameter α, and the $s_{ik}^{(t)}$ can be obtained from the amplitude values of the variable loads $p_i^{(t)}$, $X_i^{(t)}$, which depend on a parameter β.

In addition, Hooke's law is written in the form $\varDelta \sigma_{ij} = C_{ijkl} \epsilon_{kl}$, where C_{ijkl} is the elasticity tensor and ϵ_{kl} the strain tensor. Taking the symmetry properties of the elasticity tensor into account, the following equations are obtained from (2.3.108), (2.3.109):

$$\frac{\partial}{\partial x_k} \left(C_{ikmn} \frac{\partial u_m}{\partial x_n} \right) - \alpha \frac{\partial}{\partial x_k} \left(s_{jk}^{(0)} \frac{\partial u_i}{\partial x_j} \right) - \beta \Phi(t) \frac{\partial}{\partial x_k} \left(s_{jk}^{(t)} \frac{\partial u_i}{\partial x_j} \right) - \mu \frac{\partial^2 u_i}{\partial t^2} = 0,$$

$$(2.3.122)$$

$$\varDelta \sigma_{ik} n_k = \left(\alpha \frac{\partial u_i}{\partial x_j} s_{jk}^{(0)} + \beta \Phi(t) \frac{\partial u_i}{\partial x_j} s_{jk}^{(t)} \right) n_k. \quad (2.3.123)$$

The following "abbreviated" problem corresponds to (2.3.122), (2.3.113):

$$\frac{\partial}{\partial x_k} \left(C_{ikmn} \frac{\partial u_m}{\partial x_n} \right) - \mu \frac{\partial^2 u_i}{\partial t^2} = 0, \qquad \varDelta \sigma_{ik} n_k = 0, \quad (2.3.124)$$

which corresponds to the free eigenoscillations of the elastic body. The solution of (2.3.124) can be written in the form of an integral equation. The classical problem of the theory of elasticity

$$\frac{\partial}{\partial x_k} \left(C_{ikmn} \frac{\partial u_m}{\partial x_n} \right) + X_i = 0, \qquad \varDelta \sigma_{ik} n_k = p_i \quad (2.3.125)$$

allows the solution

$$u_i(P) = \int_V G_{ik}(P,Q)\,X_k(Q)\,dV_Q + \int_0 G_{ik}(P,Q)\,p_k(Q)\,dO_Q\,, \quad (2.3.126)$$

where G_{ik} is the Green tensor, V the volume, O the surface area, and P, Q are two points of the elastic body. A description of the Green tensor is given in the books by Bolotin (*1.7*, pp. 219–221).

A comparison of (2.3.124) and (2.3.125) shows that the solution (2.3.126) can be taken for (2.3.124), if the following substitution is made in (2.3.126):

$$X_i = -\mu\,\partial^2 u_i/\partial t^2, \qquad p_i = 0.$$

When $u_i(P, t) = \varphi_i(P)\,e^{i\omega t}$ is introduced and if $\omega^2 = \lambda$, the result is

$$\varphi_i(P) - \lambda \int_V \mu(Q)\,G_{ik}(P,Q)\,\varphi_k(Q)\,dV_Q = 0. \quad (2.3.127)$$

The system of integral equations (2.3.127) has the eigenvalues λ_k, $k = 1, 2, 3,\ldots$, and the eigenvector $\varphi_i^{(k)}(P)$, $i = 1, 2, 3$, corresponds to each eigenvalue λ_k. The components of the eigenvector, the eigenfunctions $\varphi_r^{(k)}$ (r fixed), are a complete and orthonormal function system. Each vector $u_i(P)$ generated by the Green tensor

$$u_i(P) = \int_V G_{ik}(P,Q)\,X_k(Q)\,dV_Q$$

can be expanded in the absolutely and uniformly convergent series

$$u_i(P) = \sum_{m=1}^{\infty} \rho_m \varphi_i^{(m)}(P), \qquad \rho_m = \int_V u_i(P)\,\varphi_i^{(m)}\,dV. \quad (2.3.128)$$

The components of the Green tensor can be written in the bilinear series

$$G_{ik}(P,Q) = \sum_{m=1}^{\infty} \frac{\varphi_i^{(m)}(P)\,\varphi_k^{(m)}(Q)}{\lambda_m}. \quad (2.3.129)$$

Just as was done in the abbreviated problem (2.3.124), the solution (2.3.126) of (2.3.125) can be applied to our problem (2.3.122), (2.3.123). For this it is necessary to replace X_i in (2.3.126) by

$$-\alpha\,\frac{\partial}{\partial x_k}\left(s_{jk}^{(0)}\,\frac{\partial u_i}{\partial x_j}\right) - \beta\Phi(t)\,\frac{\partial}{\partial x_k}\left(s_{jk}^{(t)}\,\frac{\partial u_i}{\partial x_j}\right) - \mu\,\frac{\partial^2 u_i}{\partial t^2}$$

and to replace p_i by

$$[\alpha(\partial u_i/\partial x_j)\,s_{jk}^{(0)} + \beta\Phi(t)(\partial u_i/\partial x_j)\,s_{jk}^{(t)}]\,n_k\,.$$

The following solution is the solution of (2.3.122), (2.3.123):

$$u_i(P) = - \int_V \mu(Q)\, G_{ik}(P,Q)\, \frac{\partial^2 u_k(Q,t)}{\partial t^2}\, dV_Q$$

$$- \alpha \int_V G_{ik}(P,Q)\, \frac{\partial}{\partial \xi_l} \left[s_{jl}^{(0)}\, \frac{\partial u_k(Q)}{\partial \xi_j} \right] dV_Q$$

$$- \beta \Phi(t) \int_V G_{ik}(P,Q)\, \frac{\partial}{\partial \xi_l} \left[s_{jl}^{(t)}\, \frac{\partial u_k(Q)}{\partial \xi_j} \right] dV_Q$$

$$+ \alpha \int_0 G_{ik}(P,Q)\, s_{jl}^{(0)}(Q)\, \frac{\partial u_k(Q)}{\partial \xi_j}\, n_l\, dO_Q$$

$$+ \beta \Phi(t) \int_0 G_{ik}(P,Q)\, s_{jl}^{(t)}(Q)\, \frac{\partial u_k(Q)}{\partial \xi_j}\, n_l\, dO_Q . \qquad (2.3.130)$$

For example, we have

$$G_{ik}(P,Q)\, \frac{\partial}{\partial \xi_l} \left[s_{jl}^{(0)}\, \frac{\partial u_k(Q)}{\partial \xi_j} \right] \equiv \frac{\partial}{\partial \xi_l} \left[G_{ik}(P,Q)\, s_{jl}^{(0)}(Q)\, \frac{\partial u_k(Q)}{\partial \xi_j} \right]$$

$$- \frac{\partial G_{ik}(P,Q)}{\partial \xi_l}\, s_{jl}^{(0)}(Q)\, \frac{\partial u_k(Q)}{\partial \xi_j} .$$

Applying the Gaussian formula

$$\int_V \partial F_k / \partial x_k\, dV = \int_0 F_k n_k\, dO,$$

we obtain

$$- \alpha \int_V G_{ik}(P,Q)\, \frac{\partial}{\partial \xi_l} \left[s_{jl}^{(0)}\, \frac{\partial u_k(Q)}{\partial \xi_j} \right] dV_Q$$

$$= \alpha \int_V \frac{\partial G_{ik}(P,Q)}{\partial \xi_l}\, s_{jl}^{(0)}(Q)\, \frac{\partial u_k(Q)}{\partial \xi_j}\, dV_Q$$

$$- \alpha \int_0 G_{ik}(P,Q)\, s_{jl}^{(0)}(Q)\, \frac{\partial u_k(Q)}{\partial \xi_j}\, n_l\, dO_Q . \qquad (2.3.131)$$

The third term on the right side of (2.3.130) can be transformed in a similar way.

When (2.3.131) and the transformation of the third term in (2.3.130) is substituted, the following new relationship is obtained:

$$u_i(P) = - \int_V \mu(Q)\, G_{ik}(P,Q)\, \frac{\partial^2 u_k(Q,t)}{\partial t^2}\, dV_Q$$

$$+ \alpha \int_V \frac{\partial G_{ik}(P,Q)}{\partial \xi_l}\, s_{jl}^{(0)}(Q)\, \frac{\partial u_k(Q)}{\partial \xi_j}\, dV_Q$$

$$+ \beta \Phi(t) \int_V \frac{\partial G_{ik}(P,Q)}{\partial \xi_l}\, s_{jl}^{(t)}(Q)\, \frac{\partial u_k(Q)}{\partial \xi_j}\, dV_Q , \qquad (2.3.132)$$

which will now be used as the basis of an approximate calculation. This is again justified by the certain convergence of the Galerkin method.

Let us use the trial solution

$$u_i(P, t) = \sum_{m=1}^{\infty} f_m(t)\, \varphi_i^{(m)}(P) \qquad (2.3.133)$$

of the Galerkin method for (2.3.132), where $\varphi_i^{(m)}$ are the eigenvectors of the eigenoscillation problem. The bilinear form (2.3.129) is used for the components of the Green tensor. Because of (2.3.127) we assume

$$\int_v \mu(Q)\, G_{ik}(P, Q)\, \varphi_k^{(m)}(Q)\, \frac{d^2 f_m}{dt^2}\, dV_Q = \frac{\varphi_i^{(m)}(P)}{\lambda_m}\, \frac{d^2 f_m}{dt^2}.$$

The comparison of the coefficients of the $\varphi_i^{(m)}(P)$ results in the following system of differential equations:

$$(1/\lambda_{(m)})\ddot{f}_m + f_m - \alpha a_{mn} f_n - \beta \Phi(t)\, b_{mn} f_n = 0, \qquad m = 1, 2, 3,..., \qquad (2.3.134)$$

with

$$a_{mn} = \frac{1}{\lambda_m} \int_V s_{jl}^{(0)}\, \frac{\partial \varphi_k^{(m)}}{\partial \xi_j}\, \frac{\partial \varphi_k^{(n)}}{\partial \xi_l}\, dV,$$

$$b_{mn} = \frac{1}{\lambda_m} \int_V s_{jl}^{(t)}\, \frac{\partial \varphi_k^{(m)}}{\partial \xi_j}\, \frac{\partial \varphi_k^{(n)}}{\partial \dot{\xi}_l}\, dV. \qquad (2.3.135)$$

Equations (2.3.134) are completely analogous to the system of differential equations (2.3.40) found for the theory of elastostatic stability problems. However, there is an important difference between (2.3.134) and (2.3.40). The coefficient in the last term of (2.3.134) is *variable* and usually periodic. This means that a different and more complicated method of solution is necessary than the one we found in elastostatics. Let us outline the case of periodic coefficients.

Equation (2.3.134) can also be written in the form

$$\ddot{f}_i + \Phi_{ik}(t) f_k = 0, \qquad i, k = 1, 2, 3,..., \qquad (2.3.136)$$

where the abbreviation

$$\Phi_{ik}(t) = \lambda_{(i)}[\delta_{ik} - \alpha a_{ik} - \beta \Phi(t)\, b_{ik}]$$

has been used. It is also assumed that Φ_{ik} are *periodic*, i.e., $\Phi(t + T) = \Phi(t)$.

If the new variables

$$\xi_i = f_i \quad (i = 1, 2,..., n), \qquad \xi_i = \dot{f}_{i-n} \quad (i = n+1, n+2,..., 2n)$$

are introduced, Eq. (2.3.136) is transformed into the linear system of differential equations

$$\dot{\xi}_i = \xi_{n+i}, \qquad\qquad i = 1, 2,..., n,$$

$$\dot{\xi}_i = -\Phi_{i-n,k}\xi_k \qquad i = n+1, n+2,..., 2n$$

which has periodic coefficients. This system may now be discussed as in Section 1.3.3 for linear variational equations with periodic coefficients. This means that the *characteristic exponents* or multipliers must be analyzed in order to investigate stability.

Another method will be used in practice to determine regions of instability. Use will be made of the fact that the system (2.3.136) has periodic solutions with the period T or $2T$ at the boundaries of the instability regions. Instability regions are bounded by two solutions with the same period and stability regions are bounded by two solutions with different periods. The critical value of the parameter contained in (2.3.136) is therefore easily obtained from the condition that (2.3.136) have periodic solutions.

In order to outline the calculation, let us set $\Phi(t) \equiv \cos \tau t$ in (2.3.134). When the trial solution

$$f(t) = \sum_{k=1,3,5} \left(\mathbf{A}_k \sin \frac{k\tau t}{2} + \mathbf{B}_k \cos \frac{k\tau t}{2} \right),$$

for which $\mathbf{f}(t) = [f_i(t)]$, $\mathbf{A}_k$, $\mathbf{B}_k$ are vectors, is substituted into (2.3.134), the following system of equations is obtained by a comparison of the coefficients of $\sin(k\tau t/2)$, $\cos(k\tau t/2)$:

$$(\mathbf{E} - \alpha\mathbf{A} + \tfrac{1}{2}\beta\mathbf{B} - \tfrac{1}{4}\tau^2\mathbf{C}) \mathbf{A}_1 - \tfrac{1}{2}\beta\mathbf{B}\mathbf{A}_3 = 0$$

$$(\mathbf{E} - \alpha\mathbf{A} - \tfrac{1}{4}k^2\tau^2\mathbf{C}) \mathbf{A}_k - \tfrac{1}{2}\beta\mathbf{B}(\mathbf{A}_{k-2} + \mathbf{A}_{k+2}) = 0, \qquad k = 3, 5,...,$$

$$(E - \alpha\mathbf{A} - \tfrac{1}{2}\beta\mathbf{B} - \tfrac{1}{4}\tau^2\mathbf{C}) \mathbf{B}_1 - \tfrac{1}{2}\beta\mathbf{B}\mathbf{B}_3 = 0,$$

$$(\mathbf{E} - \alpha\mathbf{A} - \tfrac{1}{4}k^2\tau^2\mathbf{C}) \mathbf{B}_k - \tfrac{1}{2}\beta\mathbf{B}(\mathbf{B}_{k-2} + \mathbf{B}_{k+2}) = 0, \qquad k = 3, 5,...,$$

with the matrices

$$\mathbf{E}, \mathbf{A} = (a_{ik}), \qquad \mathbf{B} = (b_{ik}), \qquad \mathbf{C} = (1/\lambda_i), \qquad \text{diagonal},$$

and the vectors $\mathbf{A}_k$, $\mathbf{B}_k$. Solutions having the period $4\pi/\tau$ can occur if the conditions

$$\begin{vmatrix} \mathbf{E} - \alpha\mathbf{A} \pm \dfrac{1}{2}\beta\mathbf{B} - \dfrac{1}{4}\tau^2\mathbf{C} & -\dfrac{1}{2}\beta\mathbf{B} & 0 & \cdots \\[2mm] -\dfrac{1}{2}\beta\mathbf{B} & \mathbf{E} - \alpha\mathbf{A} - \dfrac{9}{4}\tau^2\mathbf{C} & -\dfrac{1}{2}\beta\mathbf{B} & \cdots \\[2mm] 0 & -\dfrac{1}{2}\beta\mathbf{B} & \mathbf{E} - \alpha\mathbf{A} - \dfrac{25}{4}\tau^2\mathbf{C} & \cdots \\[2mm] \cdots & \cdots & \cdots & \cdots \end{vmatrix} = 0$$

$$(2.3.137)$$

are satisfied. If the trial solution

$$f(t) = \tfrac{1}{2}\mathbf{B}_0 + \sum_{k=2,4,6} \mathbf{A}_k \sin(k\tau t/2) + \mathbf{B}_k \cos(k\tau t/2)$$

is substituted into (2.3.134), the following conditions for the occurrence of solutions having the period $2\pi/\tau$ are obtained for $\Phi(t) = \cos \tau t$:

$$\begin{vmatrix} \mathbf{E} - \alpha\mathbf{A} - \tau^2\mathbf{C} & -\tfrac{1}{2}\beta\mathbf{B} & 0 & \cdots \\ -\tfrac{1}{2}\beta\mathbf{B} & \mathbf{E} - \alpha\mathbf{A} - 4\tau^2\mathbf{C} & -\tfrac{1}{2}\beta\mathbf{B} & \cdots \\ 0 & -\tfrac{1}{2}\beta\mathbf{B} & \mathbf{E} - \alpha\mathbf{A} - 16\tau^2\mathbf{C} & \cdots \\ \cdots & \cdots & \cdots & \cdots \end{vmatrix} = 0 \quad (2.3.138)$$

and

$$\begin{vmatrix} \mathbf{E} - \alpha\mathbf{A} & -\beta\mathbf{B} & 0 & 0 & \cdots \\ -\tfrac{1}{2}\beta\mathbf{B} & \mathbf{E} - \alpha\mathbf{A} - \tau^2\mathbf{C} & -\tfrac{1}{2}\beta\mathbf{B} & 0 & \cdots \\ 0 & -\tfrac{1}{2}\beta\mathbf{B} & \mathbf{E} - \alpha\mathbf{A} - 4\tau^2\mathbf{C} & -\tfrac{1}{2}\beta\mathbf{B} & \cdots \\ 0 & 0 & -\tfrac{1}{2}\beta\mathbf{B} & \mathbf{E} - \alpha\mathbf{A} - 16\tau^2\mathbf{C} & \cdots \\ \cdots & \cdots & \cdots & \cdots & \cdots \end{vmatrix} = 0.$$

$$(2.3.139)$$

The *critical parameter values* are obtained from (2.3.137), (2.3.138), and (2.3.139). In the actual calculation these infinite determinants are truncated; and finite, reduced determinants are used. This is allowable, because the infinite determinants which occur are normal and therefore *convergent*, which was proven by Bolotin (*1.7*). The reduction method is allowed for normal determinants. It is often sufficient to know the *principal instability regions* that result for $k = 1$. An approximation for them is obtained if the trial solution

$$f(t) = \mathbf{A} \sin(\tau t/2) + \mathbf{B} \cos(\tau t/2)$$

of the harmonic balance method is substituted into (2.3.134). The approximate critical parameter values then follow from the condition

$$\det(\mathbf{E} - \alpha \mathbf{A} \pm \tfrac{1}{2}\beta \mathbf{B} - \tfrac{1}{4}\tau^2 \mathbf{C}) = 0. \tag{2.3.140}$$

This condition is expanded as follows and written as two lines and two columns

$$\begin{vmatrix} 1 - \alpha a_{11} \pm \tfrac{1}{2}\beta b_{11} - (\tau^2/4\lambda_1) & -\alpha a_{12} \pm \tfrac{1}{2}\beta b_{12} \\ -\alpha a_{21} \pm \tfrac{1}{2}\beta b_{21} & 1 - \alpha a_{22} \pm \tfrac{1}{2}\beta b_{22} - (\tau^2/4\lambda_2) \end{vmatrix} = 0.$$

The book by Bolotin (*1.7*) contains an excellent and very detailed discussion of stability problems of elastokinetics as well as a large number of examples. The interested reader is referred to this book.

We will describe the previous theory by means of the following classical problem.

Example: Let us consider a straight column with a pulsating longitudinal load (Fig. 87). The differential equation for the transverse vibrations is

$$EJ \frac{\partial^4 w}{\partial x^4} + (P_0 + P_t \cos \tau t) \frac{\partial^2 w}{\partial x^2} + \mu \frac{\partial^2 w}{\partial t^2} = 0. \tag{2.3.141}$$

The trial solution $w(x, t) = f_m(t)\, \phi_m(x)$ is used, where

$$\varphi_k = \sin(k\pi x/l)$$

FIG. 87.   Buckling column with a pulsating compressive load.

are the eigenfunctions of the free column oscillations:

$$\varphi_k'' = -(k^2\pi^2/l^2)\,\varphi_k\,, \qquad \varphi_k^{IV} = (k^4\pi^4/l^4)\,\varphi_k\,. \qquad (2.3.142)$$

Equations (2.3.142) and (2.3.141) result in

$$EJ(k^4\pi^4/l^4)\,f_k(t)\,\varphi_k(x) - (P_0 + P_t\cos\tau t)(k^2\pi^2/l^2)\,f_k(t)\,\varphi_k(x) + \mu\ddot{f}_k(t)\,\varphi_k(x) = 0,$$

so that the following system of differential equations is obtained for the $f_k(t)$:

$$\mu\ddot{f}_k(t) + \left[\frac{k^4\pi^4 EJ}{l^4} - (P_0 + P_t\cos\tau t)\frac{k^2\pi^2}{l^2}\right]f_k(t) = 0, \qquad (2.3.143)$$

which corresponds to Eq. (2.3.134). A simple transformation of (2.3.143) results in

$$\ddot{f}_k(t) + \frac{k^2\pi^2(EJk^2\pi^2 - P_0 l^2)}{\mu l^4}\left(1 - \frac{l^2 P_t}{EJk^2\pi^2 - P_0 l^2}\cos\tau t\right)f_k(t) = 0, \qquad (2.3.144)$$

which is now equivalent to the system (2.3.136). A comparison of (2.3.144) and (2.3.136) shows that we must have

$$\alpha = 0, \qquad \beta = 1, \qquad \mathbf{B} = (b_m),\quad \text{diagonal,} \qquad b_m = \frac{l^2 P_t}{EJm^2\pi^2 - l^2 P_0}$$

$$(2.3.145)$$

$$\mathbf{C} = \left(\frac{1}{\lambda_m}\right), \qquad \text{diagonal,} \qquad \lambda_m = \frac{m^2\pi^2}{\mu l^4}(EJm^2\pi^2 - P_0 l^2)$$

for the rest of the calculation. The principal regions of instability are obtained from (2.3.140). When this condition is formulated for the assumptions (2.3.145) of the preceding problem, we have

$$\det(\mathbf{E} \pm \tfrac{1}{2}\mathbf{B} - (\tau^2/4)\,\mathbf{C}) = 0.$$

Since the matrices are all diagonal, we simply obtain

$$1 \pm \tfrac{1}{2}b_m - \tau^2/4\lambda_m = 0.$$

The following approximate equation for the boundaries of the principal region is obtained for $m = 1$:

$$1 \pm \frac{l^2 P_t}{2(EJ\pi^2 - l^2 P_0)} - \frac{\tau^2\mu l^4}{4\pi^2(EJ\pi^2 - P_0 l^2)} = 0,$$

which finally results in the well-known equation

$$\tau_{\text{crit}} = \frac{2\pi}{l^2} \left(\frac{EJ\pi^2 - P_0 l^2}{\mu} \right)^{1/2} \left(1 \pm \frac{P_t l^2}{2(EJ\pi^2 - P_0 l^2)} \right)^{1/2}.$$

In this simple case, it is possible to give the exact solution, because equation (2.3.144) is the Mathieu differential equation, the solution of which is known. Usually examples that occur in practice are so complicated that it is not possible to make an exact calculation, and the above approximate calculation must be carried out.

References

(*1*)   *1.* MALKIN, J. G., "Stability Theory of Motion." Oldenbourg, Munich, 1959.
      2. HAHN, W., "Theory and Application of the Direct Method of Lyapunov." Springer, Berlin, 1959.
      3. LASALLE, J., AND LEFSCHETZ, S., "Stability by Lyapunov's Direct Method with Applications." Academic Press, New York, 1961.
      4. LEFSCHETZ, S., "Stability of Nonlinear Control Systems." Academic Press, New York, 1965.
      5. BÜRGERMEISTER, G., STEUP, H., AND KRETZSCHMAR, H., "Stability Theory." Vols. I, II. Deutscher Verlag der Wissenschaften, Berlin, 1957, 1963.
      6. TIMOSHENKO, ST. P., AND GERE, J. M., "Theory of Elastic Stability." McGraw-Hill, New York, 1961.
      7. BOLOTIN, V. V., "Kinetic Stability of Elastic Systems." Deutscher Verlag der Wissenschaften, Berlin, 1961.
      8. BOLOTIN, V. V., "Nonconservative Problems of the Theory of Elastic Stability." Pergamon Press, Oxford, 1963.
      9. KOLLBRUNNER, C. F., AND MEISTER, M., "Buckling." Springer, Berlin, 1955.
    10. LETOV, A. M., "Stability in Nonlinear Control Systems." Princeton Univ. Press, Princeton, New Jersey, 1961.

(*2*)   *1.* MINORSKY, N., "Nonlinear Oscillations." Van Nostrand, Princeton, New Jersey, 1962.
      2. WHITTAKER, E. T., "Analytic Dynamics of Part a. Rigid Bodies." Cambridge Univ. Press, London and New York, 1959.
      3. KLEIN, F., AND SOMMERFELD, A., "The Theory of the Gyroscope." Johnson Reprint Corp. (Teubner), New York, Stuttgart 1965.
      4. KLOTTER, K., "Technical Vibration Theory," Vol. 2. Springer, Berlin, 1960.
      5. SOLODOVNIKOV, V. V., "Foundations of Autonomous Control." Oldenbourg, Munich, 1959.
      6. WANG, C.-T., "Applied Elasticity." McGraw-Hill, New York, 1953.
      7. BUDÓ, A., "Theoretical Mechanics." Springer, Berlin, 1963.
      8. SZABÓ, I., "Advanced Technical Mechanics." Springer, Berlin, 1960.
      9. CUNNINGHAM, W. J., "Introduction to Nonlinear Analysis." McGraw-Hill, New York, 1958.
    10. KAUDERER, H., "Non-Linear Mechanics." Springer, Berlin, 1958.
    11. OPPELT, W., "Small Handbook of Technical Control Processes." Verlag Chemie, Weinheim, 1964.
    12. POPOV, E. P., "Dynamics of Automatic Control Systems." Pergamon Press, Oxford, 1958.
    13. POPOV, E. P., AND PALTOV, J. P., "Approximate Methods for the Investigation of Non-Linear Control Systems." Akademische Verlag Gesellschaft, Leipzig, 1963.
    14. MAGNUS, K., "Vibrations." Teubner, Stuttgart, 1961.

(*3*) ZURMÜHL, R., "Matrices," p. 441. Springer, Berlin, 1958.

(4) KAMKE, E., "Differential Equations of Real Functions," pp. 120–193. Akademische Verlag Gesellschaft Geest und Portig, K.–G., Leipzig, 1952.

(5) DUSCHEK, A., "Advanced Mathematics," Vol. IV, p. 217. Springer, Wien, 1961.

(6) LANCZOS, C., "The Variational Principles of Mechanics." Univ. of Toronto Press, Toronto, 1964.

(7) FRANK, P., AND VON MISES, R., "Differential and Integral Equations of Mechanics and Physics," Vol. II, pp. 44–190. Viewig Braunschweig, New York, 1961.

(8) 1. PFLÜGER, A., "Stability Problems of Elastostatics." Springer, Berlin, 1964.
    2. BÜRGERMEISTER, G., STEUP, H. AND KRETZSCHMAR, H., "Stability Theory," Vols. I, II. Deutscher Verlag der Wissenschaften, Berlin, 1957, 1963.
    3. GERARD, G., "Introduction to Structural Stability Theory." McGraw-Hill, New York, 1962.
    4. MARGUERRE, K., "Modern Production Problems for Engineers," pp. 189–249. Springer, Berlin, 1950.
    5. LANGHAAR, H. L., "Energy Methods in Applied Mechanics," pp. 201–232. Wiley, New York, London 1962.

(9) TRICOMI, R., "Elliptical Functions," pp. 263–271. Akademische Verlag Gesellschaft Geest und Portig, K.–G., Leipzig, 1948.

(10) WANG, C.-T., "Applied Elasticity," pp. 217–222. McGraw-Hill, New York, 1953.

(11) KOITER, W. T., Elastic stability and post-buckling behavior. In "Nonlinear Problems" (R. E. Langer, ed.), pp. 257–275. Univ. of Wisconsin Press, Madison, Wisconsin, 1963.

(12) MICHLIN, S. G., "Variational Methods of Mathematical Physics," pp. 365–489. Akademie Verlag, Berlin, 1962.

(13) BISPLINGHOFF, R. L., AND ASHLEY, H., "Principles of Aeroelasticity," pp. 418–424. Wiley, New York, 1962.

Author Index

Numbers in parentheses are reference numbers and indicate that an author's work is referred to although his name is not cited in the text. Numbers in italics show the page on which the complete reference is listed.

A

Ashley, H., 253, *270*

B

Beck, M., *180*
Bisplinghoff, R. L., 253, *270*
Bolotin, V. V., 1(1), 176(1), 225, 265, 266, *269*
Borchardt, A., *210*
Budó, A., 1(2), 36(2), 176(2), *269*
Bürgermeister, G., 1(1), 176(1, 8), *269*, *270*

C

Collatz, L., *196*
Cremer, L., *29*
Cunningham, W. J., 1(2), 36(2), 176(2), *269*

D

de Bra, D. B., *12*
Duschek, A., 47(5), *270*

F

Frank, P., 125, *270*

G

Gerard, G., 176(8), *270*
Gere, J. M., 1(1), 176(1), 179(1.6), *269*
Green, A. E., *181*

H

Hahn, W., 1(1), 118, 176(1), *269*

K

Kamke, E., 40(4), *270*
Kauderer, H., 1(2), 36(2), 86, 139, 176(2), *269*
Klein, F., 1(2), *10*, 36(2), 176(2), *269*
Klotter, K., 1(2), 36(2), 139, 176(2), *269*
Koiter, W. T., 177, *270*
Kollbrunner, C. F., 1(1), 176(1), *269*
Krasnoselski, M. A., *106*
Kretzschmar, H., 1(1), 176(1, 8), *269*, *270*

L

Lanczos, C., 109, 178, *270*
Langhaar, H. L., 176(8), 177, *270*
LaSalle, J., 1(1), *71*, 176(1), *269*
Lefschetz, S., 1(1), 148, 176(1), *269*
Leipholz, H., *106*, *157*
Leonhardt, A., *29*
Leontovic, A. M., *139*
Letov, A. M., 1(1), 148, 176(1), *269*

M

Magnus, K., 1(2), 36(2), 122, 139, 144, 170, 176(2), *269*
Malkin, J. G., 1(1), 76, 79, 176(1), *269*
Marguerre, K., 176(8), *270*
Meister, M., 1(1), 176(1), *269*
Mettler, E., *256*
Michlin, S. G., 36(12), 195, *270*
Minorsky, N., 1(2), 36(2), 139, 176(2), *269*

271

Subject Index